Contents

KT-441-382

Dedication   v

Acknowledgements   vi

Preface   vii

1   Statutory regulations and organisational requirements   1

2   Applying safe working methods and emergency procedures   37

3   Effective working practices   65

4   Electrical systems and components   79

5   Electrical supply systems, protection and earthing   111

6   Electric motors   137

7   Completing electrical installations safely   163

8   Testing and commissioning electrical installation   227

9   Fault diagnosis and restoration   260

10 Restoring installations and equipment to working order   271

    Answers to self assessment questions   286

    Index   293

Electrical Installations

for Level 3

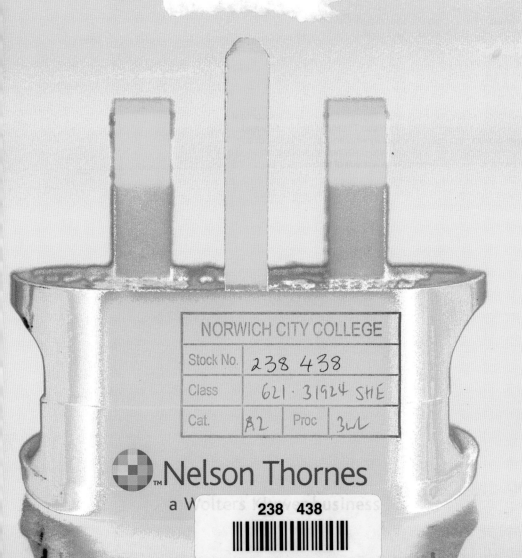

Learne

™ Nelson Thornes

a Wolters Kluwer Business

Published in 2006 by:
Nelson Thornes Ltd
Delta Place
27 Bath Road
CHELTENHAM
GL53 7TH
United Kingdom

06 07 08 09 10 / 10 9 8 7 6 5 4 3 2 1

A catalogue record for this book is available from the British Library

ISBN 0 7487 9602 9

Illustrations by Peters and Zabransky
Page make-up by Florence Production Ltd, Stoodleigh, Devon
Printed and bound in Slovenia by Korotan – Ljubljana

Dedication

Chris had over forty years experience working in the electrical installation engineering industry, both as a qualified tradesman and in managment. For a time he was also a part-time college lecturer. His first book, *Electrical Installation*, was published in 1993, followed by two updatesd editions in 1996 and 2004, the latter accompanied by a Tutor Support Pack. He also wrote *Advanced Electrical Installation* (1996) and *Electrical Option Units* (1998) for NVQ Engineering Level 2. His whole purpose in writing was to pass on the wide knowledge gained during his working life in a reader-friendly way.

Shirley Shelton

Acknowledgements

The publishers would like to thank Jon Dicken for his invaluable contribution towards this publication.

The authors and publishers gratefully acknowledge the owners of any copyright material reproduced herein.

The safety signs that appear in the book are from Instant Art/Magnum (NT).

Every effort has been made to reach the copyright holders, but the publisher would be grateful to hear from any source whose copyright they have unwittingly infringed.

Preface

This book has been derived from the March 2004 *City and Guilds* document, *Level 3 Certificate in Electrotechnical Technology*.

There are just three core units within these pages. Some of the topics you will find both interesting and absorbing, others such as the legal matters presented in Chapter 3 might be of less interest to you but please study them as the information within is vital if you wish to gain the Level 3 Certificate.

The second core unit will be far more familiar to you, dealing with the practical side of inspecting, testing and commissioning electrical installations and how best to carry out testing to *National Inspection Council* standards together with effective and efficient working practices to install electrical installations and equipment.

The last unit is all about the working practices and safety procedures you will have to put into practice when fault finding within an established installation or if, for example, you are fault diagnosing equipment within an electrical system. Two chapters serve this final unit.

This book, broken up into ten chapters, identifies with various recommended course works and underpinning knowledge requirements as laid out within the *CG 2330 Scheme Handbook*. A derived reference point is included within the introduction serving each chapter. This is to provide you with a little insight of what is to follow.

Good luck with your studies and do well in your career.

Chris Shelton

Statutory regulations and organisational requirements

Introduction

There are fifteen underpinning knowledge requirements in this chapter, all of which are compulsory.

- Safety regulations for electrotechnical activities.
- Safety – your employer's responsibilities.
- Safety and personal protective equipment.
- Safety policies, codes of practice and safe working procedures within a typical organisation.
- All about warning signs.
- Outlining the role of safety personnel.
- Accessing health and safety advice.
- Accidents within the workplace and means of control.
- Accident reporting procedures within the workplace.
- All about how to carry out a risk assessment.
- Notifying hazards to the appropriate people.
- Environmental management systems in terms of the electrotechnical industry.
- Environmental legislation as it applies to our industry.
- Changes within the electrotechnical industry – working patterns and training needs.
- The importance of constant training and maintaining good customer relations and methods to protect your customer.

Practical activities

There are four formal activities accompanying this chapter. You will, as a candidate, have to provide evidence learnt within this chapter. The activities are as follows:

- Select appropriate personal protective and safety equipment for the job you are to carry out.
- Recognise a selection of warning, prohibition and mandatory safety signs.
- Complete an accident report.
- Carry out and prepare a risk assessment programme.

Tests

You will be given a multiple choice question paper and then be assessed in accordance with City and Guilds test requirements.

Safety regulation awareness

To provide you with a complete guide to the many regulations and standards which oversee our industry would take several volumes of sheer hard work, to say the least! The paragraphs which follow are just snap-shot views to give you an awareness of some of the legal aspects which govern our industry.

The Health and Safety at Work Act [1974]

First, your legal duties and responsibilities:

1. You must cooperate fully with your employer on all aspects of health and safety.

2. Have a word with your safety officer if you have reason to believe that a health and safety problem exists at work.

3. The *Health and Safety Inspector* will provide you with good advice if a health and safety problem exists which cannot be favourably resolved at site level.

4. You must not mistreat or interfere with anything given by your employer which is for the health, safety and welfare of the company.

5. You must be responsible for your own health and also the health of others who would be affected by something you do or have forgotten to do.

6. Site accidents or any type of mishap whilst you are at work must be reported quickly and the details confidentially recorded in the accident book.

Now, your employer's responsibilities and duties:

1. He/she must prepare a written health and safety statement when five or more people are employed.

2. Your employer must make sure that the workplace is safe and without risk to health.

Figure 1.2 – *Notice warning of moving parts*

3. All personal protective clothing and equipment must be supplied free of charge. Figure 1.1 illustrates a selection.

4. Your employer must form a safety committee if two or more employee's request it in writing.

5. Adequate fire fighting equipment and means of escape must always be provided by your employer.

6. He/she must provide protection from dangerous and moving parts associated with workshop machinery. Helpful warning notices must be installed wherever possible – Figure 1.2 illustrates these.

7. Your boss is responsible for the following site facilities:
   - A place where meals can be taken. Figure 1.3 illustrates a typical temporary building.

Figure 1.1 – *Personal protective equipment*

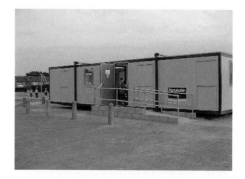

Figure 1.3 – *Site temporary building (mess room)*

- Sufficient heating arrangements.
- Drying/cloak room.
- Fresh water for drinking.
- A means to boil water.

8   Your employer must provide adequate training and supervision for apprentices and unskilled personnel before allowing them to use dangerous machinery or appliances.

9   Washing, WC facilities and clean fresh water must be provided by your employer.

10   Your employer must maintain good levels of lighting and ventilation within his/her workshop.

11   Adequate first aid equipment must be provided by your employer.

12   He/she must report dangerous incidents and certain types of diseases, for example asbestosis, to the enforcing agency.

13   The maintenance of steps and ladders are the legal responsibility of your boss.

14   Mechanical handling equipment must be available for lifting/carrying heavy loads.

15   Floors, staircases and passageways are the legal responsibility of your boss and must be properly looked after.

16   Proper safety measures must be made available for people who have to work in cramped or confined spaces.

17   Employers must ensure that all dangerous materials are recognised. Figure 1.4 shows an asbestos warning sign.

18   Your employer must take steps to protect you from the hazards of radiation, for example, certain older types of smoke detectors when stored in large quantities.

19   Workplaces must be kept free from hazards.

20   Your company has a duty to keep noise, fumes and dust under control.

These and many more regulations can be obtained as a photocopy from most public libraries; a small charge may be incurred. Alternatively, obtain or download information on-line or buy a copy of *A Guide to the Health and Safety at Work Act [1974]*.

CAUTION

ASBESTOS

Figure 1.4 – *Asbestos warning sign*

Electricity at Work Regulations [1989]

Unlike our own regulations which are now a British standard, the *Electricity at Work Regulations* are a legal requirement which affects both employer and employee alike. Should a person or employer disobey these demands, the guilty party could be heavily fined or even end up in jail! The conditions are these: a suspected wrongdoer must, through the courts of law, prove that all reasonable measures had been taken to avoid a breach of these legal requirements.

As you can imagine, there are many requirements – as examples, here are just two of them which have been summarised for an easy read:

Regulation 4: Protective equipment

Protective equipment provided must be properly used, suitably maintained and appropriate for the use intended.

Regulation 11: Means for protection against excess current

Circuits must be efficiently protected against excess current caused by short circuit or over-current and be positioned in a suitable location.

Control of Substances Hazardous to Health Regulations [1999]

This is the regulation better known to us as *COSHH*. To summarise: your employer is legally required to physically control any hazardous materials within the workplace. Hazardous substances are anything which could harm your health whilst you are at work. Consider the following examples:

- Grain dust from silos.
- Viruses and bacteria.
- PVC glue, older types of smoke detection equipment (radiation) and any substance used in work activities.
- Asbestos and lead, although highly toxic materials, are dealt with by separate laws.
- Battery acid and substances found in modern lighting.

COSHH demands that employers must deal with the following:

1 To identify workplace hazards and assess the risk factor – Figure 1.5 illustrates warning signs used in conjunction with hazardous material.
2 Decide on the precautions to be taken.
3 Choose a practical method of control.
4 Make sure that the precautionary measures in place are adhered to.

5 If necessary, monitor employees for exposure to hazardous material.
6 Keep a close watch where it has been shown that this is necessary or if COSHH make a detailed request to do this.
7 Prepare a procedure to be used in case of an emergency or workplace accident happening.
8 Your employer is obliged to notify, instruct and supervise you wherever hazardous material is in use.

To keep healthy, please obey the rules – they are there for a purpose!

Provision and use of Work Equipment Regulations [1998]

This regulation is best kept in mind as *PUWER* – it is easier to remember. To summarise:

The equipment covered by this regulation includes the following:

- Any equipment used by you in the course of your work (hacksaw, screwdrivers, hammers, etc.).
- Electric drilling machines whether 110 volt powered or battery operated.
- Ladders and steps, access equipment.
- Any tool or piece of equipment provided by you or your employer.
- Any tools or equipment provided on the construction site.

You must ensure that all your work equipment is in good working order and meets the requirements of this regulation.

- Suitable for the purpose and conditions of your job.
- Your tools and equipment must be maintained in a safe condition.
- Tools and equipment will have to be formally inspected under certain circumstances.

Risks must be eliminated wherever possible by asking your employer to provide means of protection, for example, emergency stop buttons within a workshop and personal protective equipment and by following a safe system of work.

Dangerous chemicals

Figure 1.5 – *Hazard warning sign*

These regulations are enforced by *Health and Safety inspectors*.

Further reading

The Safe Use of Work Equipment, obtainable from HMSO bookshops – ISBN number 0 7176 1626 6.

Portable appliance testing regulations

Contrary to popular belief there are no hard and fast rules laid down by the *Health and Safety Executive* concerning the frequency of the testing and inspection of portable appliances. The *Electricity at Work Regulations* puts forward that regular testing is part of any planned maintenance programme but no attempt has been made to define the word 'regular'.

The test is divided into two sections, visual and instrumentation, but please check out the Code of Practice for in-service inspection and testing.

The visual test includes the following:

1   The condition of the plug.
2   The service lead and the body of the appliance will need to be checked out.
3   The area of the lead where it enters the appliance.
4   The mechanical switching action of the appliance.

Instrumentation tests will include the following:

1   An insulation test at 500 volts DC. Values must be above 500 000 ohms (0.5 megohms).
2   An earth continuity test, ideally below 0.3 ohms.
3   A flash test but not on electronic equipment – ideally 2.5 mA maximum.
4   Load test and fuse test and inspection. For example, a washing machine would have a 13 amp fuse fitted within its plug whereas an office table lamp would have just a 3 amp fuse.
5   For your records, establish whether the appliance is either Class 1 or 2. (A Class 1 appliance is served with an earth wire whereas a Class 2 piece of equipment is double-insulated and therefore does not have an earth.)

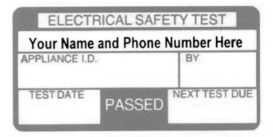

Figure 1.6 – *Portable appliance testing – passed label*

If the appliance you are testing checks out OK, stick a 'PASSED' label on it in a place where it can be seen. This type of label is illustrated as Figure 1.6. Then issue a formal certificate. Always attach a 'FAILED' label to a faulty appliance.

Frequency of testing

You must carry out tests of electrical equipment as often as necessary so as to prevent danger and to ensure that the electrical equipment is maintained in a safe condition. Hand held equipment used in garages and on site are more likely to become damaged than a printer or photocopier confined within the relative safety of a construction site office. The risk of danger arising from Class 1 equipment is the accidental disconnection of the circuit protective conductor within the plug. Figure 1.7 illustrates a typical PAT tester.

Figure 1.7 – *Portable appliance tester*

Control of Major Accident and Hazards Regulations [1999]

Better remembered as *COMAH*, this legal requirement principally affects the chemical industry and to some extent storage activities, explosives and nuclear sites. We also deal with nasty chemicals (batteries) and radioactive equipment (smoke detectors), so to some extent this regulation can be applied to us – especially if you are the main contractor of a large undertaking. (This act does not cover Northern Ireland.)

A concise summary:

- A major accident refers to a massive fire or explosion leading to serious damage to your health.
- You are legally obliged to take positive measures to prevent a major accident.
- You must inform a competent local authority if there is any significant increase in the quantity of dangerous materials you are storing.
- You, as a site operator, will have to prepare an emergency plan of action and submit it to your local authority.
- The local authority will have prepared an 'off-site' plan of action and this must be practically tested.

 All very heavy stuff, and if you are interested in reading more then log into the following government web site: www.hmso.gov.uk/si/si1999/99074302.htm#5

Noise and Statutory Nuisance Act [1993]

This is an Act of Parliament which makes it unlawful to make excessive noise in public places. If you are carrying out an outside contract, for example, within a residential street, which carries with it a fair amount of noise, then this Act will apply to you.

Briefly it covers the following:

- Loudspeakers and personal radios.
- Large machinery driven by air compressors.
- Site tools designed for chasing concrete, etc.
- Vehicles, machinery or noise generated by your own equipment. (After notifying the police, a local authority can break into a vehicle and remove it or shut down offending noisy equipment.)

You can imagine that it could become become intolerable for local residents if they had to suffer an audio diet of very loud music mixed with the general background sounds of construction throughout the day.

Noise Act [1996]

This is all about unacceptable levels of loud noise coming from your construction site or if, for example, you are rewiring an unoccupied house in a residential area of your local town. If complaints are made from your neighbours a local authority officer can, and will, issue you with a written warning notice. If the high noise level continues and the warning notice is ignored, an officer can seize and remove any offending equipment; if necessary by force! Often in cases such as these, noise pollution is caused by site radios playing at full volume. (This act is not valid in Scotland.)

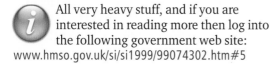 Please log onto the following web site for further details: www.hmso.gov.uk./acts/acts1996/96037-b.htm

Reportable Diseases and Dangerous Occurrences Regulations [1985]

This 1985 regulation is best remembered as *RIDDOR*. Serious accidents/occurrences under these regulations include the following:

- A site injury involving hospitalisation for more than 24 hours.
- An injury involving three or more days off from work.
- Fractured skull, pelvis or spine.
- Fractured bone within the wrist, arm, leg or ankle.
- Loss of sight in one or both eyes.
- An amputation of a toe, foot, hand, finger or thumb.
- Electric shock/burns leading to unconsciousness.
- Death.
- Suicide.

Learn to take care of your eyes; they are your livelihood – Figure 1.8 tells all! Accidents to your eyes can be caused by mechanical, chemical or thermal mishaps or a combination of all three.

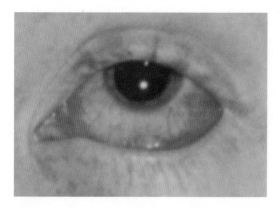

Figure 1.8 – *Take care of your eyes*

A site incident or dangerous occurrence could, for example, be the following:

- The collapse of a scaffold over five metres in height.
- The collapse of a wall or floor at your workplace.

Reportable diseases under RIDDOR which arise or are connected with work include the following:

- Asbestosis.
- Occupational asthma.
- Hepatitis and tuberculosis.
- Occupational cancer.
- Legionellosis and tetanus.
- Occupational dermatitis.

 Additional information can be obtained from the following web site: www.hse.gov.uk/pubns/hsis1.pdf

Management of Health and Safety Regulations [1998]

This legal requirement is to do with your health and safety within the workplace. For example, your employer must, by law, carry out the following requirements:

- Make a detailed risk assessment concerning all employees.
- Act upon these identified risks in a way as to reduce them, for example he/she would not ask an apprentice to carry out minor repairs on a high site crane if he/she suffered from vertigo.

- Your employer must appoint someone to administer company health and safety matters.
- Your company has a duty to inform and train employees in all health and safety matters.
- A written health and safety policy must be drawn up by your boss and be open to inspection.

Under the *1988 Work Time Regulations*, applied by two European Community Directives, all employees under the age of 18 must limit their working week to 48 hours. These restrictions, coupled to the basic working week, add a very important contribution to the health and safety of young operatives at work.

More details concerning this regulation can be obtained by logging onto the internet.

Your employer's responsibilities to maintain safety

Within the workplace

This is all about preventing accidents. Not only is it your employer's duty to provide a safe working environment but your input is equally important.

Consider the following points:

1 A first aider must be appointed by your management.
2 Necessary information and support should be provided to new working colleagues arriving within the workplace for the first time.
3 Your employer should provide assistance when you are working on live bus bars and electrical mains.
4 'STOP' buttons should be installed in your company workshop.
5 Face masks must be made available when working with certain equipment such as a masonry disc cutter or working within a farm grain silo, poultry or piggery unit.
6 Suitable ear protection must be supplied by your boss when you are working in a noisy environment.
7 Your employer must provide, free of charge, any personal protective clothing or equipment you may require – Figure 1.9 illustrates a pair of safety goggles.

Figure 1.9 – *Safety goggles*

Figure 1.10 – *It is important to place your ladder at the correct angle*

8 Fire drills and assembly areas must be organised by your management – especially on a large construction site.

9 Employers have a legal duty to ensure that their staff are physically capable of lifting and carrying heavy loads, if necessary.

10 Your employer must never allow you to use 230 volt equipment on site other than within a site office and other temporary accommodation.

11 The maintenance of step ladders and steps are the responsibility of your employer.

12 Check that your site is safe before work commences.

Plant and equipment

Plant and equipment safety is the responsibility of your employer and must be maintained by him. A few 'Do's' and 'Don'ts' for you to take onboard.

Do

- Report faulty equipment to your supervisor.
- Wear good strong protective boots or shoes whilst working on site.
- Check that your ladder is in no way damaged, especially if it is painted.
- Check out your ladder is well positioned and on a firm base at the correct angle as illustrated in Figure 1.10. The height of your ladder must be four times greater than the base (i.e. 4 up, 1 out). Position it so as to allow about *one metre* above the rung which you are to step on.

- Check out your plant regularly for defects. etc.
- Assemble access platforms on a firm level base.
- Push–pull mobile prefabricated scaffolding when moving to another location.

Don't

- Misuse and neglect plant and equipment supplied by your employer.
- Attempt to repair or meddle with plant and equipment unless you are capable of doing so.
- Use step ladders which are past their useful life.
- Obstruct passageways and staircases with plant and equipment – it could be a hazard if an emergency arose.
- Be tempted to bang hammers together – injury can be caused by splintering.
- Use a ladder with broken, damaged or missing rungs.
- Move an access tower by mechanical means – always move by hand.

Safe working systems

The best way to avoid personal injury in the workplace is to create a safe system of work before you start.

Direct contact with electricity involving serious burns and shock can also be avoided with formal training, establishing working procedures, good job planning and by use of the correct equipment and tools.

The way ahead – please consider the following:

- Predict the unforeseen.
- List potential hazards.
- Try to divide up your task into manageable sections.
- If you are the *job holder* you are duty bound to keep a friendly eye on your staff – you must be 'in-step' with one another. Team work is essential.
- Cover yourself by using formal *permit to work* authorisation procedures.
- Get into the habit of using recognised procedures as you would your tools.
- Offer each individual worker a suitable task you will know he/she can manage.
- Try to stick with your original work plan and review each step as your job progresses.

Safe working – think about the following:

- Use the correct and appropriate tool for the job you are doing – after all, you wouldn't open a tin of beans with a pair of side cutters!
- Keep your tools in good condition.
- Insist on the correct *personal protective equipment* you require for your task.
- Isolate 'live' equipment, machinery and circuits or 'lock off'. If this cannot be done, then disconnect your load conductors at your local isolating switch and make totally safe. Use 'DO NOT SWITCH ON – WORK TAKING PLACE' notices as shown in Figure 1.11 and always check out using a voltmeter. Safe isolation is important, so use safe isolating procedures at all times.
- If you are unable to completely shut down (and sometimes you *are* unable), any potential danger must be minimised – disconnect your circuit under investigation. Keep safe!
- Please remember, some de-energised equipment can become energised automatically by means of devices such as pressure switches, thermostats, light-sensitive

Figure 1.11 – *A 'Do not switch on – work taking place' notice*

and movement switches. Keep this possibility in mind.

- It is a wise precaution to work in pairs on 'live' mains – if one of you gets into trouble the other is available to help.
- Get to know your first aider.
- Check out where your nearest medical facility/burns unit is located together with their telephone number.
- Keep a land-line telephone or cell phone near to you whilst working on 'live' electrical mains equipment.

Providing a safe working environment

It is essential to have a safe working environment for all concerned. There are many ways to achieve this goal and listed in concise form below are a few – first we will deal with workplace electrical matters.

1 Do you know how to contact your electrical supplier in case an emergency arises? Log in their phone number into your mobile phone.

2 Do you know where your main switch is located? Always isolate 'live' equipment before attempting to work on it.

3 Don't overload site transformers – the voltage will drop and site 110 volt fluorescent lighting arrangements will stop working as intended.

4 Always use the correct size fuse or select the right MCB for your over-current device.

5 Never use faulty equipment – always report it to your supervisor.

6 Carry out portable appliance testing on site equipment regularly.

7 Replace any site-damaged cables and flexes serving equipment.

8 Keep diffusers firmly on temporary site-fitted fluorescent lighting. The chemical elements within the tube are toxic.

9 Provide emergency lighting to serve escape routes and stair wells.

10 A clean and tidy site is often a safe site!

Now, let us consider non-electrical matters.

1 A good working space is essential – avoid cramped or poky areas if this can be avoided.

2 Never overcrowd your working space – move unessential items to one side.

3 Keep all passageways and staircases free from trash in case an emergency arises – remove and skip regularly.

4 Good levels of light are essential when working under site conditions during the winter months.

5 Install exit signs, fire fighting signs and general information signs within the site – Figure 1.12 illustrates these.

6 Place guards around hoists and other moving site machinery parts.

7 Warn of dangers such as laser light or asbestos – Figure 1.13 illustrates this.

8 Independent, clean and well stocked site toilets are essential for both sexes. Hot and cold water must be available.

9 Make sure that your meal break facilities are clean, tidy, well lit and heated and provide a means in which to boil water for hot drinks, etc.

10 Personal protective equipment such as a hard hat, high visibility waist coat and toe-protective boots or shoes must be worn when working on site. Figure 1.14 illustrates these garments.

11 Provide a site drying room.

12 A simple site fire alarm system should be considered.

13 Make available in-house site training if this is felt necessary.

Figure 1.12 – *General information signs*

Figure 1.13 – *Laser light and asbestos warning signs*

Figure 1.15 – *Pedestrian walkways*

Figure 1.14 – *Personal protect equipment – hard hat and high visibility waistcoat*

14 Free up ground to make available a safe vehicle parking area for the work force. A safe pedestrian walkway should also be provided and indicated as shown in Figure 1.15.

I am sure you can list further ways in which to create a safe working site environment – treat this challenge as an in-house exercise among yourselves!

Safe methods of handling, storing and transporting goods and materials

Prior to moving material items you must consider the risk involved in such an undertaking – especially if you have to move bulky electrical items from storage to site. Follow your in-house procedures and rules when handling and positioning goods under your care and record any problem areas which crop up. Check that the materials to be delivered are not damaged, for example it would be easy to break one or two glass lamp shades loosely placed in a damaged cardboard box.

Consider the following points:

■ Wear suitable protective clothing and a high visibility waist-coat.

■ Store as recommended by the manufacturer.

■ Whilst in store, place your equipment so that it is accessible for when you want to move it.

■ Label goods which are unsuitable or cannot be issued for a particular reason.

■ Dispose of accumulated waste – place it in your skip.

■ Keep records and dates of material movements and damages, etc.

It is best to keep your material goods off the ground on wooden or plastic pallets (plastic pallets are stronger and bigger). This way they are easier to handle – especially if you are a trained fork-lift-truck driver or have access to one.

Store your materials in a cool dry place as moisture can cause a great deal of damage. Sheeted goods, plywood for example, for making construction site electrical cabinets should be stored flat on the ground, preferably under a suitable PVC sheet as this will stop them warping. Breakable items can be stored on a wooden or steel storage rack placed at eye level so they can be seen easily – Figure 1.16 illustrates this.

Conduit is best kept on steel racks whilst conduit inspection boxes can be placed within wooden boxes and kept on the floor. Keep your electrical accessories within their cardboard boxes and store them within labelled pigeon-holes – this way they will be safer.

Safe transporting materials to site

The best way is to get your electrical wholesaler to deliver your goods – this way, if anything is damaged, the loss is not yours!

If this is not possible, please follow the following guidelines.

- Get assistance to load-up your van or truck and, if necessary, to unload upon delivery.

Figure 1.16 – *Storage racks for material items*

- Take care not to damage the edges nor scratch the surface of your load – it might be a costly mistake!
- When placing materials on a roof-rack, secure with nylon rope whilst conduit can be placed and secured with an expandable bungee cord. It is wise to tie a piece of red cloth on the end of the conduit to keep within the traffic rules and regulations. It is not a good idea to drive with your rear van door half open in order to accommodate materials within – it is far more sensible to use a bigger vehicle. Drive safely!

Reporting an accident

Within the workplace

If you have an accident at work you are obliged to enter all details onto a '*privacy-act*' form. After returning the document to your site manager, it is filed with others in the site accident book. As your account of the accident is considered 'confidential' this database has restricted viewing rights and should be placed under lock and key within the site office.

If, for example, you should lose your footing and cut your head your site manager is obliged to ferry you safely to the nearest Accident and Emergency Unit and wait with you until you have been accepted for treatment by a doctor.

Figure 1.17 illustrates a typical page of a site accident book. This document varies in shape, size and format and is confidential.

Road traffic accident (RTA)

Providing no one is hurt, exchange names, addresses, vehicle index numbers, insurance company details and telephone numbers, etc. In the case of a company van, the telephone number will be that of your company.

Do not make any comment as to who was responsible for the RTA – let the insurance company or police deal will this matter. If safe to do so, make an on-site rough sketch as to how the accident took place, entering data guidelines and measurements as appropriate.

You will then be asked to fill out an *accident investigation form* in which you will have to provide a detailed and accurate account of what happened and draw a site diagram of the lead-up

ACCIDENT BOOK

1 About the person who had the accident	2 About you, the person filling in this book	3 About the accident
▼ Give full name ▼ Give the home address ▼ Give the occupation	▼ Please sign the book and date it ▼ If you did not have the accident write your address and occupation	▼ When it happened ▼ Where it happened

Name _____
Address _____
_____
_____
_____ Postcode _____
Occupation _____

Your signature   Date
[   ] /   /
Address _____
_____
_____ Postcode _____
Occupation _____

Date   Time
[/   /   ]
In what room or place did
the accident happen? ____
_____
_____
_____

Name _____
Address _____
_____
_____
_____ Postcode _____
Occupation _____

Your signature   Date
[   ] /   /
Address _____
_____
_____ Postcode _____
Occupation _____

Date   Time
[/   /   ]
In what room or place did
the accident happen? ____
_____
_____
_____

Name _____
Address _____
_____
_____
_____ Postcode _____
Occupation _____

Your signature   Date
[   ] /   /
Address _____
_____
_____ Postcode _____
Occupation _____

Date   Time
[/   /   ]
In what room or place did
the accident happen? ____
_____
_____
_____

Name _____
Address _____
_____
_____
_____ Postcode _____
Occupation _____

Your signature   Date
[   ] /   /
Address _____
_____
_____ Postcode _____
Occupation _____

Date   Time
[/   /   ]
In what room or place did
the accident happen? ____
_____
_____
_____

Figure 1.17 – *A typical page from a site accident book*

to your mishap. At times a representative of your firm's insurance company will do it for you at your company's offices.

Instruction, training and supervision of employees

An employer is obliged by law to inform, instruct, train and supervise his/her workforce as necessary. This is to ensure that their health and safety during work time activities is fully considered. Whether the staff are older and well established, young school leavers or people who have been newly recruited, all are subjected to this ruling.

Other legal regulations which target this area are:

■ *The Manual Handing and Lifting Operations Regulations 1992.*

- *The Personal Protective Equipment Regulations 1992.*
- *The Management of Health and Safety at Work Regulations 1992.*

What sort of training is required?

In general terms and relevant to our industry, please think about the following points.

1 Your company's health and safety policy and procedures.
2 First aid measures.
3 Accident reporting.
4 Site work fire hazards and drill.
5 Site evacuation procedures – normally carried out during a site-induction programme.
6 New equipment, test instruments and site machinery.
7 Specific health and safety risks which need to be highlighted together with procedures which need to be taken.
8 The responsibilities of individuals in respect to work time health and safety matters.
9 The need to advise both site visitors and members of the public of dangerous activities. This is best publicised by means of a site 'Hazardous Activity' notice board as shown in Figure 1.18.
10 Any changes which could affect the health and safety of the work force.

Newly recruited staff and young people straight from school must be provided with training before they start work and introduced to harsh site conditions.

Refresher training is equally important. This will help to reinforce past training and hopefully create good working practices within our industry. The effect of training can be evaluated by observation, asking the right questions and by gaining general feedback from others.

On-site training

Accidents at work are costly and can be quite upsetting for the person(s) concerned so a site safety *induction programme* is important for all newcomers. It provides an excellent opportunity to mould and influence new operatives, whether they are casual, part-time workers or are part of a larger organisation. Induction provides them with a positive attitude to site health and safety. All contractors need to go through this procedure to provide them with the skills to carry out their jobs safely.

The Tara electrical and mechanical engineering company

Hazardous Activities - data

Date: *27th November*
Activity: *High voltage cables spanning the site will be taken down and redirected underground. It has been programmed to last between 0900 to 1400 hours. The site supply will not be affected by this work. Your contact for this operation is Andrea Schaffer.*

Figure 1.18 – *Site notice board warning of hazardous activities*

Induction techniques vary from site to site. Some will provide a brief lecture followed by an instructional safety video; others will have an informal chat coupled with browsing through a well worn, large print/illustrative safety folder.

A form is then filled in providing a summary and snap-shot view of the induction procedure. This you are asked to sign.

The main contractor will, more often than not, offer you a weather-proof *induction badge* to stick on your hard hat – Figure 1.19 illustrates one such.

All about safety policies

If your company employs five or more people they are required, by law, to provide a *safety policy statement*.

This document consists of three parts. Briefly they are:

1 *A general statement of intent* – this is your company's attitude towards the management of health and safety matters.

2 *Organisational matters* – briefly, this outlines the command structure within your company in terms of health and safety matters:

   – who in your firm is responsible for what,

   – who is accountable for what,

   – the responsibilities people carry,

   – your internal monitoring procedures.

3 *Arrangements* – this is a practical schedule of the type of policy applied. An example follows:

   – safety training,

   – safe working systems,

   – noise control,

   – accident reporting,

   – fire safety prevention,

   – good site housekeeping,

   – environmental control,

   – safe working environment.

Additional information, can be quickly obtained by logging onto the following web site: www.healthandsafety.co.uk/SafPolnfo.html

TARA ELECTRICAL

Site induction at: *Whiteparish ghb*
Date: *16th October*
Trade: *Electrician (JIB approved)*
Company name: *Fire alarms UK plc*

Figure 1.19 – *After a site induction session you will be given an 'induction sticker'*

Operator safety requirements

Personal protective equipment (PPE)

This is legally defined as all types of equipment, which includes clothing, designed to protect you against all weathers and conditions. PPE is only intended to be worn by people at work which will protect against one or more risks to their health or safety. Examples are:

- Hard hat.

- Gloves – Figure 1.20 shows a pair of electrician's gloves.
- Eye protection – see Figure 1.21.
- High visibility clothing.
- Safety footwear – Figure 1.22 illustrates this.
- Safety harness for working in dangerous locations – Figure 1.23 illustrates this.
- Water/weatherproof clothing – see Figure 1.24.

Figure 1.22 – *Only wear safety footwear under site conditions*

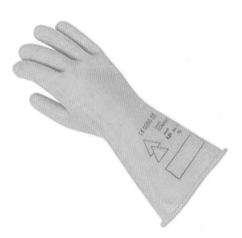

Figure 1.20 – *An electrician's pair of safety gloves are often made from latex*

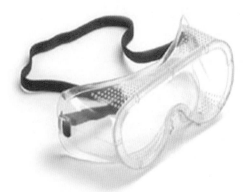

Figure 1.21 – *Wear the correct eye protection when required*

Figure 1.23 – *Wear safety harnesses for hazardous high work*

Figure 1.24 – *Site weatherproof clothing*

The requirements of these regulations

Briefly, they are as follows.

1 It must be appropriate for what it is intended for, the risks involved and the conditions within the place you are working.

2 It has got to adequately control the risks involved and not add to them!

3 Your PPE must be completely adjustable – for example, you must be able to wear your hard hat comfortably without it falling below your ears.

4 Your state of health must be taken into consideration – it wouldn't be wise to wear an insulated jacket if you were running a temperature.

5 You must be able to carry out your job satisfactorily without discomfort and be able to clearly see and communicate with others.

6 If you are wearing more than one PPE item they must all be well-matched and suited for use with each other.

7 Common sense maintenance is required – store in a dry, cool place and, if damaged, replace or get professionally repaired.

Respiratory protective equipment (RPE)

RPE is no substitute for the proper control of industrial dust and fumes which enter our workplace. Many RPE products will not provide protection against all gases and toxic vapours so you must be certain that the product you have selected is suitable for the job you are doing.

The general safety requirements for respiratory protective equipment are as follows:

- Your equipment must be suitable for the purpose for which it is used.
- RPE products must fit the wearer properly.
- It must be used properly in accordance with the manufacturer's instructions.
- If your respirator filter is not the disposable type, it must be cleaned using warm water and disinfected daily.
- All RPE equipment must be 'CE' marked for your safety.
- Your protective equipment must have a good seal – no leakage must occur.

Never try to modify your equipment for it could prove to do more harm than good and never wear protective equipment in a warm confined space where the oxygen level could be reduced.

Secure areas

Secure areas can be created under construction site conditions with the use of modular barriers. These are designed to positively mesh together providing security within. There are many different types of barrier for all different industrial purposes; Figure 1.25 provides an example. In secure areas or hazardous places notices must accompany any site erected barrier as illustrated in Figure 1.26.

Barrier requirements are:

- That they are strong and durable and do as intended.
- They are styled so as to provide visual alerts to site operatives.

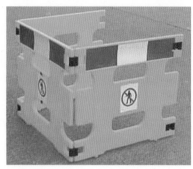

Figure 1.25 – *Modular barriers are useful to cordon off working areas from others*

Figure 1.26 – *Hazardous area warning notice*

Barriers can be constructed to seal off prohibited areas of sites from unwanted intervention or meddling from other trades or members or the public. The more hazardous your task, the heavier your barrier system will be.

Carrying out policies, practices and procedures

Code of practice

A code of practice can originate from a legal document, such as for example *The Electrical Safety Act* of 2002. It provides a great deal of practical common sense and advice to manage electrical risk in your workplace. The 2002 Act requires us to practice our skill safely and to ensure that all electrical equipment and cabling installed is done in accordance with the manufacturer's instructions and BS 7671.

The Act places a responsibility on your employer to guarantee that work is carried out to the highest standards and that all electrical equipment is safe.

What you are obliged to do

Please browse through the following check list; responsibilities fall both on you and your employer.

- Your employer is accountable for your personal safety whilst at work.
- A duty is placed on the manufacturer, supplier and designer of an installation.
- You, as an electrical worker, must make sure your installation is safe, that it has been tested and that you have followed the spirit of BS 7671.
- You have an obligation to carry out repairs and remedial work safely and that the equipment you use is also safe.
- If you are the job holder you are legally responsible for the safety of your installation.
- You must respect anything which is provided for your safety by your employer and not place anyone at risk whilst carrying out your duties within the workplace.
- Your employer is duty bound to instruct you in the use of personal protective equipment.
- Risk management must be used but please remember hazards and risks are *not* the same thing:
  - hazard, a potential to cause harm,
  - risk, the likelihood that death, injury or illness might arise.
- There are three ways in which to meet your commitments:
  - by statutory (legal) regulations,
  - ministerial notes and memos,
  - codes of practice.

Safe working practices and procedures

Safe working practices and procedure are prepared to assist you to work safely under harsh site and workplace conditions. Hopefully, this guidance will help prevent accidents and injuries from occurring and will make your time at work a lot safer.

Please take time to look through the following guidelines:

1 *Communications* – be clear and precise between yourselves, your client and other trades working alongside of you.

2 *Risk assessment procedures* – consider all the work to be done and calculate the risks involved.

3 *Method statement* – how are you to carry out your work safely and show how identified risks can be controlled.

4 *When work starts, have*:
 – a safety survey,
 – an adequately trained workforce,
 – a workforce competent to carry out the task in hand.

5 *Dangerous/forbidden areas* – these you will be advised about. Notices and warning signs should be fitted.

6 *Scaffolding requirements* – this, other than modular (DIY) scaffolding, has to be carried out by a specialised company.

7 *Confined spaces* – care must be taken. It is wise to work in pairs for your own safety.

8 *Personal protective equipment and clothing* – wear these as necessary and when required.

9 *The control of pollution* – this is your responsibility and your client may require risk assessment details from you. If you find asbestos, identify the pollution with warning notices, see Figure 1.27, and ask a specialised company to deal with it.

10 *Road safety* – a site vehicle park must be made available as there is no safe place on site for cars and vans. Remember that pedestrian safe walkways are also needed, especially within the boundaries of the site.

11 *Emergency procedures must be identified and tested*:
 – alarm points,
 – assembly points (after a building has been vacated),
 – fire extinguishers, Figure 1.28 illustrates this,
 – notices, Figure 1.29 illustrates this,
 – record and file as necessary.

Figure 1.27 – *Asbestos alert warning sign*

Figure 1.28 – *Typical site fire extinguisher*

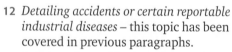

Figure 1.29 – *Fire safety notices*

Figure 1.30 – *Hazardous and harmful material signs*

12 *Detailing accidents or certain reportable industrial diseases* – this topic has been covered in previous paragraphs.

13 *Safety and warning signs* – use and position where they will prove to have the greatest effect on the workforce.

14 *Hazardous and harmful substances and materials* – see Figure 1.30:
   – Use in accordance with the manufacture's instructions and guidelines.
   – Your client may ask you for a risk assessment – this you will have to provide.
   – Keep an inventory of your stock.
   – Maintain records and data concerning issued material, dates and to whom issued.
   – Keep all harmful and hazardous materials under lock and key.

15 *Concerning your welfare:*
   – office,
   – the use of a telephone,
   – male and female toilet blocks,
   – canteen,
   – drying room/cloak room.

Safety policies

Let us imagine we have to produce a concise safety policy for your college or learning centre. The contents could be similar to the following snap-shot view which follows.

■ This concise safety policy statement will summarise the schedules which are to be in force throughout your college. They have been written to ensure you have a safe and healthy working environment. This will apply to students, staff and contractors alike. This safety policy document will be available to everyone in some shape or form.

■ Heads of departments or people who are responsible for the practical safety of a department must produce a formal written *risk assessment* as to where these risks have been identified, for example, the boiler house or an electrical laboratory, etc.

■ The operation of this policy will cover several legal documents (*The Health and Safety at Work Act*, various Codes of Practice and so on) and is required to be monitored.

- College *high risk areas*, where for example radiation is used or welding is carried out, must produce their own departmental safety handbooks. These are usually drawn up by your sectional safety office (if your college is large enough to have one) and updated annually. This will provide general health and safety procedures and arrangements for dealing with emergencies, etc.

- Each college *high risk areas* must be identified and made known. Do this by publicising, writing articles within your college in-house magazine, the use of clear notices, e.g. Figure 1.31, posters and handouts.

- Your college safety policy will also include the following issues which are reviewed in full within the pages of the policy document. This is particularly important to new students or members of staff:
  - on the job training in health and safety matters will be discussed,
  - basic skills training (how best to operate fire extinguishers and other emergency equipment),

  - emergency procedures adopted by the college,
  - health and safety concerns,
  - basic first aid treatment in case of an emergency,
  - internal/external emergency telephone numbers,
  - fire condition and evacuation procedures.

- Your college safety policy will also outline the need to give particular attention to young people and those with special needs. These will include students in wheel chairs, diabetics, and those who show a mild degree of dyslexia, etc, or are slightly handicapped in any way.

- Safety training in practical terms is usually the responsibility of the department head although larger departments will normally have their own safety committees.

- Often a college safety policy will dictate that safety training should start during the first few weeks of the academic year and frequently will form part of your course.

Figure 1.31 – *High risk area sign*

Safety signs

Warning signs

These signs, mainly triangular in shape, are colour-coded bright yellow supporting a black border and symbol. They are placed in dangerous or hazardous places to instruct, advise and alert people of perils ahead. Figure 1.32 illustrates a selection.

Prohibition signs

These are usually oblong in some shape or form. The message part of the sign is always presented as a four-sided red backdrop figure with white lettering. The pictorial logo is constantly black on a white background. Prohibition signs advise staff, workers and visitors of what is not allowed. Figure 1.33 provides a selection for you to browse over.

Figure 1.32 – *Typical site warning signs*

Figure 1.33 – *Prohibition signs*

Mandatory signs

These signs are usually rectangular in shape, although at times you may come across square ones. They have blue and white symbols and backgrounds with white text. This family of signs is to warn and inform workers and visitors alike of actions which must be carried out to provide a safer workplace environment. Figure 1.34 provides a selection of examples.

Fire equipment signs

These can be in all shapes and sizes. Fire signs require a white border around the edge of the safety sign. These signs have a red backdrop with white text and symbols and are generally positioned near the location of fire fighting equipment. Sometimes arrowed signs are used to point the way to fire fighting equipment – Figure 1.35 illustrates this.

Figure 1.34 – *Mandatory signs*

Roles, responsibilities and powers

Safety officer

A safety officer can either be a pain in the neck or add a positive safety contribution to the workplace. A large electrical or construction company will appoint a *qualified safety officer* to monitor matters of safety within the workplace.

Figure 1.35 – *Fire equipment signs*

You might well have one attached to your own company or technical college.

A safety officer holds down a very important and responsible job – they work hard to protect and promote the health and safety of the workforce. They are there to give advice, information and guidance.

The role they play is very varied, but in general terms, please consider the following:

- To ensure that the workplace and the people who work there have been risk assessed.
- To carry out a programme of regular safety inspections either by request or as a spot check.
- To ensure there is a system in place for the continual monitoring of company safety matters.
- To make, receive and read reports and comment accordingly.
- To monitor the company's safety policy between major inspections.
- To make sure that recommended remedial work is carried out to an acceptable safety standard.

The duties and responsibilities of a safety officer

As you can imagine there are many – so please browse through the following simplified checklist and feel free to ask your course tutor should anything puzzle you.

1 To advise on matters of health and safety to individuals and management.
2 To maintain records of accidents within the workplace, reports and investigations for a minimum of seven years.
3 To carry out yearly reviews and appraisals on all matters of health and safety.
4 To produce risk assessments concerning specified site work.
5 To check that all electrical equipment is portable appliance tested and documented regularly.
6 To see that new staff are adequately trained in health and safety aspects.

The powers of a safety officer

A health and safety officer can enter your workplace at any time during the working day or after a major accident caused by work time tasks.

After an accident he/she will need to investigate the possible cause of the accident and to establish whether a violation of the Health and Safety Act has been committed. Advice as to how to avoid a similar incident will also be given.

Other powers of the safety officer include the following:

- Taking measurements, samples and photographs if required.
- Removing articles, dismantling equipment and arranging for the equipment to be tested.
- Taking away anything which could be harmful, if left at the scene of an accident.
- Interviewing people who might prove helpful in establishing the cause of the accident.
- Checking, looking through and photocopying relevant documents.

If the safety office seriously considers you have broken the law other measures can be brought into force. These briefly include the following:

- He/she can issue an *informal warning* (this amounts to a slap on the wrists). This will be delivered either by word of mouth or, in more serious cases, it will be in writing.
- An *improvement notice* can be issued to the person responsible.
- More seriously, a *prohibition notice* can be given to the proprietor. This will mean total or partial close down of all commercial or industrial activities.
- Lastly, the offending company or individual can be prosecuted under the Health and Safety at Work Act.

Safety representatives

Unlike a safety officer, a *safety representative* has no legal duties whatsoever, other than those of an employee. He/she is often appointed by a trade union movement which is recognised within your workplace.

The only legal rights they have are within the workplace and are mainly confined to the following roles:

1 They can represent employees in discussions with the employer on welfare, health and safety matters.
2 Investigate hazardous or dangerous incidents within their place of work.
3 Probe into complaints.

4 Attend safety committees, on or off their premises.

5 Carry out workplace inspections and look at documents applicable to health, safety and welfare.

6 Be paid for time spent on tuition and functions, etc.

Health and safety executive inspectors

The role of a health and safety inspector amounts to the following:

- The enforcement of rules and regulations which relate to health and safety maters.
- To act as a mediator with trade union representatives.
- To monitor safety performance in the workplace.
- To monitor the standard of newly acquired machinery.
- To advise employers and employees on health and safety matters.
- To keep a professional eye on potential risk areas.
- To investigate and gather information concerning accidents and dangerous events.
- To encourage health and safety activities.
- To talk to employers' organisations regarding workplace health and safety topics.
- The health and safety inspector can conduct legal proceedings against a company or individual who is in breach of the Health and Safety Act.
- The inspector is able to liaise freely with the European Community in respect to all health and safety directives.

The powers of a health and safety executive inspector

Section 20 of the Health and Safety at Work Act gives an inspector far-reaching powers of enforcement. Briefly, these are:

- Right of entry (day or night), and accompanied by a policeman or any person who is approved by the HSE if it is felt necessary.
- Equipment, for testing purposes, can be taken into premises and used accordingly.
- The inspector can ask for certain parts of a building to be left undisturbed, for practical, or any other, reasons.
- He/she has a right to take photographs, video footage and make sound recordings as required.
- Section 20 of the *HASAWA* allows an inspector to acquire samples of articles or substances.
- He/she is able to order both plant and equipment to be dismantled or destroyed if they are an immediate danger to others.
- Relevant documents and books can be photocopied.
- He/she can conduct interviews with individuals and request a written statement from them.
- A health and safety inspector can start legal proceedings against a company or individual.

Environmental health officer (EHO)

An EHO does a fine job in protecting the general public and construction site workers alike from environmental hazards.

Construction sites are, at times, problem areas from smoke, fumes, dust and general industrial noise or from radios being played extra loud. An EHO will look into any objection made by a member of the public concerning this type of activity. Advice is offered on how to carry out the programme of work without causing a problem to others.

Environmental health officers have a wide range of duties, much of which extends beyond policing construction sites and industrial premises. These duties are linked to housing problems, *HASAW* and environmental protection.

Other duties include:

- Hygiene and safety issues.
- Food poisoning, both at home and with the workplace.
- Infectious diseases within the workplace.
- Accident investigation.
- Keeping records and writing reports.

The powers of an environmental health officer

Their powers are far reaching and include the following selection:

- Inspection of construction sites.

- A full investigation into assumed misconduct within a site environment.
- To issue warning notices on offending companies and individuals.
- To take legal proceedings.

Accessing advice and literature on health and safety matters

There are several places where you can obtain knowledge and advice on maters of health and safety:

- Your college library.
- Any public library (for this you will have to pay a small fee for research, photocopying and handling).
- Your college (or work) health and safety officer.
- The internet. The following two sites will be helpful to you: www.Incs.ac.uk/depts/safety/policy/appexdix.htm www.Iancs.ac.uk/depts/safety/policy/appendix.htm

Causes of accidents

Studies have shown that the most common causes of an accident are:

1 carelessness,
2 lack of knowledge,
3 human limitations,
4 fatigue and listlessness,
5 horse play,
6 drug taking and drinking alcohol,
7 faulty or unguarded machinery,
8 a badly ventilated workplace,
9 a dirty or overcrowded workplace,
10 a poorly-lit workplace.

One of the major causes of accidents is carelessness. It can stem from an innocent act of day-dreaming or being preoccupied with personal concerns through to more serious problems such as tiredness, the 'morning after the night before' feeling, or just assuming an off-hand and casual attitude to the work in hand because the boss has upset you.

Accidents can also be caused by a lack of experience or technical knowledge and often happens when you are told to carry out a task for which you are not qualified to do or don't know much about.

Strains and muscular disorders are also responsible for many missing man-hours in the electrical industry – time that need not have been lost if we had been more sensible.

Long hours, working at weekends and insufficient sleep are the major factors which are responsible for weariness and listlessness. Try to achieve a sensible balance between work and play. This will help to reduce day-time tiredness and stress levels which often accompany an extended overtime policy.

Don't be tempted when bored to engage in high jinks with workmates or apprentices, such behaviour may result in accidents, serious or otherwise.

It is often said that a tidy site is a safe site. It's surprising what you can trip over within your workplace. Please remove all your off-cuts of cable and material wrappings from the floor and place them in a suitable container for collection at a later date. Keep passageways and staircases free from material equipment and large tools, just in case there is an emergency and people are panicking to get out of the building.

Overalls and industrial work wear not only shield your everyday working cloths from unnecessary wear and site contamination but will provide a reasonable degree of protection from many hazardous working locations we find ourselves in from time to time. Wear also the correct protective clothing for the work you are doing.

Preventing accidents

Preventing accidents on site really amounts to applying good common sense. Here are a few ideas.

- Place a guard or barrier around the hazard – for example, an open lift shaft or moving machinery.

- Replace the hazard with something less dangerous – for example, an open fronted floodlight with a high-impact glass/pvc protected floodlight, see Figure 1.36, or by changing the lighting arrangement to a 110 volt fluorescent fitting with a diffuser fitted.

Figure 1.36 – *High impact, glass protected flood light*

■ Do away with the hazard all together – for example, equipment and material items left within a passageway leading to an exit.

■ Wear personal protective equipment – a high visibility waistcoat for example when working in areas exposed to traffic, or a harness, when working at heights.

■ Safety induction courses, education and site publicity, etc.

Reporting accidents

Before the *Data Protection Act of 1998* was put into operation it was commonplace to have an open site-accident book which anyone could browse to their heart's content. Now you are given a blank *Accident Report Form* which you are asked to fill in.

Once completed, the form is returned to your site manager. They will then file it and lock it away in a secure cabinet.

This procedure is always carried out after you have had your injuries attended to and the area where your accident happened has been made completely safe.

Accident books vary as to the amount of data you will need to supply but to give you an idea of what is expected please browse through the

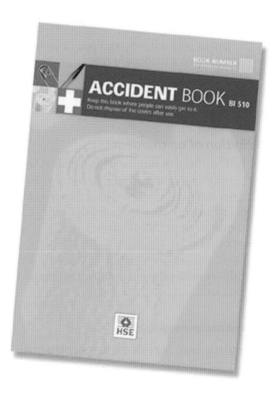

Figure 1.37 – *Typical accident book binder*

following list. Figure 1.37 illustrates a typical accident book binder.

1 Summary of the accident and where the accident happened.

2 The full name and trade of injured party – possibly your age and the name of your company will also be asked for.

3 A brief description of events leading up to the accident.

4 Details of witnesses involved.

5 Details of injury or loss suffered.

6 Conclusions drawn.

7 Managerial recommendations and suggestions.

8 Supporting material could include the following:
   – photographs – this could be done using your mobile
   – video – ideal if you have a camera handy or when the accident is serious enough to warrant it,

– diagrams and measurements of the incident – this is a good option as most of us have a pen or pencil handy and you can always tear open a cardboard box for writing material.

9 The date and time of your mishap.

10 Your signature.

Definition of an accident

The HSE's official definition of an accident is:

'. . . any unplanned event that resulted in injury or ill health of people or damage or loss to property, plant, materials or the environment or loss of business opportunity.'

Accidents at work don't just involve people – you must also consider the environment, material and business losses.

Carrying out a risk assessment

A risk assessment is nothing to worry about; it's just a complicated term we use to describe an in-depth examination of what, within your workplace, could cause you harm. Real life examples include the following:

■ Disused, uncovered wells.

■ Rats – these carry the *Hanta virus*, a deadly disease known as *HPS* which is transmitted to us through their droppings, urine or saliva. If you have rats – please get professional help.

■ Rotten or missing timbers.

■ Unsafe walls and ceilings.

■ Unlit and potential hazardous areas.

A site risk assessment is to ensure that nobody working on site will get hurt and is a legal requirement you must carry out before any work is started on site. Your hardest task will be to decide whether the hazard you have picked is important enough to include in your report.

The risks within your workplace can be measured by following these five simple steps.

The 1st step – Spend time having a good look around where you are to work and identify any potential hazards.

The 2nd step – Decide which of your staff might be harmed and why.

The 3rd step – Calculate all the risks involved and then decide whether more needs to be done to ensure all-round safety.

The 4th step – Make an entry in your site diary what you have discovered. This might be required in the future.

The 5th step – Check out your assessment of the risks involved but don't over-complicate things and never be afraid to revise if necessary.

Remember, other trades and members of the public who work alongside of you could be hurt by your activities.

Preparing a report

This is a brief but imaginary report concerning an ex-factory which is to be redeveloped into a commercial warehouse.

Tara Electrical Services plc
Whiteparish
Wiltshire – UK
Risk assessment – The old body spray factory

Survey and assessment carried out by Kris Zimmermann 27th May:

1 Most floors are slippery (oil/grease) or uneven.

2 Several floorboards are rotten in the old office area by the main door.

3 There are oil spillages throughout the boiler house.

4 There is waste material, boxes and old documents, etc, in every area of the old factory.

5 There are three 50 litre containers of inflammable liquid (paraffin) in the boiler house store room.

6 Some machine guards are missing from movable parts.

7 Several electrical circuits are physically damaged – TPN circuits 3, 8, 10, 11 and 12. Many industrial fluorescent fittings are hanging from their flexes without mechanical support.

8 Many broken windows – dust and fumes are unpleasant when windy.

9 There are nine containers of paint products stored within the old materials store – some have been opened and the paint has been spilt on the floor.

10 There is just one area within the old factory which can be classed as high (8.2 metre) – this is attached to the old boiler house – access equipment will be required to serve these heights.

Assessment of risks

a) The existing electrical installation will have to be made completely safe and dismantled by Tara Electrical Services plc.

b) Two 10 kVA 110 volt site transformers will be temporarily hard wired to serve the ground and first floor areas. Each site transformer will be served with a TPN supply protected by a 40 amp Type 'C' or 'D' MCB and a 30 milliamp RCD.

c) Access will be made by modular aluminium scaffolding.

d) The remainder of the existing factory has far too many hazards to enable work to progress safely. Wilts Internal Demolition Company will be asked to submit a price for carrying out this work to bring the proposed workplace to an acceptable safety standard.

Report compiled by Kris Zimmermann – 27 May

Reporting procedures – hazards

To your supervisor

Just informally tell your supervisor of a site or workplace hazard you are concerned about. It is a good idea to cover yourself by jotting down the main core of your conversation in your personal day diary. If your request for help or guidance is ignored or forgotten, then formally draft your unease on paper and keep a copy for yourself, but always remember to date your correspondence – Figure 1.38 illustrates this.

Safety officer or representative

If you think your boss is exposing you to unnecessary risks, for example, not providing the correct personal protective equipment whilst working at heights (see Figure 1.39) or near to water, you can contact your safety representative to arrange a meeting. He/she is obliged to listen to your concerns concerning matters of health and safety. If the matter is still

INTERNAL MEMO - TARA ELECTRICAL COMPANY   Number...00468

Date: *17th January*
From: *Dean Linkardson*
To: *H. Truden Area 2 - Supervisor*
Re: Site: *Whiteparish MgbH - Phase 1*
Subject matter: *Health and safety - site hazard*

The deep hole near to the electrical mains still hasn't been made safe.
Storm water has leaked through a hole in the factory roof and has has frozen around the hole in the floor.
This problem has been reported twice before in conversation with my supervisor.

D. Linkardson

Figure 1.38 – *Hazard reporting – a memo to your supervisor*

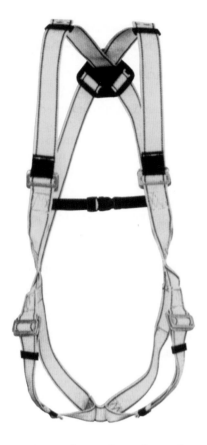

Figure 1.39 – *Personal protective equipment for working at heights*

not resolved, you can get in touch with the HSE who will deal with your concern sympathetically and in strict confidence. The UK HSE Info-line is on 08701 545500.

Fire officer

A hazard can be described as anything which could cause an accident or create danger – flammable liquid stored within a wooden site store where fire has broken out in an adjacent building, for example. Site and workplace hazards can be prevented from turning into accidents by reporting relative information and data to an appropriate fire officer attending the incident. This is best done verbally – but be precise.

Works rescue team

This is very much a 'now' situation. If you can provide any facts which would help your works search and rescue team, especially where people are concerned, you must offer this information to the *team leader*. He/she will always be dressed differently to the other crew members and will probably have their team position printed on the back of their PPE garment.

Environmental management systems – BSEN ISO 14001

Briefly, to carry out the wishes of BSEN ISO 14001 an environmental management system (EMS) must be formed. This is to provide a framework to control the impact a company's commercial and industrial activities could have on the environment. This code of practice does not in itself state any particular work related standards and conditions but it does seek to ensure that an organisation will continue to improve its environmental performance.

Here is a selection of working activities which need to be controlled within an electrical installation engineering company:

- The emission of dust and fumes.
- The control of industrial site/workplace noise levels.
- The burning of rubbish and cable installation.
- Spillage of harmful substances – transformer oil, PVC adhesive, etc.
- The release of contaminates and irritants by means of the public sewage system.
- The dumping of electrical rubbish and unwanted site materials.
- The disposal of toxic material such as battery acid and disused radioactive smoke detectors. etc.
- The masking of arc-weld emissions – ultraviolet light is dangerous.

Acts and regulations briefly explained

The Environmental Protection Act of 1990

The aim of this Act of Parliament is to control pollution and the knock-on effect it would have on the environment. It covers many of our shared industrial interests including the following:

- Air quality – good clean air, free from industrial dust and fumes, etc.
- Construction site noise levels – they must never become a nuisance to the general public.
- Vehicle and industrial plant emissions – this refers to company trucks, vans, motorised cutting equipment and site generators, for example.
- Water pollution in the forms of toxic or harmful waste disposed of by means of the site toilet or just thrown on the ground.

Pollution Prevention and Control Act of 1999

This Act requires companies to apply for a permit to carry out their legally recognised work. Only certain polluting industries are affected. To summarise, they are:

- The paint spraying industry (cars, vans and trucks and so forth).
- Metal finishing industries which use solvents.
- Large scale industrial cleaning companies.
- Companies which treat wood with chemicals as a preservative.
- Processes which create and liberate vast amounts of dust and other fine particulates into the atmosphere.

The Clean Air Act of 1993

Cable burning is an offence unless authorised, which will upset many an electrician! Under this Act it is a crime to cause or permit the release of *dark smoke* from a building site or an industrial trade premise. Local authorities are able to turn any part of the district they serve into a *smoke controlled area*. You may also be surprised to learn that owners of some tall furnace chimneys need to seek approval to operate.

Radioactive Substances Act of 1993

Most of us use or have handled equipment which incorporates a radioactive material – the common ionisation smoke detector, for example. All radioactive substances are the responsibility of the Environment Agency. Manufacturers within the UK who make radioactive devices are required to register with the Environment Agency for inclusion on a public register.

Controlled Waste Regulations of 1998

This Act of Parliament stops us from disposing of waste in a way which is convenient for us. It was drafted to regulate the dumping of all kinds of waste from old fridges to the type of trash we accumulate at the end of our job. This law prohibits waste emissions to the atmosphere, the ground and all waterways.

A special permit will be required if you want to discharge industrial waste such as oil, etc. – this is obtainable from your local water provider.

If you store oil on site to run machinery you will require a double-skinned storage vessel and a secondary containment system such as a drip tray to catch any oil spillage.

Dangerous Substances and Preparations and Chemical Regulations of 2000

This Act of Parliament is about restricting certain dangerous substances and preparations used in industry. It prohibits the supply to the general public of substances listed in Schedule 2 of these regulations – some examples are:

- chloroform – a colourless volatile, sweet smelling liquid,
- carbon tetrachloride – a colourless, non-flammable liquid used as a solvent.

The two listed are just a passing snap-shot; there are many others to choose from! Certain products marketed must have a warning label saying 'Restricted to professional users'. These are often cancer-producing chemicals and substances which can cause deformities in humans.

Factors within our industry causing change

Changing demand

Our industry is constantly adjusting to the demands of others and policies made by government. New national housing targets require additional skills to meet the demands made by our national leaders. Recession has the opposite effect and many of us are plunged out of work overnight, finding it difficult to find the most humble of jobs.

An increased scale of competition

Here are a few suggestions to sustain adequate business arrangements.

- By the use of promotions – for example, half-price PAT testing, a free extended warranty on all work carried out, free visual checks on older properties, etc.
- By increasing office and workplace efficiency and reducing price levels on work set aside for tendering.
- By price, by negotiated settlement with your client.
- By promoting attractive package deals with your potential client.

In competitive terms

1 *Marketing* – Advertise well and try to make an impression. A five minute promotional slot on your local television channel as a news item is worth a great deal of advertising space. Consider changing the design of your company vehicles to be more up to date

2 *Productivity* – Provide a bonus scheme, offer a price for certain jobs to be carried out. Start a profit sharing scheme. Give a small percentage of profit back to an operative who brings in additional work for your company.

3 *Quality* – Buy-in middle priced goods you can rely on. Low-priced electrical accessories may seem attractive but when they break down before your warranty runs out and you have to travel a couple of hours to site to replace them, it really does reduce your expected profit margin. Regular inspection by management is essential to ensure that the installation is being carried out as intended. The site electrical foreman must be made more responsible.

When new technologies are introduced

This can be quite distressing if you are in your 40s and you have set your mind to thinking you have learnt all there is to known about the work you do.

Don't worry! Broad based skills will still be required within the workplace but to stay afloat you will have to retrain and update a few times. Do this, either on the job, guided by someone who has had previous experience, or book-in a return visit to your old college for a short course.

As your work becomes more and more technically orientated, new skills such as electronics will have to be studied. There will be a positive demand for inter-related electrical skills within the workplace.

We could experience both changeable and flexible working arrangements on some high profile jobs – shift work, flexi-hours, night shifts and split shifts, for example. Be prepared in the future, especially if you work for a large organisation, who only knows you by a payroll number, to expect far more frequent job changes.

My advice to you is to work hard both practically and academically; and with the awards you achieve, develop new and better career patterns for the future.

Meeting your customer's expectations

Is the job you have done fit for the purpose intended? Your professional livelihood really hinges on this and whether you have met your customer's expectations. If you fail you would have lost a most valuable customer – a customer who could recommend you to others.

Your performance and workplace attitudes will also have been measured against those who have come before you. If you have used your client's tea, coffee and cookies without consent it will not have created a good impression – nor will having to return on numerous occasions to put right things which were found to be wrong. This too, will be remembered for quite a long time to come.

Keeping the cost down before you start work is equally important. Unless specified, buy-in middle priced wiring accessories. Never go down-market – it will do you far more harm than good!

On a small job, an extension to a house, for example, it is best to offer your customer a price. By doing this he will not have any nasty shocks when you present him with the bill. If the work you are carrying out is based on time and material, try to keep an eye on costs and the travelling time involved to your job for whatever you charge is bound to be too much as far as your customer is concerned!

Your agreed completion dates must be kept. Your customer will be relying on you, especially if what you have done has to be opened for business on time. Reliability is another key factor for the continuation of your work and to promote a good trading image for your business.

The legal obligations of a sales contract

A sales contract is a legally binding written agreement between two people.

Take time to draft a really good sales contract between you and your customer; use your own headed sales contract form, see Figure 1.40 for an example, and tailor-make its legal contents to your proposed new job, making sure that the contract is more favourable to you!

TARA ELECTRICAL COMPANY, WHITEPARISH SQ1 1PQ, ENGLAND

Equipment Sales Agreement

Contract Number: ADG 0985   **Date: 25th December**

1. This Sales Agreement is entered into by and between *Klein College* **and** *Tara Electrical Company, Whiteparish SQ1 1PQ England*

2. The product: *To supply and replace two hundred and fifty six ionisation smoke detectors, as SE1749 and sixteen heat detectors as HD1748 within the college between 1st January and 21st February*

3. The cost: *Five thousand and forty eight pound, thirty two pence plus VAT (£5,048.32 + VAT)*

4. The time for payment for the equipment is as follows:

    The first instalment: *30th January*

    The second instalment: *28th February*

    The final instalment: *30th March*

5. Commission charge: *Eight percent of net price*

6. Warranties: *Two years from completion*

Disclaimer: *NO OTHER WARRANTIES EXIST*

Signed on behalf of: *Tara Electrical Company*   **Date:**

....

Signed on behalf of: *Klein College*   **Date:**

Figure 1.40 – *A typical sales contract*

What are the contents of a sales agreement?

To provide a snap-shot view, please consider the following points:

- Correctly state the price of the contract – remember to spell out any discounts or additional charges (for example, commissioning charges by others).

- State the start and finishing dates of the work to be carried out.

- Indicate whether there will be any price adjustments during the course of the contract – important when your contract is long term.

- Payment and credit terms – how and when payment will be due or whether payment will be based on productivity.

- Warranties – the length of time your warranty will last after the work is completed.

- Disclaimers – which other warranties exist and so forth. This is best spelt out in BOLD TYPE.

- Liability limitations – this can mean, for example, the maximum amount you are liable for and limited to the purchase price of any particular item or piece of electrical equipment that has failed. It can also assume that your company is in no way responsible for loss of your customer's profits whilst work is carried out.

A sales contract is a legal obligation which binds both you and your customer to the letter of the contract. Should any party pull out or default from this agreement, legal proceedings can be started against the offender.

Handy hints

- If you have to document any important information always remember to keep a copy filed safely away.

- The following regulations apply to modular aluminium scaffolding systems:
  - The Construction (Working Places) Regulations 1966,
  - The Health and Safety at Work Act 1974.

- Some religious orders forbid the use of radios, televisions and mobile phones on site. Jewish customers would not allow food to be consumed within the workplace – please respect their rules.

- There is no formal time-span for testing portable appliances but it is worth checking them out regularly under site conditions.

- Remove the plaster off the insulation serving your conductors – it will help to provide you with good insulation test values.

- Floor joists are best drilled when the floor above hasn't been laid. If practical, keep your holes horizontal to the floor below – it's easier to wire this way. Other joists are fitted with weakened knockouts.

- Cables serving future outside lighting fittings should have their conductors separated from each other and their ends well insulated to avoid problems occurring before the light is fitted.

- Tungsten halogen lamps will eventually blacken if a reduced voltage is applied by means of a dimmer switch.

- The efficiency of a hot water immersion heater may be taken as: (a) unlagged: from 77–85%, (b) insulation fabric lagged: from 85–95%, (c) factory moulded lagged: from 85–98%.

- The *efficacy* of a lamp is the ratio of light output in lumens to the lamp power in watts.

- Fluorescent tubes can be aligned with precision by positioning the small indents formed in the aluminium end cap to face the floor.

- A high frequency fluorescent fitting can trigger a residual current device into fault condition. This type of fitting has a suppressor connected between the phase, neutral and earthing terminals.

- Keep telephone and data cables well away from low voltage mains installations. Never mix audio

Handy hints

- cables with telephone cables if sound induction is to be avoided.
- Take care when cutting small solid conductors with sharp wire cutters. Target the waste copper to the ground as damage to the eye could result from sloppy cutting.

- Digital meters are easy to read but take care to read correctly. Observe the position of the decimal place.

Summary so far . . .

- *All personal protective clothing must be supplied free of charge by your employer*
- *Your employer is legally required to control dust and other hazardous materials within the workplace.*
- *Reportable accidents include the following: death, fractured skull, loss of sight, an amputation or any injury involving more than three days off from work. This is not a complete list.*
- *Provide a safe site working environment – illuminate passageways, stairs and escape routes well.*
- *An employer is legally bound to instruct, train and supervise his employees.*
- *Respiratory protective equipment (RPE) provides protection against vapours, gases and toxic fumes but they must fit the wearer properly as intended.*
- *Site information signs: warning (yellow), prohibition (red), mandatory (blue) and fire equipment (red with white logos and text).*
- *Safe working practices and procedures are for your safety. They assist you to carry out your task securely under harsh site or workplace conditions.*
- *A safety policy simply lays out a series of safety schedules which are to be in force throughout your workplace.*
- *A safety representative has no legal duties other than those of an employee. He/she is appointed to represent the employees.*
- *An environmental health officer's (EHO) job is to protect the general public and workplace operatives from environmental hazards.*

- *Places where you can obtain both knowledge and advice on matters of health and safety are as follows:*
  - *large public libraries,*
  - *your own college library,*
  - *site or works safety officer,*
  - *the internet.*
- *Carelessness, fatigue and lack of knowledge are three major causes of accidents in the workplace.*
- *Accidents within the workplace don't just involve people – you must also consider the environment, material and business losses.*
- *A formal risk assessment is an in-depth examination of what, in your workplace, could cause you harm.*
- *Should you have to report a hazard to your supervisor, remember to record written details within your day diary – it might prove useful to you in the future.*
- *An environmental management system provides a framework to control the impact your company's commercial and industrial activities could have on the environment.*
- *The Controlled Waste Regulations of 1998 stop us from disposing of waste in a way which could harm the environment.*
- *When new technologies are introduced within the workplace, take time to retrain. Do this by means of 'on the job experience' guided by a knowledgeable colleague or attend a short course at your local college.*
- *A sales contract is legally binding. If any party pulls out or defaults from this agreement, legal proceedings can be started against the offender.*

Review questions

1. To whom would you speak if you felt that a health and safety problem existed within your workplace?

2. What site facilities is your boss responsible for?

3. Site portable appliance testing must be carried out:
   (a) every three months,
   (b) once every year,
   (c) there is no required length of time between tests,
   (d) tests are carried out every other month

4. List two reportable diseases which arise or are connected with work.

5. Describe a safe method of handling and transporting material items

6. When, by law, is a company required to provide a safety policy statement?

7. Briefly describe a site induction course.

8. Briefly describe a safety policy statement.

9. Describe the principal role of an environmental health officer.

10. Provide two ideas of how we can prevent accidents occurring at work.

11. If you store oil on site to run a machine, what type of storage vessel do you require?

12. The need to advise site visitors and members of the public of dangerous activities is best made known by:
    (a) by using a yellow warning sign advising of danger,
    (b) by displaying a red 'Danger men at work' sign,
    (c) by having someone to advise of the activity in progress,
    (d) by publicising them by means of a 'Hazardous activity' notice board.

13. Briefly describe the use for respiratory protective equipment within the workplace.

14. In your own words, explain the formal definition of an accident.

15. Describe what is meant by a legal sales contract.

Applying safe working methods and emergency procedures

Introduction

Derived from Outcome 2, this chapter forms part of your Unit 1 Element (Application of healthy safety and electrical principles). The aim of this chapter is to encourage you to work safely and efficiently and to understand common applications of health and safety.

The supporting knowledge contained within this chapter comprises the following items:

- Permits to work and where it is considered unadvisable to work alone.
- First aid and ways of preventing injury.
- Actions to take if a workmate has a bad electric shock and precautions against contact with electricity.
- The causes of asphyxiation (preventing oxygen from reaching the lungs and other bodily tissues) and the appropriate emergency action to take if called upon to do something.
- The procedures to be taken in event of a site emergency occurring.
- Methods of fire prevention and means of controlling a fire.
- The physical conditions which cause both combustion and extinction.
- Fire procedures.

For formal details please refer to the CG 2330 Scheme Hand Book.

Practical activities

You will be expected to know and to carry out the following practical activities:

- To recognise probable safety hazards within the workplace.
- Know the most sensible place to keep site first aid equipment.
- Show your assessor you are able to outline emergency and evacuation procedures within the workplace.
- Be capable of identifying likely causes of fire within the workplace.
- To be able to recognise different types of fire extinguisher commonly used on site and match them with different forms of fire – paper, oil, electrical and chemical, etc.

Tests

The supporting knowledge requirements will be assessed by the use of a multiple choice question paper produced by City and Guilds in accordance with the test specifications.

Permit to work/enter at a named location

This is a formal written system used to control certain types of work that could be potentially dangerous. You will find this scheme adopted on a large construction site, for example when working within a nuclear power station or hired to carry out electrical work within a submarine, tunnelling project or hospital. Permit-to-work systems can arise where you least expect them and, in theory, can be used on *any* site where it is felt additional health and safety control is essential. You will find that a *Permit-to-Work* is necessary when the work you have been asked to carry out is only achievable if the usual safeguards are overlooked or dropped altogether. Similarly, you will have to gain written permission if new working hazards have been brought in.

Numerous construction-site related injuries could have been avoided if a *Permit-to-Work* had been issued and lives, undoubtedly, would have been saved.

The system is designed to look after your health, safety and welfare where site working conditions are harsh, dangerous or have an element of risk attached to them and can be issued by a resident engineer. Examples of harsh site working follow.

- Working underground.
- Bridge work.
- Working in an explosive atmosphere (battery charging room, live gas installations, etc).
- High risk areas near to the construction and assembly of heavy infrastructure.
- Work near or within the path of cranes, heavy plant, etc.
- Areas where a potential danger exists.
- Where the normal safeguards have to be dropped in order to carry out the work required.

A *Permit-to-Work* system will ensure that only authorised and properly trained personnel are able to carry out the intended tasks. With a thorough induction as to the nature of the work involved and its location, an operative will feel far more confident about undertaking their task.

Permits will vary from company to company. Some will use specially manufactured forms, whilst others will choose the option of producing their own in-house pro-forma. Styles and formats will obviously differ but you could expect some or all of the following considerations:

- Your name, trade and the company you are working for.
- Where you are allowed to work.
- Identified hazards and risk factors within the work area.
- The precautions you must take.
- Whether you are permitted to use standard power tools.
- Whether your 110 volt power tools have to be intrinsically spark-proof safe.
- The type of personal protective clothing you must wear.
- A list of tools you are *not* permitted to use.
- Your responsibilities whilst within the permit-to-work area.
- The need for fire prevention.
- The type of atmosphere you will be working in (explosive, oppressive, restrictive, hot, cold, etc).
- Whether smoking is allowed.
- Whether you are permitted to work alone.
- A place for your signature.
- A box to tick indicating that you have been induced.

These are just some of many – it all depends on the type of risk factor involved and where you are to carry out your duties. Figure 2.1 illustrates a typical *Permit-to-Work* pro-forma.

Lock-off procedures

Lock-off procedures are essential for your well being and those of others who work alongside of you. It provides peace of mind that the circuit you are to work on will remain isolated from the electrical mains until you are ready to bring it back on-line once again. Imagine, for an example, that your job is fault-finding a three-phase electric motor problem – it could be highly dangerous if the circuit were switched 'ON' in error. Once it was considered sufficient to remove the supply fuses but this is no longer an option.

Glamco Electrical Engineering Company &

In house permit to work   Number 009875

Name..

Your trade...

Company...

Details of work to be carried out...............................

....

Name............ (Date) from .... to .... (Time)

Precautions to be taken..

PPE to have..

Tools not permitted..

Special requirements...

Known hazards...

Signed.. (Site safety officer)........................(Dated)

Signed.. (Trades person)...........................(Dated)

Figure 2.1 – *Permit-to-work certificate*

Most EU/UK switch gear now has a facility to physically lock the supply source in the *off mode* or prohibit access to a distribution centre by means of a *lock and key.* Work can then be carried out in safety. The lock is either an integral feature within a distribution unit or can be site-fitted to the unit. Some types of switch fuses have holes within the 'ON/OFF' handle movement through which a padlock can be slipped. Once locked off, *recheck your circuit* then inform any body else who could be involved at a non-technical level.

Once this has been done, hang a notice over your switch-gear or distribution centre advising that electrical work is being carried out. This type of advisory notice is illustrated in Figure 2.2. Sometimes it might be advisable to scribble the date work commenced on the notice as additional information for all to see. Do this by using white PVC tape and a permanent marker pen.

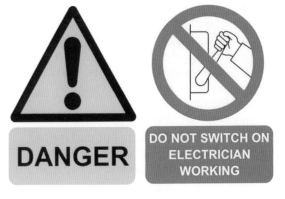

Figure 2.2 – *Advisory/warning notice that work is being carried out on a circuit*

When not to work on your own

There will be many situations when, as an electrician or a senior apprentice, you will feel it is unadvisable or just plain unsafe to work in isolation. Your boss, no doubt, would prefer if you worked on your own – the rewards would be far greater. Do not be pressurised – remember that your health and safety at work must always come first before profit. Listed in random order are a few circumstances where it would be best *not* to work unaccompanied:

1. In restricted and cramped spaces – can cause blood circulatory problems and lack of fresh breathable air will cause drowsiness.
2. Within deep trenches outside – the trench could collapse from pressure exerted from ground level by heavy machinery.
3. Near unguarded moving machinery – personal accident and involvement with the machinery.
4. Where a fire risk exists – a site fire; risk of personal injury.
5. When working with or near to toxic or corrosive material – could be overcome with fumes or gases; a health hazard.
6. Working from tall ladders – could get into difficulties; slip or slide factors. You would also need help in erecting and transporting.
7. In a grain silo – the windpipe (trachea) will become restricted due to fine dust if a bad fitting face mask is used or if the mask is clogged with contaminants.
8. Near to a 'live' bus-bar chamber – fear of direct contact with electricity.
9. Carrying out work from a cherry picker access machine – it is possible to get stuck in an elevated position; especially if the machine has not been kept in good working order and the emergency lowering lever fails to operate as intended.
10. Working from a prefabricated mobile access tower – two operatives must manhandle all physical tower moves. Two people are also needed when building or dismantling a mobile tower.
11. Wiring a large steel or PVC-u conduit installation – it is very difficult to carry out this operation on your own; two people are always required.
12. Installing long lengths of heavy steel wire armoured cable – two or more people are required if personal injury is to be avoided (back strain, hernia, pulled muscles, etc.).

The need for the provision of first aid treatment

Where you should keep your first aid box

The need for first aid facilities on *any* site is important for the health and safety of you and your workmates. It is not intended to replace the professional skill provided by a medical doctor or trained paramedic but to act as short term remedial solution until professional help arrives.

Keep your first aid box in your *site office* in a position where all can see. Never place it under lock and key or tucked away somewhere out of sight. Remember too, all accidents at work, however small, must be reported and logged into your site's *accident book*. It is wise to place a green and white 'First Aid' sign, as shown in Figure 2.3, on or by your site office door. This will indicate to others you are competent to carry out this site service to those who require it.

Location of your qualified first-aider

Every one of us owes it to his or herself to acquire a basic knowledge of first aid so we are able to do the right thing should an emergency arise.

Figure 2.3 – *A 'FIRST AID POST' advisory sign*

Figure 2.4 – *First aiders have specially designed hard hats to indicate who they are*

However, *Health and Safety at Work* demands that we appoint a *first-aider* to deal with our minor medical matters. On well organised sites your first-aider will have a white cross on a green background on his/her hard hat, as illustrated in Figure 2.4.

Get to know your first-aider, where he/she is working and where your site medical centre is located if you are unable to carry this facility yourself. When working on small sites these might turn out to be your responsibility!

Preventing injury to your eyes, hands, limbs and skin

Eyes

It really does not take too much effort to slip on a pair of safety goggles when carrying out work which has the potential of damaging your eyes. Apprentices and younger electricians seem to think that eye accidents only happen to other people. One day that other person might be you!

Always wear eye protection when carrying out the following work:

- Drilling walls and concrete ceilings.
- When using a rotary angle grinder.
- During battery maintenance work.

- Chasing walls and concrete ceilings.
- Forming air ducting holes for extract fans.
- Making off epoxy-resin steel wire armoured cable joints.
- When forming a ducting within a concrete floor.
- Hole cutting in steel and plastic trunking installations and electrical accessory boxes or switch-gear.
- Welding and brazing.
- Corrosive or hostile atmospheres – use a suitable respirator for this task as shown in Figure 2.5.

Hands

Site washroom hygiene and effective skin care ranks top in importance with numerous large building contractors today. Many larger construction sites provide, free of charge, a variety of hand-care creams, via dispensers, which can be applied before, during or after work. A barrier cream rubbed into your hands before starting work acts as a protector against grime and oil and will be washed off before each meal break, leaving your hands in good condition. Before leaving site a special cream

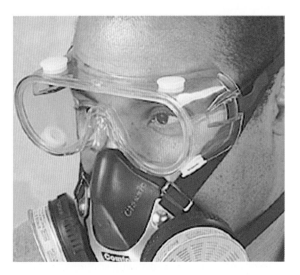

Figure 2.5 – *Choose your safety goggles and respirator with care*

can be applied to your hands which adds natural oils to your skin, depleted during the working day.

Remember to apply a suitable plaster to an open wound or abrasion. Your injury may not look much to fuss about but viruses and bacteria are microscopic and will invade your body through any open wound. Play safe and stick a plaster on!

There are many different types of industrial gloves to suit most site work applications. Always wear gloves when conditions become hostile.

Arms and legs

When carrying out your electrical duties within a factory or an industrial building which has moving machinery it is essential to wear the correct protective clothing and footwear. Be dressed in well fitting overalls and suitable toe-protective footwear; if necessary, wear a hair net and head protection even if the area you are working in is not judged to be a hard hat zone. Never work wearing jeans and trainers. You may look better in them but they will not protect you if loose clothing is caught in moving machinery. Ideally, moving machinery should be confined within safety guards, or screens and fences should be employed, but if you are carrying out maintenance work this may not be at all practical.

Site work can also have its dangers if you decide to abandon your overalls for a pullover and jeans. Even a battery operated drill can cause injury or damage if entangled within a woollen pullover.

Be sensible – dress for the part; wear protective clothing. Figure 2.6 illustrates some.

Coping with someone who has had a bad electric shock

I hope you are never in a position to have to deal with an electric shock victim – it is not the sort of incident you would want to deal with regularly. In case this duty falls on you one day, please remember these seven essential considerations. It might go a long way in saving someone's life!

Figure 2.6 – *Get kitted up for the job in hand – PPE is essential*

- Firstly, stay calm and do not panic.
- Next, if practical, isolate the supply of electricity from the victim.
- If impractical, move by dragging or pulling the casualty away from the source of current. Do this by using anything which is insulated, such as a dry wooden broom, nylon rope or even a length of dry hose pipe. Keep away from construction site timbers – they could be damp and able to conduct electricity.
- Keep your hands off the injured person. If you touch their bare skin, you too will receive an electric shock. If it will help, slip on a pair of electrician's insulated gloves. These will protect you.
- Be very careful what type of fabric you use to separate the victim from the source of electricity. Keep well away from marginally damp material such as a towel or a discarded waste rag.

- If the person is unconscious but breathing and severely burnt, get in touch with the emergency services on your mobile phone or summons your site's first aid services. Alternatively, drive your patient to the local accident and emergency unit. Whatever you decide, always seek advice and tell one of your team what has happened and where you are going.

- Depending on what you decide, ask for assistance if you choose to drive your patient to the local A & E unit. Stay with them until they are seen by a professional medic.

Mouth to mouth resuscitation

Breathing can stop as a result from direct contact with electricity. A sobering thought! If this happens to one of your workmates, brain damage will most likely take place within about three minutes. With this in mind, it is *crucial* that air is forced into the victim's lungs at a controlled rate until he/she starts to breath normally again.

Please familiarise yourself with these seven important stages whenever applying mouth to mouth resuscitation – it just might help to save someone's life!

1. Arrange your casualty on his or her back. Place their head to one side and remove anything from their mouth which should not be there (chewing gum, sweets, a half eaten sandwich, etc.). This is illustrated in Figure 2.7.

2. Gently lean the victim's head backwards. Do this by placing one of your hands on their forehead and your other hand on the back of their neck. This will allow their mouth to open and provide a clear airway passage to their lungs, see Figure 2.8. Now check that they have a pulse at their neck.

3. If you can detect a pulse, apply a little pressure to their chin (this will open their mouth even further) then pinch their nostrils together and blow two speedy but generous breaths by covering the victim's mouth with your own, see Figure 2.9.

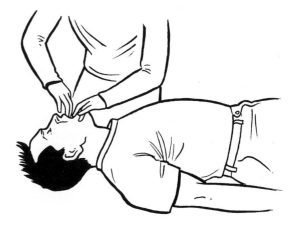

Figure 2.7 – *Mouth to mouth – lay the victim on their back*

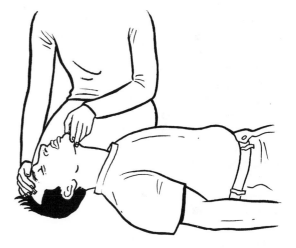

Figure 2.8 – *Clear the air passage*

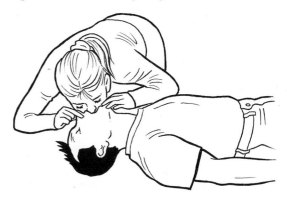

Figure 2.9 – *Cover the victim's mouth with your own*

4. After your second ventilation make sure that their chest responds by rising and falling. If there is no movement, tilt their head well back and blow through their nose but make sure their mouth is closed using your hand.

5. Carry on to breath progressively into your casualty's mouth at a normal breathing rate, keeping an eye on the rise and fall of their chest each time you artificially inflate it.

6. Once their breathing is re-established and considered normal, recheck the victim's pulse and, if all is well, place them in the *recovery position* as shown in Figure 2.10. Doing this will stop their tongue or body fluids blocking the entry to their windpipe. If your patient is too heavy to move, get help but be aware of a possible relapse.

7. Call the ambulance. If you are on a large construction site, get in touch with your medical centre but *always* stay with your patient until professional help arrives.

Figure 2.10 – *The recovery position*

Safety measures against electric shock

Risks arising from the use of electrical equipment

There are many but here are just a handful of examples:

- Electric shock caused by fatigue due to a long working day.
- Shock due to wrongly assuming that a conductor is dead.
- Damaged or strained back caused by lifting heavy electrical equipment.
- Death or injury caused by faulty electrical appliances.
- Wrist strain caused by hand-held drilling machines seizing in stubborn masonry. This

happens when a machine is not served with a built-in clutch.

- Explosion or fire when smoking in an unventilated battery charging room – batteries give off hydrogen whilst charging.
- Electric burns caused by flash-over conditions. This could happen whilst using installation test equipment.
- Health problems when electrical work is undertaken in a working grain storage silo, poultry farm or piggery. Keep to the rules and wear a face mask.
- Lung damage caused by airborne asbestos dust whilst drilling unrecognised building material.
- Mild but stabbing shocks caused through capacitance stored within long cable runs.
- Crushed feet caused by the misuse of large wooden cable drums.
- Electric shock caused by touching rain soaked wooden distribution poles supporting uninsulated conductors.
- Shock caused by mishandling charged capacitors.
- Respiratory (lung) problems caused when work is carried out within loft spaces lined with decaying fibreglass thermal insulation.
- Flashovers and burns to the limbs caused by wristwatches and bracelets when worn when work is carried out on live equipment or heavy secondary cells (batteries).
- Shock caused by frayed or damage flexes serving power drills.
- Wood splinters, cuts and scratches brought about by handling large wooden cable drums.
- Accidents caused by the delay of repairing damaged plant and site equipment.
- Electric shock caused by the damage of temporary site cabling and equipment.

Ways to reduce the risks

The following is a random list suggesting ways to minimise dangers and hazards when working with electrical equipment:

1. Never fool around with 500–1000 volt insulation testers – their voltage output can upset the electrical rhythm of the heart.

2. Wear the correct type of protective clothing.

3. Be sure you are able to manage heavy electrical plant and equipment. If not, get help!

4. Always work with someone you can trust on large, live, electrical main distribution centres. Not a very nice sight if you trip or fall into a bus bar chamber.

5. Never be tempted to use 230 volt hand-held power tools under site conditions.

6. Do not day dream. It could lead to an accident, especially if you were working from a ladder or on live equipment.

7. Remove all your personal jewellery when working on live equipment or heavy secondary cells (rechargeable batteries). It could snag on live conductive parts.

8. After handing lead acid secondary cells, always wash your hands thoroughly. Both lead and acid are toxic materials and can harm you. Wear gloves!

9. Wear eye protection when hammering home masonry nails. Although most will bend some will snap.

10. Regularly check over the plugs which serve your site power equipment. Many an accident has been caused by a loose or disconnected circuit protective conductor.

11. Regularly sharpen your screwdrivers and trim any steel over-hang from your bolster or masonry cold chisel, as shown in Figure 2.11.

12. When working in a workshop, protect yourself against the dangers of earth leakage currents by using a plug-in *residual current device (RCD)* which has a tripping current of 30 milliamps.

13. Never over-stretch using an electric power drill if muscular strain is to be avoided.

14. Refractory bricks (the type of brick which is placed within a storage heater) contain masses of iron filings. Wear industrial gloves when working with these items – they could save your hands from damage. Never be tempted to carry any more than you can comfortable manage if a hernia (a tear within the stomach lining) is to be avoided.

15. Treat fully charged capacitors with respect. A charge from a large capacitor will provide an unpleasant stabbing shock resulting in flesh burns to the hands.

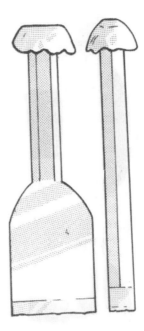

Figure 2.11 – *Trim dangerous overhang from your bolster*

16. Use the correct protective clothing when handing secondary cells.

17. Keep battery rooms well ventilated to prevent a build up of hydrogen gas.

18. When insulation testing very long runs of *steel wire armoured* or *mineral insulated cable (MI)*, always discharge each conductor to earth and to each other to avoid capacitance within the conductors and protective sheath. This can be dangerous if working from a stepladder and when totally unaware of what can happen.

19. Use the correct type of protective equipment if forced to dismantle asbestos from old electrical equipment. Better still, get a specialised company to do it for you.

Why we should carry out regular site portable appliance testing (PAT)

A PAT programme of work is a vital component of any site prepared health and safety policy. Electricity can kill. Each year about 1000 accidents involving direct contact with electricity are reported to the *Health and Safety Executive [HSE]* – roughly thirty are fatal.

Ot the thousand or so accidents which are reported annually, 25% involve portable appliances and that is 25% too many!

The *Electricity at Work Regulations 1989* places a legal responsibility on an employer to act in accordance with these regulations. This means they must take realistic practical steps to make sure that no danger can result from using any portable appliance within the site.

The H&SE does not put forward any timescale as to the frequency of the testing and inspection of portable appliances. The *Memorandum of Guidance* concerning the *Electricity at Work Regulations* only suggests 'regular inspection of equipment' – so these recommendations are not legal requirements. Putting into practice this set of proposals falls with BS 7671 – *Requirements for Electrical Installations*. Our regulations are supported by a series of *Guidance Notes*, of which the following is applicable to portable appliance testing.

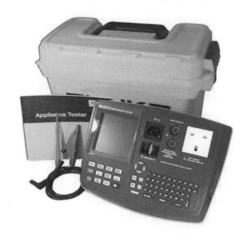

Figure 2.12 – *A PAT tester suitable for site work*

Code of Practice for In-service Inspection and Testing of Electrical Equipment

Your college database would probably have a copy of this publication. If not, feel free to order a copy from your local public library.

It is recommended that portable appliance testing and inspection should be carried out regularly every three months on a construction site. By doing this we are able to keep a check on the condition of the equipment and appliances used. Electrical devices served by 230 volt, AC, may be used within the safety zone of temporary accommodation but all power tools and lighting arrangements on site must only be supplied by 110 volts.

Figure 2.12 illustrates a typical PAT tester, suitable for both 230 and 110 volt appliances.

Portable appliances – general safety rules

Don't wait for the next PAT inspection to come around – check your power tools yourself, regularly. Listed below are a few guidelines to consider:

- Remove any plastic insulation tape around your flexible lead and rewire your appliance using the correct size, colour and voltage rating of cable. Site-used appliances must be wired using yellow sheathed cable. If the appliance is served with 230 volt and used within a safe zone (for example, site office), then rewire using a blue outer sheathed cable.

- Inspect the complete length of the flexible lead serving your portable appliance. Look for exposed but insulated conductors which have been separated from the sheath. This can happen at both ends of the cable. Make sure there is no exposed copper cable to be seen or snagged outer sheath along its route.

- Check for exposed or broken casing. This is potentially dangerous to the user. The appliance must be handed back to your company for a new one. Never put up with sub-standard equipment. Check out the *Code of Practice for In-service Inspection and Testing of Electrical Equipment*.

- Physically inspect the condition of each plug serving each of your portable appliances. Make sure the cord grip device is sound and the plug isn't due for retirement.

- A safety guard must always accompany a wandering lead light.

- High impact diffusers must always be fitted to 110 volt temporary lighting arrangements.

- Make certain the green/yellow *circuit protective conductor (cpc)* is sound and securely terminated to the correct plug pin. This is marked either with a capital letter E or the correct BS EN symbol, as illustrated in Figure 2.13.

Figure 2.13 – *This is the BSEN symbol for an electrical earth*

- Confirm that the connections serving your appliance are all terminated as intended and that each conductor is secure and not broken internally.
- If you have to install a semi-permanent appliance under site conditions, for example a large humidifier, wire using a suitable size of steel wire armoured cable. This will minimise damage caused by people walking on the service lead and heavy plant being dragged across it.

Problems associated with the use of reduced voltage portable equipment

Catalogued in random order are a selection of problem areas linked to reduced voltage portable equipment:

1. The wrong size cable can be used when rewiring heavy heating appliances. A 3000 watt, 110 volt heater will draw 27.27 amps. It would be wrong to rewire using 1.5 mm² conductors as within a 230 volt installation.
2. Repeated use of your appliance can snap a conductor within the flexible service lead. This can happen at any end. If the *phase* or *neutral* are affected you will know immediately. Your appliance will stop working! A dangerous situation will arise when the earth wire (cpc) is broken. You will not be aware of the condition of the cable until a fault to earth materialises and your appliance becomes 'live'.
3. Leads can become snagged, exposing live conductors.
4. It is easy to crack or break the plastic housing serving a site power tool. They often have to

put up with harsh treatment and, although they will tolerate a few knocks, eventually they will break.
5. Dust and general fine building debris can enter your machine and cause premature wear and tear on the bearings and other moving parts. Switches can also break down.
6. Conductors fitted to plugs can work loose. Usually it is the cord-grip which is responsible – screws become loose and are never tightened.
7. They are attractive items to steal. Always take your reduced voltage appliances with you when stopping for tea or lunch breaks. It is a small insurance premium!

The source of reduced voltage

A reduced voltage installation serving a construction site takes the form of one or more 10 kVA transformers. They can be larger than this figure, or some can be smaller but 10 kVA is usually what to expect. They are available as single- or three-phase machines.

KVA stands for kilovolt-amp and is the product of the voltage in kilovolts and the current in amps. Let us suppose a single-phase transformer draws approximately 43.49 amps when connected to a single-phase supply. Given the following expression, we will calculate the size of the transformer:

$$kV \times A = \text{size of transformer} \qquad (2.1)$$
$$0.23 \text{ kilovolts} \times 43.49 \text{ amps} = 10.001$$

This would be classed as a 10 kVA transformer and would be supplied with 10 mm² three-core steel wire armoured cable from the site's electrical distribution centre. Figure 2.14 illustrates a typical site transformer.

A brief description

- A steel body made weather proof.
- Often painted red and yellow.
- Incorporating a 100 amp isolator which also act as a 30 milliamp residual current device.
- The sockets are formed from PVC-u and colour-coded yellow.
- Most are equipped with a mix of 16 and 32 amp sockets.

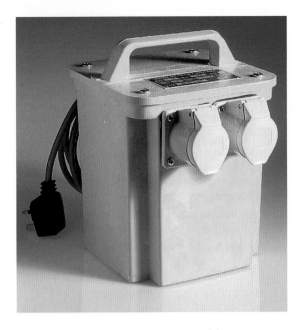

Figure 2.14 – A typical site transformer delivering 110 volts, AC

- The 16 amp sockets tend to outnumber the larger ones.
- Some older models had an independent thermal cutout button which was designed to be pushed if the transformer over-heated during use.

Single-phase site transformers are available in three kVA ratings: 3, 5 and 10. On a large construction site there would be many, all rated at 10 kVA whereas smaller sites (for example, when building a single house) the builder would prefer a 3 or 5 kVA transformer.

Power requirements – single-phase transformers at 230 volts

- A 10 kVA single transformer must be protected by a 50 amp, Type D or C miniature circuit breaker. Remember only to install a Type D or C. If a Type B is employed it will auto-switch to 'OFF' as soon as the transformer is switched 'ON'. This is because the initial rush of current is in excess of the rated miniature circuit breaker. A Type D's and C's characteristics are somewhat slower.
- Protect a 5 kVA transformer using either a 20 amp Type D circuit or C breaker or a 32 amp breaker. Some 5 kVA transformers can operate successfully protected by a Type B breaker. This is something you will have to find out yourself! If a 20 amp breaker is used, wire with 2.5 mm^2 three-core, steel wire armoured cable was used. When protecting with a 32 amp breaker then you must use a 6 mm^2 three-core armoured cable.
- A small 3 kVA transformer can be protected by a 16 amp Type B miniature circuit breaker. These transformers are easy enough to carry around with you and plug into the nearest 13 amp power socket outlet. They are usually coloured yellow but can be obtained in red or green depending on the manufacturer. A useful tool to add to an apprentice's tool kit.

Reduced voltage site equipment available

As you can imagine, there are many to choose from but here is a random selection.

1. Power tools and hand inspection lamps.
2. Twin fluorescent lighting arrangement on a stand as illustrated in Figure 2.15.
3. SON T lighting at 70, 250 and 400 watt.

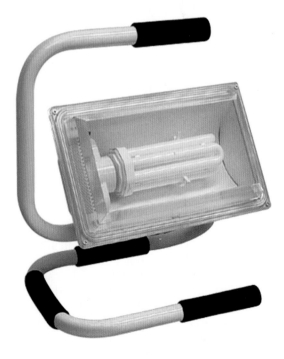

Figure 2.15 – Portable fluorescent lighting arrangements

4. General fluorescent lighting (5 foot and IP65 rated).
5. Infrared light heaters.
6. Four-way extension boxes, known also as splitter boxes.
7. De-humidifier.
8. Three metre spaced festoon lighting cabling arrangements.
9. Tungsten-halogen lights on a stand at 500 and 300 watts.
10. Reduced voltage (110 volts) ventilation fans.

Do remember that the current flowing in a circuit served by 110 volts will always be greater than for a 230 volt appliance with the same power rating. As an example, consider 29, 60 watt, 110 volt lamps served by a festoon lighting arrangement. The current drawn from this circuit would be 15.81 amps. This arrangement would be connected into a 16 amp circuit on the site transformer.

Safety training

This is important for our health and safety whilst working under site conditions. It is known generally as a *site-safety induction procedure* and can be carried out in a number of ways. The three most popular are as follows:

- A video presentation coupled with a brief lecture from a member of the construction company's management team.
- A group discussion around an 'A3' full colour presentation book where the leaves are slowly turned to relative items and snippets of information are usually remarked upon.
- A one-to-one discussion where the site manager discusses company policy matters and briefly guides you through basic health and safety issues. Some of the topics which could be covered include the following:
  - The construction company's health and safety policy (you will most likely be given a 'guidance booklet' free of charge from the principal contractors).
  - The layout of the site and traffic routes to expect (you might be offered paperwork in support of this item).
  - Where to park and where parking is forbidden (often a site-plan will be given to you as a guide).
  - The days the site is open and the hours of operation.
  - Where the first aid facilities are located; how to recognise the site first-aider and where the accident book is kept.
  - Welfare amenities – canteen, toilet block and drying room, etc. (if the site is large, you will be given a site-plan).
  - Where the health and safety plan together with the site risk assessment files are located (usually kept within the general office).
  - Fire procedures and where the site extinguishers are sited.
  - A study and understanding of the construction company's rules and regulations.

When all is done you will be given an *induction sticker* for your hard hat then asked to sign a site-safety induction form, of which the top copy will be given to you. The carbon copy is filed. Figure 2.16 illustrates the type of form you can expect to sign whilst Figure 2.17 provides an insight of how your newly acquired induction sticker might be designed.

If the site is small, most of this will fall by the way-side. It is a great pity as health and safety at work is really *everyone's* concern.

Warning signs and notices

Signs

There are many, of which most are designed as a yellow triangle supporting a black border in which a black pictogram or symbol has been added. Any text accompanying this category of sign is also black.

Site warning signs are placed in positions where people can see them easily. They are there to instruct, advise and forewarn site personnel and visitors of possible dangers within the workplace. They should never be ignored.

Figure 2.18 illustrates a small selection of signs, some of which you may have seen; others will be new to you.

Notices

Warning notices are also yellow in colour but are designed as a square or four-sided figure. Some warning notices are a mixture of both designs and

**Ishkian Mybll
Electrical Engineers
Newerton Village**

Safety is our first concern
SITE SAFETY INDUCTION

Our site manager is: David Sandingham

Our site supervisor is: Michael Gauntly

Our plant/materials controller is: Andrea King

Our site first aider is: Scott Pankerton

Our fire marshal is: Emma Salterton

Remember! A tidy site has the makings of a safe site

For your works you (or your firm) have completed the relevant:		Signed by operative ....
Company Method Statement	Yes/No	**Name (Block Capitals)**
Risk assessment	Yes/No	**of (Company)**
Permits (young operatives etc.)	Yes/No	**Dated**
Issuing of personal protective clothing	Yes/No	**Inducted by**

TOPICS TO BE COVERED IN THIS INDUCTION

Site Manager (or his represenrive) should ensure all new arrivals to this construction site are advised concerning the following:

- ☑ Layout of site and traffic routes
- ☑ Site hours of operation
- ☑ Location of first aid box, trained first aider and accident book
- ☑ Arrangements for welfare facilities
- ☑ Fire procedure and fire extinguishers
- ☑ Location of the health and safety plan and risk assessments
- ☑ Company's health and safety policy statement guidance booklet for contractors
- ☑ **Study and understand the company's rules and regulations**

In addition the Site Manager should cover the following points of general guidance:

✓ **DO** Make sure you are aware of the Health and Safety plan.

✓ **DO** Make sure you have copies and have read the Method Statement and Risk Assessment relating to your works (if applicable).

✓ **DO** Wear helmets safety footwear and hi-vis jackets at all times; gloves, ear defenders and goggles when working conditions dictate.

✓ **DO** Ensure all materials and plants are always stored safely.

✓ **DO** Ensure power cable are free from damage, dry and are not a trip hazard.

✓ **DO** Ensure adequate fire extinguishers are within reach if using blow torches or similar.

✓ **DO** Report anything unsafe to the Site Manager immediately.

✓ **DO** Keep your work place tidy.

✗ **DON'T** Alter or remove scaffolding or safety equipment provided for your, and other workers' benefit.

✗ **DON'T** Consume food and drink at your work place - use the canteen.

✗ **DON'T** Use toilets in the building, use the facilities provided.

✗ **DON'T** Use basins or sinks within the building, use the facilities provided.

✗ **DON'T** Use 240v appliances. **Only 110 volt appliances.**

✗ **DON'T** Use uninspected, incomplete or unsafe scaffolding.

✗ **DON'T** Use any plant unless you can prove you are properly qualified.

✗ **DON'T** Enter areas restricted during certain operations.

Figure 2.16 – *A site induction form*

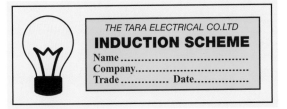

Figure 2.17 – *An induction sticker to place on your hard hat*

have incorporated within their message both a four-sided figure and yellow triangle as shown in Figure 2.19.

Isolation procedures – how to avoid getting a shock!

Accidents which involve direct contact with live conductors can be minimised if you make a routine examination of your circuit before you carry out any work on it. This type of inspection is particularly essential if, for example, you are to carry out maintenance work on a machine within a factory. Figure 2.20 suggests a practical

way in which this could be achieved.

This type of inspection will not take up much of your time and could save your life. Once your circuit has been safely disconnected, remember to lock off and hang a yellow warning notice on your switch-gear advising that the circuit is 'dead' and being worked on. Figure 2.21 illustrates this.

Figure 2.19 – *Some warning signs have a notice underneath them*

Figure 2.18 – *BSEN warning signs*

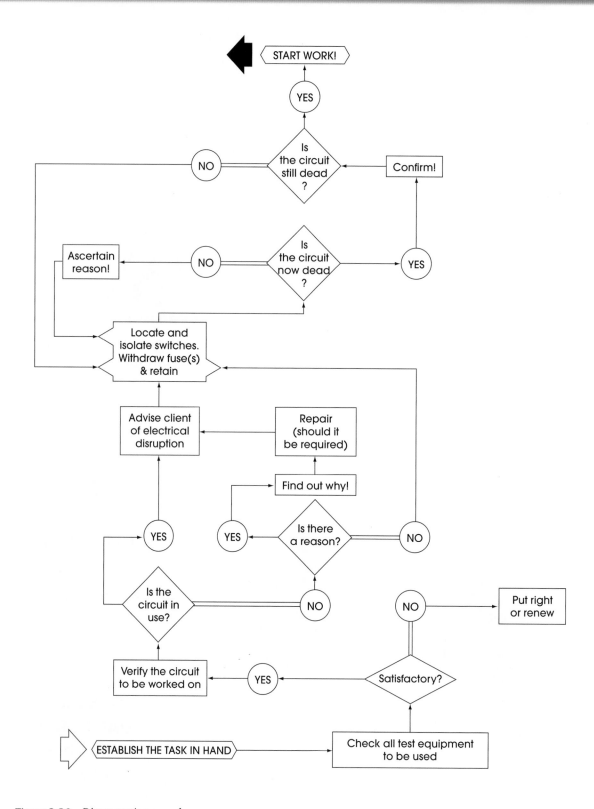

Figure 2.20 – *Disconnection procedure*

Figure 2.21 – *A warning notice that a circuit is being worked on*

Ways you can protect against direct contact with electricity

1. Insulation of all live parts (i.e. the phase and neutral conductors).
2. Obstacles placed in the way.
3. Out of reach (for example, power lines).
4. Residual current devices trip rated at 30 milliamp (ideal for supplementary protection).
5. Placing within enclosures, etc.

Asphyxia

What is asphyxia?

As humans we rely on a constant supply of oxygen for our very existence. Its importance cannot be over-stated. If we get into a position where our body is starved of oxygen, medical problems will occur in less than one minute; our body tissues will rapidly deteriorate and die. This is known as asphyxia and it is dangerous.

How is it caused?

There are many causes – some of which might appear a little fanciful when related to a construction site but here is the list for good measure.

- Choking on food within the site canteen.
- Drowning if work is carried out near to a canal or river.
- Suffocation as a result of working in a hot/confined place.
- Strangulation.
- Paralysis due to a fall resulting in an accident to the back.
- Injury to a person's lungs resulting from an accident.
- Injury to the chest cavity resulting from an impact accident.
- General lack of oxygen in the air due to inadequate ventilation.

The symptoms

- Unnatural breathing – laboured or noisy.
- The victim's lips have a bluish ting to them.
- Can be frothing at the mouth.
- Not breathing.
- Could be unconscious.

What to do

- Act quickly – but do not panic.
- Get help if you can.
- Check out what is causing the problem. Remove, if necessary, any food within your casualty's mouth which could be blocking their windpipe (the trachea), etc.
- Move him/her into the fresh air.
- Loosen the clothing around their neck.
- Start *artificial respiration* (mouth to mouth or mechanical ventilation) as described in previous paragraphs.
- Once you have re-established their breathing, place your patient in the *recovery position* making sure that their tongue will not fall to the back of their mouth to obstruct their windpipe.
- Do not leave your patient but carefully monitor their breathing and well being until professional help arrives.
- Telephone for an ambulance if you were unable to get anyone else to help out – but be sure you remember your mobile telephone

number, as the operator will want to confirm it after you have supplied all the information. Asphyxia is a very serious condition which could lead to death if not acted on without delay. Do not undervalue its importance.

Defining terms of reference

Dangerous occurrence

Any incident within your workplace which would give rise to serious concern – a site fire, an explosion or toxic atmosphere are examples.

If necessary, get outside help by calling the emergency services but please remember to be precise when using the telephone. Accurate details and your own cellphone number must be offered to the emergency service you are calling.

Hazardous malfunction

A dangerous breakdown can occur at site level any time when risk assessment has been taken lightly or not at all.

The term *hazardous malfunction* doesn't only mean an unsafe electrical or mechanical breakdown, but can relate to a multitude of other risky occurrences within the workplace.

Here are some of them for your records.

- Walls and ceilings becoming unsafe after local demolition.
- Rotten flooring and tripping hazards found where not expected.
- Unsafe roof spaces and staircases caused by new building works or alterations.
- The discovery of a disused well within or near an old house.
- A mechanical crane failure or other mechanical problem causing it to be unsafe.
- A buried power cable which has been exposed and severely damaged through excavation.
- A temporary lighting failure in an area served with poor natural lighting.

You should be able to cope with a hazardous malfunction at site level. However, if you do run into trouble never be too proud to seek the advice and help of others.

When the site emergency alarms sound

A brief introduction

A fire will develop whenever heat, fuel and oxygen are present. Under site conditions a fire can be caused by one of the following general reasons:

- Electrical faults.
- Cigarette ends careless thrown to the floor.
- A slipshod approach when carrying out gas or electric welding.
- Deliberate fire-raising (arson).
- Fault conditions within power tools and other electrical equipment.
- Too much current within a cable (for example, a small cable serving a large electrical load whose over-current protection is set too high).
- Temporary halogen lighting knocked to the floor to rest among combustible material.
- Loose terminals serving your temporary power and lighting arrangements. When arcing occurs it generates heat – heat can cause fire!

On discovering a fire

There are three simple rules:

1. Sound the alarm but if this is not practicable tell someone and shout 'FIRE' at the top of your voice repeatedly.
2. Try to put the fire out with an appropriate fire extinguisher but do not endanger yourself.
3. If you are unable to control the fire, get out of the building fast along the quickest and approved escape route. Advise co-workers along the way and call the emergency services. When you give the fire brigade your site address, do not switch off your mobile phone until it is repeated back to you.

Site evacuation

Site evacuation should be known to all working on site. This is because it forms part of an *induction process* which we are all subjected to when first reporting for work. This emergency practice is not only for fires but for any type of crisis which encompasses the workplace. The site evacuation procedure should be rehearsed

Figure 2.22 – *Fire exit signs must be displayed on construction sites*

Tara Electrical Engineering Company GmbH

Procedure to adopt in case of an emergency
1. On discovering a fire
2. Sound the alarm & try to extinguish the flames
3. DO NOT put your self at risk
4. If unable to put out, make your way out of the building both quickly and safely
5. All routes are identified
6. Make your way to the approved **ASSEMBLY POINT** at **LINGBRIDGE CLOSE**
7. There you will be checked in/debriefed

H.T. SCHIFFER (Safety Officer)

Figure 2.23 – *This type of fire evacuation notice is often found in site canteens*

formally every so often so that it is always fresh in our minds.

Emergency evacuation notices must be processed in clear readable text. Directional signs are required to be clear in their meaning and placed where they can be easily seen, see Figure 2.22.

'On discovering a fire' notices must be placed within areas where people can see and read them easily, for example:

■ In the canteen at eye level whilst sitting down.
■ Within the site office.
■ In the WC cubicles and hand washing areas provided.
■ Near signing in/clocking off points.

Figure 2.23 illustrates such a notice.

When the emergency alarm sounds

■ Stop what you are doing at once.
■ If it is safe to do so isolate all your power tools from the site's temporary reduced voltage power system.
■ Again, if safe, isolate the temporary power serving the site at the source of supply. After doing this the site's emergency lighting units will take over.
■ Do not stop to gather up your personal belongings – just focus your thoughts on your own safety by getting out of the building as quickly as possible; be single minded!
■ Do not run (this can cause bottlenecks) but make your way out of the building quickly, taking the official escape routes. If you are working in an occupied building, *do not use the lifts*. When alternative routes are open to you (for example, through an undeveloped door or window, etc.) – take them if safe to do so. It could mean escaping from the building in seconds, rather than minutes. Gather at the official *assembly point* – do *not* enter the building again unless you are told you can do so by the *Fire Marshal* who is in a position to assist you. Once safely at the assembly point your name together with the company you work for could be taken and listed. Assembly point signs are square or oblong in shape and green in colour. White lettering and logos are used to convey the message.
■ If you are in charge of the job (the job holder), please check that all your staff and any

visitors you might have had are safe and sound.

Site assembly points

Construction site assembly points are important as a means of recognising safety areas where workers can gather in the event of a crisis. These areas are not usually within the working boundaries but are chosen well away from any potential harm. Often they are sited on grassed areas or nearby car parks. You will be told where your assembly points are situated when you complete your *site-safety induction course* upon reporting to the site manager for work.

Preventing and controlling fires

Causes of fire

It might surprise you to read that about 500 people die and some 11 000 people are injured from fires within the workplace and home. We can all help a little to provide a safer environment for each and every one of us. Here are a few guidelines which will help within the workplace.

■ Many construction site fires occur within the works canteen whilst cooking. Watch this point and be careful, especially if you are not used to preparing your own meals.

■ Be careful if smoking is allowed on site and observe 'NO SMOKING' notices.

■ Faulty or sub-standard electrical gear can be responsible for fires within a construction site. Remember the current is much higher when equipment operates from a 110 volt system.

■ Welding and the use of an angle grinder can create a fire hazard if not appropriately managed.

■ Misuse or careless use of site heating appliances can cause fires – these appliances are not for drying your overalls out!

Controlling the spread of fire

The four stages in the life of a fire are as follows:

■ *Ignition* – a cigarette end, a lighted match or concentrated heat source such as a halogen lamp.

■ *Primary fire growth period* – flashover is between 500–600 degrees Celsius.

■ *Fully developed fire stage* – where the fire will spread to other areas.

■ *Decay* – when all the combustible material or oxygen levels are spent and the fire dies down to smoulder.

Controlling the spread of fire on a building site is difficult as automatic equipment such as fire and smoke dampers together with permanent electromagnetically controlled fire doors needed to carry out this process are not yet installed or, at best, not commissioned.

The law now requires that smoke and fumes must be *contained within an area* or *compartment* of a building. It is the thick heavy smoke which kills most victims of fire – breathing in vast quantities of smoke is usually fatal.

Listed are a few methods of containing both smoke and fumes which will help control the spread of fire throughout a building under construction.

■ A reduction or removal of all non-essential combustible material and site litter throughout.

■ A *ban on all flammable liquids* and *low pressure gas (LPG)* canisters stored within the building. Materials such as this can be stored safely within a steel lock-up suitably signed/labelled as shown in Figure 2.24.

■ A temporary plumbing installation serving a *quick response water sprinkler system* – an effective tool in fire fighting in the early stages.

■ Temporary but internal separation of construction areas but served by site assembled *fire doors*.

■ Possibly, large auto-controlled site extraction fans suitably positioned in selected corridors to swiftly remove deadly smoke and fumes.

Figure 2.24 – *Danger of flammable liquids sign*

Fire prevention

You all have a common responsibility to one another not to put others at risk from fire. Fire can damage, it can suffocate – and fire can KILL!

Most construction site fires have simple causes and can be dealt with at the site level quite easily.

Listed are fourteen suggestions how you can create a safer workplace. Sorry the list is long – but it is important for our safety.

■ Site fire doors must never be wedged 'open'.

■ Test your temporary fire alarm system regularly – but always announce when the alarm is to be sounded.

■ Store low pressure gas (LPG) cylinders outside in a locked container; this also includes flammable solvents and adhesives.

■ Place appropriate *fire fighting equipment* (extinguishers and blankets, etc.) at either end of a corridor or where they can be seen easily. Fire fighting equipment is always accompanied by an appropriate *fire equipment sign* as illustrated in Figure 2.25. These are either oblong or square in shape with white lettering and a pictogram on a red background.

■ Ensure that all escape routes are easily recognised and are free from tools, equipment and debris. Figure 2.26 shows a selection of fire exit signs. These are either square or

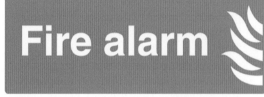

Figure 2.25 – *Fire equipment signs which are found on construction sites*

Figure 2.26 – *Fire exit signs must be displayed where most needed*

rectangular in shape and have a base colour of green. The added pictograms and accompanying text are white.

■ Check that all emergency exits are unlocked or are not permanently shut.

■ Never over-fill a chip pan with oil when preparing a meal in the site canteen.

■ Use only intrinsically safe power tools (this also includes battery operated equipment) in places where there may be flammable vapours.

■ Always keep a suitable fire extinguisher at hand when carrying out 'hot-work' such as welding, braising, or when using a gas blow lamp.

■ Never burn your waste material on site.

■ Keep a fire *blanket* in your works canteen and use it if your chip pan catches fire – never try to extinguish with water! A fire blanket sign, red in colour with white lettering, is illustrated in Figure 2.27.

■ Stub your cigarette end out sensibly – don't just throw them away; it could just cause a fire!

■ Keep your damp overalls and other *personal protective equipment clothing* off temporary halogen heaters. They are very hot and must not be used to dry clothing. Use your drying room facilities!

■ Remove any faulty electrical equipment from the site.

■ Do not over-load multi-socket adaptors with appliances within in your works canteen. Canteens are often poorly served with sockets – but be sensible.

■ Use the correct size fuse in 13 amp plugs within the safety zone.

■ Always use wall mounted convector heaters for heating accommodation blocks.

■ Finally, regularly test the site's temporary electrical installation together with any PAT work which has to be done. Please provide formal written data and give to your site manager, as described in Chapter 8 of this book.

Workplace fire drills

If you have in your custody a fire certificate issued by your local fire and rescue service it will usually specify the frequency you should carry out site fire drills. There are no hard and fast rules concerning this matter but on a large construction site it would be wise to have a formal fire drill every four months. This would take on board trades' people who join the site later within the development period.

Each drill must be formally logged as illustrated in Figure 2.28 and entered into your health and safety at work records. The following should be taken into account:

1. Date/time of fire drill.
2. Weather – fine, windy, stormy, etc.
3. The approximate number of people taking part in the fire drill.
4. The total time taken to evacuate everyone from the building to the safety of the assembly point from the start of the alarm.
5. The interval from sounding the alarm to the last person *leaving* the building.
6. Relative comments and patterns to learn from.

It might be best to let the local fire and rescue service know you are having a site evacuation drill, if for example a large number of people are to assemble in public areas such as a green or car park.

Design your fire drill to the practical needs and requirements of your site; publish and hand out to construction workers and trades' people after the formal site induction procedures have taken place. This way they will always have something to refer to should ever an emergency situation arise. One point to consider: check out, under normal site working conditions, that the alarm bells are clearly heard, even over the loudest background noise.

Figure 2.27 – *This type of sign is for a fire blanket*

```
Date/Time ...............................Practice/fire incident [please state]
Weather.................................................................................
Approximate number of people involved .......................................
Evacuation time from sounding the alarm to assembly point......................
.............................................................................................
If an actual fire incident - state brief details........................................
.............................................................................................
Relative comments and learning points .......................................
.............................................................................................
.............................................................................................

Signature of Site Safety Office ................................. Date.................
```

Figure 2.28 – *A page from a site fire drill log*

Fire fighting equipment for different types of fires

In January 1997 all newly manufactured fire extinguishers were colour-coded red. The contents of each type of extinguisher can be easily recognised by a secondary colour in the form of a wide band totalling just 5% of the red base colour. Table 2.1 provides concise details of how and where to use them when tackling a fire in its early stages. We have also included Halon 1211, a fire extinguishing gas which was withdrawn throughout the EU in 2003. Some smaller sites may still carry this type of extinguisher – they should not be used.

Automatic fire fighting equipment and detection

Listed are a few of the many types of fire fighting equipment of which some of you will be familiar with:

1. Battery operated smoke detector and alarm – it is important to change the battery regularly; once every year.

2. Electrical mains-operated ionisation smoke detector – can incorporate an integral audio alarm or be part of a larger fire detection and alarm system automated from a control panel.

3. Stand-alone or control panel served optical fire alarm detection units.

4. Fixed temperature heat detectors – ideally suited where a variation in temperature is common. Unsuitable for damp areas.

5. Rate of rise heat detector – operates at a fixed top temperature of 60 degrees Celsius.

6. Audio alarms used in conjunction with a construction site fire alarm panel.

Other automatic devices which can be used to some degree or another include the following:

■ High temperature heat detectors which operate at 90°C.

■ Duct detectors (photocell type).

■ Beam detector (gallium arsenide, infrared light type).

■ Automatic fire doors (electromagnetic type) – these operate in conjunction with an auxiliary relay connected to the fire alarm panel, which when energised, isolates power to all the electromagnets serving the fire doors. Once de-magnetised, the fire doors close.

■ Automatic roller shutter doors activated by an alarm status. These are often to be found in warehouses.

■ Mechanical sprinkler head – activated by heat from a fire.

Table 2.1 – *Fire extinguishers and their use*

Contents	Used for	Comments	How to use	How it works
Water (Painted all red)	Timber, plastic, cloth, etc	Not to be used on oil/fat fires nor electrical fires	Aim the jet at the base and keep moving across the target area	By cooling the burning material
Powder (has a blue band)	Solid fires, liquid oil and grease and electrical fires	See Figure 2.29. Do not use on chip pan fires	Point discharge horn at the base of the fire. Rapidly sweep from side to side until out.	Knocks down the flames and forms a smothering skin. Also has a cooling effect.
AFF* foam (has a cream band)	Solid fuels, liquids, grease and oils	Do not use on chip pan or electrical fires	Aim at the base of solids but apply to the inside edge of a container containing a liquid base fire.	Forms a film across the surface of the burning liquid. Has a cooling effect.
Carbon dioxide, CO_2 (has a black band)	Liquid, solid and electrically fuelled fires.	Not to be used on chip pan and fires. The fumes are harmful.	Aim at the base of the fire and keep moving across the target area.	Displaces the oxygen in the air and suffocates the flames.
Halon 1211 (has a green band) Now replaced with FM200®	Flammable liquids and solids	Dangerous fumes given off. *Now withdrawn from use within the EU.* Can still be found on smaller sites	Direct at the base of the fire and sweep from side to side. Do not use in confined spaces.	Suffocates
Fire blanket (has a red container)	Solid and liquid fires. Ideal for clothing and chip-pan fires.	Be sure that all your target area is covered.	Place very carefully over the fire. Take care not to burn your hands.	Smothers and deprives the fire from oxygen. Made from woven glass fibre fabric.

*Aqueous film-forming foam.

Figure 2.29 – *A notice which accompanies a powder based fire extinguisher*

Fire protection rules and regulations

A brief insight

The Health and Safety Executive have published that, on average, there are *eleven construction site fires* throughout the UK daily. Not only are people killed or badly injured but there is a heavy financial burden to acknowledge.

The *Construction (Health, Safety and Welfare) Regulations 1996* require a site manager or job holder to take positive and constructive measures to avoid a fire from starting and to ensure that all building workers and visitors are fully protected from the outbreak of a fire.

The following regulations have been written in a concise form to enable a clearer understanding.

Prevention of risk from fire – Regulation 18

You must take both suitable and sufficient steps to prevent the risk of injury to anyone within the construction site arising from:

a) explosion or fire,

b) any construction material or location liable to cause asphyxiation,

c) flooding.

Emergency routes and exits – Regulation 19

a) You must ensure that there are sufficient emergency exit routes provided and take into consideration the number of people who will be working on site at any given time.

b) All your emergency exits must end within a 'safe area'.

c) Any doors you have along the exit route must be free from obstruction.

d) One hundred and ten volt emergency lighting units can be provided along an emergency exit route if considered necessary.

e) Routes leading to a 'safe area' must be indicated by the use of suitable signs.

Emergency procedures – Regulation 20

a) A site management must draw up a plan for dealing with any foreseeable emergency and practical measures for the evacuation of the site, or part of it, from workers.

b) All site workers must become familiar with emergency procedures. (This could be introduced during the site-induction period).

c) The site management must arrange for fire-drills to be carried out and practiced at suitable intervals.

Fire detection and fire-fighting – Regulation 21

In the interest of health and safety, the following must be provided within a construction site:

a) Suitable and sufficient fire-fighting equipment. These could include fire blankets, extinguishers, sand buckets and live hose pipes. These utensils must be clearly indicated by the use of red fire equipment signs.

b) Suitably located detectors and fire alarm break-glass units should be installed throughout the site.

c) All fire alarm equipment must be properly maintained and tested/inspected at regular intervals.

d) Any equipment *not* designed to come into use automatically, for example a hand-cranked alarm bell or a fire alarm break-glass unit, must be totally accessible.

e) It is important that all site workers know how to use and operate the fire fighting equipment available in case an emergency arises.

f) In high fire-risk areas permission must first be given by a senior member of staff before any electrical installation work is carried out.

g) All site fire-fighting equipment must be accompanied by fixed signs.

Further reading

Please refer to the *Construction (Health, Safety and Welfare) Regulations 1996* or visit their web-site or, if you prefer a hard copy, the following book from your college library: *A Guide to Managing Health and Safety in Construction (HSE book 1995)*; the ISBN is 0 7176 0755 0.

Figure 2.30 – *A safe area sign*

Handy hints

- Do wear your hard hat as intended.

- Lock your battery drill when not in use to help prevent accidents from occurring whilst carrying.

- Properly maintained tools and plant will provide years of reliable service.

- Wear eye protection when hammering home plastic cable clips – targeting incorrectly can cause both clip and nail to fly and it's usually your face which gets hit!

- It is wise not to attach any type of discharge light to a lead. Many have power factor correcting capacitors in circuit and handling the plug-pins after removal from a socket will give you a very unpleasant and painful electric shock.

- Site emergency lighting units operated from a 110 volt power source can safely be fitted with long nylon cable straps if traditional fitting is a problem.

- Due to the 'funnelling effect' of wind within an open ended building it is best to securely fix mobile towers, ladders and steps to the nearby infrastructure if an accident is to be avoided.

- Never run during a site emergency evacuation – you could trip over or, if several of you are running, it could cause a bottleneck at doors or exit points.

- Unlike BS 7671, which is a voluntary code of practice, *The Electricity at Work Regulations (1989)* is legally binding.

- Avoid using a neon test screw driver – it's unprofessional.

- Allow a jigsaw to stop before removing – haste can lead to the blade shattering.

- Take care when handling bricks which serve night storage heaters. They contain masses of iron filings and can cause problems if your skin is damaged through handling them.

- Keep telephone installations well away from mains electricity cables.

- If you are confused or just want reassurance please don't be afraid to ask someone you trust for the advice you need.

Summary so far . . .

- *A 'permit-to-work' is designed to look after your health, safety and welfare where site conditions are harsh, dangerous or if there is an element of risk attached to the work you proposed to do.*

- *A 'lock-off procedure' provides you with reassurance that the circuit you are to work on will remain isolated from the electrical supply.*

- *There will be times when it will impossible for you to work by yourself. Always consider both the practical and safety aspects of what you are about to do.*

- *Be always on guard against the possibility of an electric shock. There are many reasons why we*

have them. These range from carelessness brought on by tiredness to bad workmanship and faulty equipment.

- *A portable appliance testing programme is a very important factor of your site's Health and Safety policy.*

- *There are many site problems associated with reduced voltage appliances from the accidental removal of conductors from plugs, etc., to wrongly wired power tools and dangerously cracked safety insulation serving equipment.*

- *The source of site power is from a reduced voltage (step-down) transformer. Its size is measured in kVA and is the product of the*

Summary so far . . .

potential current delivered by the transformer within the secondary winding and the supply voltage in kilovolts. A 230 volt supply would be 0.23 kV.

- Site safety training is known as 'induction' or sometimes recognised as the site safety induction procedure. This initiation is important for our health and safety whilst working under site conditions. Your induction can be presented as a video or in a one-to-one chat over a cup of coffee.

- Most warning signs are designed as a yellow, black bordered triangle in which a sign or symbol has been placed. They are placed around the site to instruct, guide and forewarn people of possible dangers within the workplace.

- Reduce your chance of getting an electric shock from direct contact by making a sensible inspection of the circuit you are to work on. Do this before any work is carried out.

- If we get into a position where our body is starved of oxygen, asphyxia will rapidly follow in less than a minute. At its worst, this can lead to death!

- Any incident or event within your workplace which would give rise to serious concern, such as a fire or explosion, etc., is classed as a dangerous occurrence.

- Be familiar with your site's fire drill procedure. Never run, as bottlenecks can form at fire doors and staircases. Panic can easily result from this.

- The four stages of a fire are: ignition, growth period, fully developed stage and decay. Only tackle a fire in its very early stages and never

put yourself at risk. People are more important than property!

- You can do much to halt the spread of fire, such as not wedging the temporary site fire doors open or storing low pressure gas bottles within the building. Paper, wood and other flammable materials must be cleared from the site regularly. If you have the luxury of an automatic electronic fire detection and alarm system, make sure that it is tested regularly.

- There are no hard and fast rules concerning site fire drills but after training is over, please remember to formally log them as described and illustrated in Figure 2.28.

- Water must never be used on an electrical or canteen chip-pan fire. It is fine to tackle timber, plastics and cloth fuelled blazes, but always use carbon dioxide (CO_2) to extinguish an electrical fire. Powder is ideal for use on liquid oil and grease based fires whilst a fire blanket is very efficient at smothering a fire in a chip pan.

- Regulation 19 of the Fire Protection Rules and Regulations demands that your site provides sufficient emergency exit routes whilst Regulation 18 states that sufficient steps must be made to protect from the risk of injury arising from fire, explosion, flooding and asphyxiation.

- Site automatic fire detection and alarm equipment can include smoke detectors and alarms, heat detectors and magnetic-catch fire doors.

- The extinguishing gas 'Halon 1211' has now been withdrawn from construction sites within the EU and is now replaced by an agent called FM200®.

Review questions

1. Briefly describe why we have 'permits to work' on larger building sites.

2. Compose a few lines regarding 'lock-off procedures'.

3. Give a minimum of three examples where it is best not to work by yourself.

4. Where is it best to report an accident at work:
   - the site manager,
   - your company accident book,
   - the local Health and Safety people,
   - your company office.
   Please write down your option.

5. Provide four practical examples where it would be wise to wear eye protection.

6. How would you move a casualty of electric shock away from live conductive parts?

7. List five practical safety measures to guard against electric shock.

8. List five sensible ways to limit the risk of electric shock.

9. The HSE advises that construction site portable appliance testing should be carried out as follows:
   - every three months,
   - at the beginning of construction work then every six months,
   - no recommendations offered by the HSE,
   - once every four months during construction work.
   Please write your option

10. A single phase site transformer will deliver a total of 87 amps. Its practical working kVA rating is as follows:
    - 10 kVA,
    - 20 kVA,
    - 30 kVA,
    - 50 kVA.
    Please write your option

11. Briefly outline five items discussed during a site induction session.

12. Summarise the symptoms of site related asphyxia.

13. List at least five means of fire prevention.

14. How may fire detection and fire fighting be improved at site level?

15. A fire started in a chip-pan is best put out by one of the following:
    - a red water base extinguisher,
    - carbon dioxide (CO_2) extinguisher,
    - a fire blanket made from glass fibre fabric,
    - Halon 1211 extinguisher.

Effective working practices

Introduction

There are six underpinning knowledge requirements serving this chapter, all of which are compulsory. In summary form they are as follows:

- Learning and personal performance, career structure and training.
- The features of team working.
- Employment rules and regulations within our industry.
- Processes to carry out electrical work safely.
- Standards for assessing working practices and procedures.
- The benefits of improving working practices and procedures.

Practical activities

There are no practical activities associated with this chapter.

Tests

You will be given a multiple choice question paper and then be assessed in accordance with City and Guilds test requirements.

Learning related to personal performance

Learning new skills, techniques and procedures within the electrical installation industry will help to promote personal confidence and greatly increase your chances of promotion – especially if you are technically *unqualified,* working as an electrician and desperately needing to move away from your present employment. Make learning a lifetime habit – be prepared for change, no matter in what form it comes.

We have seen many changes within our industry throughout the decades including two changes of colours relating to cables, three systems of sizing conductors, the introduction of electronic appliances within the workplace and many more.

How may I step on the learning ladder?

There are many ways; here are just a few:

- A City and Guilds or NVQ evening or day release college course.

- Enrol with a *Direct Learning* organisation.
- Take-up distance learning (a correspondence course, for example).
- Computer *'on line'* courses – available on the net.
- Being shown and taught within your workplace.
- By enrolling into an evening NVQ college course whilst carrying out practical applications within your workplace. This is ideal if you are not qualified as a practicing electrician but asked to carry out the duties of one.

Career patterns within the electrical installation engineering industry

The *Joint Industry Board* is an organisation which exists to help us by acting as a coordinating agency for the electrical installation engineering industry throughout most of the UK.

The grading system

Electrical workers are graded according to experience and qualifications. In summary the grades are as follows:

- electrical labourer,
- apprentice,
- electrician,
- approved electrician,
- technician.

Electrical labourer – No technical or academic qualifications are required by a person with this job title as all work carried out is unskilled but supervised.

Apprentice – This is a person who is bound by a formal agreement to learn a craft, skill or trade coupled with formal technical education. Once qualified, he or she is graded as an electrician.

Electrician – To be graded as an electrician the following criteria must be met:

- Must be adequately trained and served a recognised apprenticeship.
- Must have been awarded the current NVQ certificate in electrical installation engineering.
- Must demonstrate he/she has had experience relative to age.
- Must be able to have the ability to carry out electrical installation work efficiently.
- A working knowledge of BS 7671 is essential.

Approved electrician – To be graded as an approved electrician a person must hold the following qualifications:

- To have undertaken an formal registered apprenticeship with on-the-job training.
- A current and appropriate NVQ award at level 3.
- At least 22 years of age and a minimum of two years' experience working as an electrician.
- He/she must be able to work without supervision.
- A good working knowledge of BS 7671 and other British Standards together with electrical Codes of Practice serving our industry.
- Has gained the City and Guilds 2391 Certificate, *Inspection and Testing*.

Technician – To qualify you will require the following:

- A registered apprenticeship with practical training.
- A relevant NVQ in electrical installation engineering at level 3.
- At least 27 years of age.
- A minimum of five years' experience as a site supervisor, a job holder or as an approved electrician.
- At least three years' experience as a supervisor.
- Knowledge of design and laying out electrical installations cost-effectively.
- Knowledge of BS 7671, other British Standards and Codes of Practice.
- Has gained the City and Guilds 2391 Certificate, *Inspection and Testing*.

Career options

There are many career opportunities which you can consider once you have fully qualified but unless you have made your mind up beforehand, the choice can be difficult. Often the job you would like to do will require additional qualifications so it's wise to decide which career option you wish to follow as soon as possible so that the appropriate path may be taken.

Here is just a handful you can browse through:

Contracting – Working for an employer in the electrical contracting industry involves a great deal of installation work and possibly long hours and staying away from home. Alternatively, if you wish, you can branch out on your own and form your own electrical company but if you feel this is too big of a step to take in one go, you can always work as an *agency electrician* where the time you spend on the job can be very flexible.

Maintenance – Another alternative is to become a maintenance electrician working within a factory or hospital, an experimental station or perhaps in television, etc – the opportunities are unlimited.

Specialised electrical engineering – Specialised skills are required in many areas such as radio, the theatre, mining and within the offshore oil/gas industry, just to name a few. These are all highly skilled jobs so it's well worth putting some additional effort in to gain extra qualifications if this sort of work appeals to you.

Management – If management appeals to you there are many opportunities you can take but be prepared for career moves in order to get the position you finally want. Additional vocational qualifications will be required if you are survive this change – an electrical installations office can be quite a jungle if you are not use to it!

Technical education – Teaching is another option to be considered. You could take the post of a tutor at your local technical college, training centre or a manufacturer's school. Again, additional vocational qualifications will be needed if you take this path.

Technical consultancy – This position involves providing technical advice and helping to develop new ideas and concepts within the field of electrical installation engineering. This post is often established within a larger, nationally based electrical wholesaler.

Methods of team working

There are four stages of team working based on management methodology.

1. Storming
This is the stage which develops ideas, theoretical concept and collects together necessary relevant information.

2. Forming
The forming part of 'team working' checks out, amends and corrects the ideas and plans which have been made into suitable designs and applications.

3. Norming
Norming is the term used when you set up the ideas thrashed out to work within the limitations of your company.

4. Performing
Performing is the act of carrying out the ideas you have developed in a practical location.

Employment regulations – your responsibilities

Employment Rights Act (1996)
This section might prove to be a useful friend for it's all about your rights as an employee. There

are, regrettably, many small electrical companies who seem to be a law unto themselves who will try to get away with avoiding huge slices of legislation which other larger companies cannot.

As the employment regulations are fairly wide-ranging I have abridged them to include points of interest which are appropriate to our industry.

Under this Act you have a right to:

- A written contract/statement of employment. This will include:
  - Your employer's name and address.
  - The date you started with your company.
  - Details of pay rates (normal working and overtime).
  - When you are paid (weekly, monthly, etc.).
  - Your hours of work.
  - Paid annual leave and public holidays.
  - Sick pay (state or company).
  - Company pension details.
  - Notice of the termination of employment – either way.
  - Your job title.
  - If your job is not permanent, then the period of employment.
- An itemised pay statement – hours worked, rate of pay, overtime hours worked, gross and net wages, tax deducted, bonus payment and expenses, etc.
- Not to be harassed from unauthorised deduction from your wages – deductions can be made but you must sign a written statement authorising it. This signed declaration must state the amount to be deducted, and whether it is weekly, monthly or quarterly, etc.
- A minimum period of notice on the termination of your employment.
- Redundancy pay after a minimum of two years' service.
- Not to be unfairly dismissed – for example, if you are unable to work due to an accident or if you are forced to have time off for your dependents.
- Have time off to care for, or arrange care for, a dependent.
- Have four weeks' paid annual leave and be paid for any untaken leave when leaving the company

Other rights include the following:

- Should you accidentally lose or damage anything, your employer cannot insist in being refunded to the full value. There is a limit.
- Your employer has to guarantee payment of wages.
- On religious grounds you can object to working on a Sunday if you choose.
- You have an absolute right to time off for public duties – jury services, for instance, or representing employees.
- You have a right to arrange for additional training or courses without fear of being sacked.

 For additional information log on to the following web site:
http://www.john.antell.co.uk

The Data Protection Act of 1998

This act is to protect us against the misuse of personal data for commercial and other undercover reasons. These are the rules which a *data controller* must keep.

- The data obtained from a person must be processed both fairly and lawfully. If the data to be stored is to protect the subject, then the person's permission need not be required.
- The data you obtain may only be used for lawful purposes – it must not be reprocessed, sold to someone else or used for any other purposes other than for which it was intended.
- The personal data you process must be sufficient for your needs and not excessive in relation to why it is required. When collecting data, you must state the purpose of why it is required.
- Your data records must be both correct and up-to-date. Often, college students are asked to take their personal data home to check they haven't made any mistakes whilst processing their data forms.
- Data must not be kept any longer than is required. If, for example, you have collected personal information concerning several job applicants, your records must only be stored for the actual period of assessment. They must then be destroyed.

- All data must be processed in accordance with the rights of the *data subject* – the person who you are collecting data about. If that person wishes know their personal records, they can make a written request to you, the *data controller*, to see them. You will then have to reply to this request within forty days.
- Data agencies have a legal right to protect subjects against unauthorised and unlawful processing of personal data and against accidental loss or damage of documents or electronic filing. All data processed must be both private and secure.
- Data must not be transferred anywhere outside the EC – unless that country can ensure complete protection of the data supplied. Data must not be sold on to anyone else unless the subject of the data gives written consent.

 Additional information may be gained by logging into the Government's official web site – just type in Data Protection Act 1998 into a search engine – it should be among the first two or three entries.

Disability Discrimination Act (1995)

A person is said to have a disability if he/she has a permanent/semi-permanent physical or mental problem which has a bad effect on their ability to succeed with ordinary, regular daily living or if he/she has had such a disability in the past.

If you have a physical or mental problem that is not short term such as arthritis, multiple sclerosis or paranoia schizophrenia, all three of which can be controlled by drugs, then you will be sheltered under the umbrella of the Disability Discrimination Act of 1995. Approximately 20% of the working population have a disability and that is about *seven million* people! Many have fallen victims of industrial accidents and have been forced to take to a wheelchair.

This Act of Parliament was the very first anti-discrimination regulation to be passed which targets discrimination on the grounds of disability and is law throughout England, Scotland and Wales.

Challenging decisions

You can challenge decisions made within the workplace and take them to court concerning any possible discrimination on the grounds of disability. For example:

- If an employer or potential employer has treated a disabled person less favourably due to his/her physical or mental disability.
- If an employer has preyed on, or bullied, an employee with a disability, he/she will have good grounds to prosecute their employer within a court of law.
- If an employer has failed to make any modifications or practical adjustments to accommodate a person's disability the disabled person can take legal action to put right what is obviously wrong. An example would be wheelchair access.

The Race Relations Act of 1976 (amended in 2000)

The spirit of this Act is to uphold equal opportunities and maintain good relations between people of different ethnic groups – this is something we all have to work hard with.

There a few point in law which apply to the workplace:

- Direct discrimination within the workplace – less favourable treatment than offered to other operatives
- Indirect discrimination – failing to notice or deliberately overlooking a talented operative for site promotion or to work within a managerial level because he/she is racially different.
- Victimisation – preying on and hunting out with racial harassment, abuse and ethnic name calling.
- Vicarious liability (i.e. second-hand or indirect liability) – this is a legal term to describe the racial acts of an employee or a group of employees in which all knowledge is denied by his/her company, or if known would disapprove of them.

The 2000 (Amendment) Act places a legal responsibility on your company to encourage equal opportunity, promote good race relations among different ethnic groups of workers and, above all, to do away with illegal racial intolerance within the workplace. Your company is also obliged to put together and offer a suitable proposal of how this might be put into practice within the workplace.

Workers within the electrical installation industry must be treated fairly regardless of the following points:

- Whether they are men or women.
- No matter what faith they adopt.
- Their ethnic origin.
- Their nationality.
- Their race.
- The colour of their skin.
- Their age.

Sex Discrimination Act of 1975

The Sex Discrimination Act applies to the whole of Great Britain other than Northern Ireland, making it unlawful to sexually discriminate in matters of employment, training and technical education. The Act of Parliament provides the workforce with a direct path to a civil court or industrial tribunal where their case may be impartially heard and solutions offered for alleged sex discrimination. Subcontractors are also covered by this Act.

What is sex discrimination?

This Act of Parliament applies equally to both men and women – either can be discriminated against. Basically, there are two kinds which we must consider:

- direct discrimination,
- indirect discrimination.

Direct – This is when a women is treated on the grounds of her sex – for example, your boss might feel that a fully qualified female electrician would not fit in with the rest of his all-male company.

This is a another example amounting to direct sex discrimination. If an employer treats a married person of either sex in a less favourable way than a single person he would be guilty of direct sex discrimination.

Indirect – This applies equally to both men and woman.

Another illegal condition of employment would be to insist that all recruited personal must be a minimum of 56 kg and be at least 1.8 metres tall, or that the only way to secure a job would be to be single.

Complaints concerning sex discrimination are dealt with within an industrial tribunal.

Sexual harassment

Sexual harassment within the workplace is a criminal offence and, if convicted, can result in a very heavy fine, imprisonment or even both.

Where can I obtain further information?

 Log onto the following web site where you will find an easy to read guide: http://www.womenandequalityunit.gov.uk/legislation/discrimination_act.htm

Human Rights Act 1998

This is a complicated legal document which has now been incorporated into our domestic law. It is now illegal for a public authority, an individual or a company to act in any way which opposes this Act of Parliament. In summary form, please consider the following, which states that everyone has a right to:

■ Life – this is protected by law.

■ A fair trial – you are entitled to a fair trial and a public hearing within a reasonable span of time. This will be channelled through the medium of the courts, industrial tribunals, inquiries and by governmental announcement.

■ To respect privacy and family life – your private life will remain private with no prying from administrative systems other than in accordance with our domestic law.

■ Freedom of thought and expression – you have the absolute right to freedom of opinion, and to receive and give to others both information and ideas.

■ Not to be discriminated against – you cannot be discriminated against on the grounds of the following:
  – language,
  – race,
  – your sex,
  – religion,
  – the colour of your skin,
  – where you were born,
  – your political opinion.

■ The protection of their property – every one of us is entitled to the peaceful enjoyment of our possessions and property; except subject to the conditions which are provided for by our laws.

Your rights as a resident within the EU may need to be balanced against the rights of others or for the good of everyone.

Carrying out electrical work safely and efficiently

For our own sakes and for the safety of others we must try to work both safely and efficiently during our day to day activities. Please browse through the following table, Table 3.1 – you could find a few points which might help your day along a little easier.

Standards by which we assess both working practices and procedures

First let us take a look at ISO 9000

ISO stands for the *International Organisation for Standardisation* – it is in effect a world-wide association of national standards: BS, for example, represents the United Kingdom and ANSI serves the United States of America. To date it can boast of over 40 000 member companies spanning some 100 countries adopting ISO 9000 standards. The ISO 9000 standard, revised in December 2000, is just a common collective name which can be applied to a whole set of values we adopt within our day to day working practices and procedures.

ISO 9000 incorporates three standards:

■ ISO 9001 – looks into business activity, materials, product design, production/distribution and areas of customer service.

■ ISO 9002 – This is about the same as the previous standard except that the sections which cover both design and customer services are excluded.

■ ISO 9003 – This standard deals with inspection and testing procedures.

In addition to these three standards there are a few advisory documents which can be browsed

Table 3.1– *Carrying out your work both safely and efficiently.*

Your installation task	Relevant comment
Checking your instructions and specifications	This is most important as if you get it wrong it could cost your company a lot of money. Read the small print and read it carefully!
Checking your work for suitability	Is what you are installing suitable? You could wire a fire alarm installation using flexible cable. Yes, it would work – but it certainly would not be a suitable choice.
Deciding your sequence of work	Be practical – don't make work for yourself. If, as an example, you are carrying out a conduit installation, first check over the route, move obstructions which can be moved, calculate the size of conduit you will need, mark the position of your inspection boxes, then mount your saddles – install your conduit and then wire. Getting the sequence right will save you time and your company money.
Determining what tools and equipment you should use	It's a good idea just to select the tools and equipment you will require for your day's work – the rest may be safely locked away until needed.
Checking out the skills you will require	Look into the skills that are called for in order to do your work. If there is an area where you will need to brush up on, browse through an up-to-date electrical text book just to re-familiarise yourself with the subject once more. We all tend to forget things at some time or another.
Checking out what other trades are doing where you are working	Inquire, or ask the site manager whether other trades will be working in the same location as you will. The last thing you will need is for a carpet fitter to lay a floor covering when, for example, you are carrying out a formal test and inspection or doing a third-fix installation within a small flat.
Observing safety procedures and practices	Do remember to carry out safety practices and procedures. The more familiar you are with your work the easier it is to get distracted and have an accident. Don't end up as a statistic of industrial misfortune.
Making sure that your completed job meets with the specification and there are no departures from BS 7671	Read every word of your specification – it is important to get it right first time if waste is to be avoided. Not only may specific accessories be asked for (make, model, colour, etc.) but you may even be asked to supply calculations you have worked to determine the size of conductors when laid within thermal insulation – correction factors. Labelling, and the material your label is made from, the size of the lettering and the style of lettering (font) can also appear within a specification. Any departures from BS 7671 must be detailed when drawing up your installation test certificate and 'as-fitted' drawings. Test certificates are debated within Chapter 8.

through. As an example, please consider the following:

■ ISO 9000 – Guidelines for Selection and Use.
■ ISO 9004 – Quality Management and Quality System Elements – Guidelines
■ ISO 9000–3 – Guidelines for the Application of ISO 9001 to the Development, Supply and Maintenance of Software.

To play an active part within the International Organisation for Standards your company must register and then meet the terms of the organisation. There are several different decisive factors you will have to meet of which some are listed below.

■ To have a good reliable *quality system* – this means that good team leaders together with dedicated team members are required.

- Good record keeping and methods of identifying and tracking products – this will be one of your team member's responsibilities.
- Good training among employees – BS 7671 and beyond for technical staff.
- An 'out of house' audit system incorporating evidence of compliance – your Insulation test and inspection certificate will help your auditor.
- Good measurement and testing procedures and calibrated test equipment – as called for by BS 7671.
- In-house statistical procedures – your office and team leader will deal with this area.
- A well established method of handling, storing, packaging and delivering goods and commodities – in-house procedures.
- A good procedure for testing and inspecting your product – this will be associated with the electrical work you do.

How the rules are observed

The observance of these rules is confirmed by setting up a 'quality system' which acknowledges some twenty elements. Documenting the data associated with these elements is the formal responsibility of your company. As an example, here are a few:

- design control,
- testing,
- trade trading,
- inspection procedures,
- buying department,
- the responsibilities of your management team.

Each main route in all twenty areas will be described and formally recorded in a simple, easy to follow language code stating 'who is responsible for what' and 'what is carried out by whom'.

A team leader will be responsible for a number of team associates. The team members will take care of all documentation that relates to the *quality system* for their group.

Different groups within your company, such as your electrical installation engineering or test and inspection department, will have their own relative *quality systems* – so each will deal with different parts/sections of the whole standard.

After your 'quality system' has been formed

Once your company's *quality system* has been set up and is working as intended, an auditor/examiner will have to be chosen to review and assess the system to ensure that it meets the terms of the published ISO 9000 standards.

If problems are discovered during the auditing period your company will have time to put things right, after which a certificate will be issued recording the level of compliance you have reached.

Investors in People

This is a standard, promoted by Investors in People UK, which is a public body whose stakeholder (that is, someone who has a share or interest in an organisation) is the Department of Education and Skills.

I feel you will agree that an electrical installation company's greatest asset is the people they employ to carry out their tasks and projects. In order to make money and succeed, where others might have failed, the workforce has to be well trained, knowledgeable regarding their level of skill and be able to work unsupervised and resourcefully.

The Investors in People Standard advances and cultivates workable schemes which are both practical and sound in order to develop both productivity and performance within the workplace.

What is it all about?

The Investors in People Standard is a way to provide a schedule so you will be able to bring about your ambitions. You will find it will greatly help you to contact these people. It could considerably change the way you carry out your everyday business activities; but because different companies operate in different ways, Investors in People will not adopt any single method in order to get their message across. All they will do is to provide a support framework to help both you and your company get the most of what you have and to achieve success by means of your workforce.

The cornerstone of this Standard rests on four basic people-development principles:

- Commitment – to achieve business goals and targets by developing your in-house workforce.
- Planning – to review your in-house technical training and development needs.
- Action – making sure what you have decided at management level is put into practice.
- Evaluation – assessing the result of technical training and progress made by individuals within the workplace. Measuring the overall effect the Standard has produced within your company

Towards achieving the Standard – the way ahead

There are ten outlines or 'points' which must be considered:

1. First, a scheme must be developed which will radically improve the performance of your company by the use of its workforce.
2. Both train and develop to a higher potential within your workforce.
3. Plan to promote quality of opportunity within your staff. Everyone must have an equal chance to develop their skills within the workplace.
4. Managers must not only lead but develop their staff to achieve a higher effectiveness – for example, some people are better at doing certain things than others or 'horses for courses'.
5. Managers must be efficient in leading their staff, directing and encouraging their workforce to further development.
6. Your workforce's input must always be acknowledged and appreciated. Don't take them for granted. Praise them – pay them!
7. Staff must be encouraged to be responsible by involving them in decision-making.
8. Encourage your workforce to learn and develop both efficiently and professionally.

9. By doing what is required, Investors in People will improve your company's overall performance rating.
10. Constantly monitor your scheme and add on-going improvements as to the way people are managed whenever it is felt necessary.

 The complete standard can be obtained, at cost, by e-mailing the following address: iipuk@tsc.co.uk

As well as the obvious benefits of the Standard, plus the Guidance Notes to serve the above ten outlines and how these demands can be met, the package also includes information concerning additional ways your organisation can be helped by Investors in People.

How can I tell if any of this has had any effect?

A good way is to informally talk to your immediate staff and outside workforce and find out, by just by asking the right questions – whether, for example the Standard has made any difference to them personally or within the company generally. This would be a good start. Alternatively, publish your own questionnaire using a multiple choice question method – this way it would be reasonably straightforward for you to assess the findings.

The benefits of improving working practices and procedures

Customer satisfaction

You might be surprised to learn that the outlay involved in finding and keeping a new client is between five to seven times greater than keeping an existing one! This is clearly a serious message – a point we should all take note of.

Customer service satisfaction should be a high priority for an electrical installation or engineering company. Measure your *customer satisfaction factor* by learning more about your client and from the informational feedback you will receive from them. This will help to pave your way towards improved business efficiency practices and will have a positive knock-on effect by expanding your profits.

How can I gauge customer satisfaction?

There are many ways – here are just a few of them to consider and talk about among yourselves:

- questionnaires (multiple choice are easier to process) sent to your customers,
- telephone interviews with people who matter,
- e-mail surveys directed at certain individuals within your clients' organisation,
- an old fashioned 'face to face' interview,
- employing a specialised company who will do all the hard work and research for you. The results obtained will be very professional – but at a cost!
- by means of your own web site.

How can I change my working practices and procedures to produce customer satisfaction?

Here are ten suggestions you might like to think about:

1. Target areas where you feel your company needs to improve. Obtain this data from past feedback sources.
2. Quickly respond to enquires or emergency call out requests.
3. Development training, from management to in-house employees, must be considered.
4. Aim for a clearer command of communication between all concerned from management and employees to client.
5. Keep all your customer data records up to date – important contact names and telephone numbers are the lifeline of your company.
6. Back up, whether it is information or operatives, must be delivered quickly and efficiently to where it is needed.
7. Possibly *monetary incentives* could be considered for outstanding improvements to customer relations, or if customer service targets have been met.
8. Try to build long term customer relationships – it will help your business prosper.
9. Identify aspects and issues which could be responsible for lost sales, productivity and customer confidence.
10. Operatives who are performing well in your customer's eyes should be financially rewarded by their company.

If you do well and you have plenty of customer satisfaction your clients will come back to you time and time again. It's worthwhile to work hard at it!

Improved productivity

An improvement in site production does not happen overnight, or by accident. It has to be planned and planned well in advance of any work carried out at site level. Here is a practical list of ways to improve productivity at the site level. Please feel free to use this listing as a discussion point among yourselves:

1. Better off-site planning is needed before your project begins.
2. The way you are to do things must be well organised before work actually starts.
3. A greater input and commitment from the company's management team is required.
4. Create productivity targets together with incentives.
5. Provide a team bonus scheme based on genuine productivity.
6. By employing the same team to work on a similar project will significantly increase productivity within the workplace.
7. Sending information electronically rather than by conventional means or by word of mouth is more efficient.
8. Direct labour costs about one quarter of your project costs – use your own staff; avoid agency operatives.
9. Fit global tracking devices to company transport to eliminate 'time-stealing' and journey time wasting. This will not be popular!
10. Avoid road traffic delays – start early; work flexi-time.
11. On-site teamwork is most important between different trades and site management.
12. Research your project at the design stage and turn up with a sensible programme of work and the time it will take to complete.
13. Productivity can be increased by avoiding site-related stress and keeping accidents low.
14. Insist that your deliveries arrive on time and at the right location.

15. On very large sites (the size of the Euro-Tunnel for example), employ a specialist logistics contractor to distribute material in and around the site. Productivity will rise if tradesmen don't have to collect materials from their company store or local wholesaler where time is wasted.

16. Use the right people for the right job – don't, for example, ask an electrician who is highly efficient at wiring houses or old people's homes to carry out an electrical installation within a new supermarket project; it will not work!

17. During the early stages of any project, your design team need to work closely with the main contractor. Ask questions and get involved!

18. Improved 'site-housekeeping' within your immediate working area will help towards additional productivity and keep accidents at bay.

19. Good site welfare facilities are most helpful.

20. An experienced job holder is essential to improve site productivity.

If taken seriously, these points will lead to significant improvements in site productivity. This is something we all have to work on.

More efficient use of resources

We are all guilty of waste and not being very efficient with the resources we have available to us. Here are a few ideas that will help at site level:

- Look after your assets – they are there to help you.
- Identify risks and hazards which will affect your plant and equipment.
- Choose appropriate working practices to make more efficient use of the resources available to you.
- Try not to waste energy such as gas in portable cylinders serving site heating or cooking equipment – it can be used for another day.
- Try to make certain that efficient management practices are carried out at company level.
- Recycle your site waste, if possible.
- Avoid using a ladder where it is better to use steps.

- Use a multimeter instead of several different instruments.
- Don't waste your company's petty cash on unnecessary purchases.
- Avoid unnecessary journey in company vehicles – fuel is expensive.
- Use the right power tool for the job. Don't, for example drill out a square hole to serve a steel accessory box where a chasing machine would be a far better choice.

There are many more points which could be considered, as you can well imagine – but space is limited. Aim for continuous improvements in this field; it will be an advantage for you in the long run and could make your job that much easier.

Increased profitability

There are several ways in which your company can increase their profitability and that doesn't mean a reduction in pay or doing away with travelling time! Here are a few suggestions to think about. Would any of these proposals fit comfortably into your company?

1. Team up with a successful building organisation. You will not win all of their contracts but you will be added to their 'Preferred Electrical Contractors List' – and that is worth a lot of business in the long run.

2. Specialise in carrying out Building Regulations – Part 'P' work. Part 'P' came into force in January 2005.

3. Become a member of the National Inspection Council for Electrical Installation Contracting. This will help you win local government contracts.

4. Create a test department to carry out electrical installation tests and inspections to National Inspection Council standards.

5. Contribute to portable appliance testing work serving industrial, commercial and building organisations.

6. Offer electrical surveys, periodic testing and inspection work to estate agents, industry and commerce and to building companies.

7. Become a 'Kitemark' electrical installer – details from the British Standards Institute, London.

8. Build up an electrical maintenance department to serve the needs of industry, commerce, domestic sites and agriculture within your area.

9. Advertise your specialities in your local newspaper – be bold!

10. Form a design and control department for specialised electrical work focusing on industry and commerce.

Try to develop these ideas from a theoretical to a practical level among yourselves or as a discussion group. Talk about the advantages and disadvantages of such an untried venture. How, for example, would any of the ten new enterprises fit within your company?

Handy hints

- Provide a laminated floor-plan layout attached to each landing wall for all to see. Reinforce the plan by temporary labelling areas of importance or flat numbers. This will help people who are new to the site to settle in quicker.

- Indicate all site fire escape routes and exit points throughout a large construction site.

- If you are forced to leave unconnected cable ends – terminate them safely and if it is possible they could ever become 'LIVE', highlight the cable with white or yellow PVC insulation tape and write 'LIVE' using a dark coloured permanent marker pen.

- Part 'P' of the Building Regulations came into force on 1 January 2005. It applies to both houses, flats, all types of dwellings and businesses which are attached to dwellings which share a common metered supply.

- Never even try to test an installation without first disconnecting all RCBO's in circuit. You will record values of just below half of one megaohm from your neutral conductor to earth – test properly!

- The HSE estimates there are around 36 000 people with an advanced stage of vibration white finger (VWF). Don't let this happen to you – select vibrating hand tools, such as drilling or chasing machines, with care. Adopt a low vibration purchasing policy. Manufacturers of vibrating tools and equipment identify vibration levels in units of metres per second squared. Mathematically this is written as m/s^2.

- Quickly check out the internal wiring arrangements of a combination distribution centre served by two independently fitted residual current devices. Manufacturers have been known to make mistakes.

- If you are in doubt always have a word with an experienced colleague you can trust.

- Bending springs used for heavy gauge PVC-u conduits are colour coded green whilst those used for the lighter gauge are coloured white.

- Obtain all your site variation orders in writing before work is carried out. Never rely on trust – people have short memories!

- Use only the correct intrinsically-safe power tools and equipment in laminable areas. Standard low voltage 110 volt gear offers no protection against igniting laminable dust or vapour.

- Take care when cutting single solid conductors (1.0 mm^2, for example), with sharp side cutters. Target the waste copper to the ground as serious damage to your eye could result from careless cutting.

- Digital test meters are easy to read but mistakes can be made when targeting the positing of the decimal point,

- A high frequency fluorescent light can trigger a residual current device into thinking that a fault condition has occurred. Many of these fittings have a suppressor fitted between the phase, neutral and earth.

Summary so far . . .

- JIB member companies grade their electrical workers as follows:
  - electrical labourer,
  - apprentice,
  - electrician,
  - approved electrician,
  - technician

- There are many career paths you can choose, once qualified. Management, technical education and installation are just three of many.

- Methods of team working are: storming, forming, norming and performing.

- You have many rights under the Employment Rights Act of 1996. These include the following: a formal job title, an itemised pay statement, and a minimum period of notice in which to terminate your employment. You are also entitled to four weeks' paid holiday annually.

- The Data Protection Act defends you against the misuse of personal data for commercial and other reasons.

- A person is said to have a disability if they have a permanent or semi-permanent physical or mental problem which has a bad effect on their ability to succeed with an ordinary job or day-to-day living – or if he/she has had a disability in the past.

- The Race Relations Act of 1976 (amended in 2000) stems racial discrimination in job opportunity programming and employment within the workplace.

- The Sex Discrimination Act makes it unlawful to sexually discriminate in matters of employment, training and technical education. (Applies to the UK other than Northern Ireland.)

- The Human Rights Act makes it illegal for a public authority, a company, or an individual to act in any way which is against the spirit of this Act of Parliament.

- Select only the tools and equipment you will need for your day's work – the rest are best safely locked away until required.

- The International Organisation for Standardisation is a world-wide association of national standards.

- Investors in People UK is a public body whose stakeholder is the Department of Education and Skills. Their aim is to create and cultivate workable schemes which are both practical and sound in order to develop both productivity and performance within the workplace.

- Customer satisfaction is an important factor to consider in any business. The all-round outlay involved of finding and keeping a new customer is between five to seven times greater than keeping an existing one.

- An improvement in site production requires planning at every level. Taken seriously, your productivity planning will lead to considerable advances within the workplace.

- Use the resources available to you more efficiently. Avoid wasting bottled gas serving site heating appliances – it can be used for another day!

- There are several ways listed within this chapter to increase company profitability. One way is to create a department to deal with all aspects of portable appliance testing (PAT) work.

- Learning new skills, techniques and procedures will help to provide you with additional confidence and greatly increase your chances of a better job.

- Your employment rights state that you cannot be unfairly dismissed, if for example, you are unable to work due to an accident or you are forced to have time off because of your dependants.

- The Employment Rights Act of 1996 states that you will be able to claim redundancy pay after you have been with your company for two years. How much you will receive will depend on your age and how long you have been with your company.

- A simple way to gauge your customer's satisfaction in to send a simple-to-fill-in questionnaire a few weeks after the job has finished.

Review questions

1. Provide three practical examples of how you may increase your learning potential.

2. What qualifications must you have in order to be graded as an approved electrician?

3. What are the four stages of team working?

4. Under the Employment Regulations we are all allowed the following amount of paid leave annually: (a) four weeks, (b) three weeks, (c) twenty-one days, (d) up to five weeks.

5. Data stored within a company's computer must keep to one of the following requirements: (a) only very personal data is permitted, (b) data can be freely reprocessed or given to others, (c) must only be sufficient for your company's needs, (d) information, under the Act, may be kept as long as required

6. An employer does not have to make any modifications or practical adjustments to accommodate a person's disability. TRUE/FALSE

7. Describe, in terms of the Race Relations Act of 1996, what is meant by 'direct discrimination' within the workplace.

8. Provide a brief example of direct sex discrimination at work.

9. List four of the 'rights' we all enjoy under the Human Rights Act of 1998.

10. Why is it sometimes wise to check out with your site manager the activities of other trades where you plan to work next?

11. ISO 9000 can be loosely described as one of the following: (a) a standard which only deals with testing and inspection, (b) a UK government body responsible for electrical standards, (c) an international association of national standards organisations, (d) a standard originated from the British Standards Institute.

12. Describe briefly the aims of the Investors in People Standard

13. Recommend three ways of researching customer satisfaction.

14. Suggest a way of reducing 'time stealing' and 'journey-wasting' within your company.

15. Summarise in simple terms why teaming up with a successful building company could increase your own organisation's profitability.

Introduction

There are seven underpinning knowledge requirements in this chapter, all of which are compulsory. In summary form, they are as follows:

- ☐ Resistors and Ohm's law.
- ☐ Magnetism and magnetic circuits.
- ☐ Inductance and inductive components.
- ☐ Capacitors.
- ☐ The effects of resistance, inductance, capacitance and impedance in an AC circuit.
- ☐ Semiconductor devices in simple rectified circuits.
- ☐ Basic electronic circuits and components.

Practical activities

There are no practical activities accompanying this chapter.

Tests

Questions relating to the contents of this chapter will be included within your paper.

Resistors

What is electrical resistance?

All materials have a natural intrinsic (built-in) resistance to the passage of electricity which can be formally defined as:

'The ratio of a potential difference, in volts, placed across an electrical component to the current in amps which is passing through it.'

The symbols we use are R and W. In general terms most conductors of electricity will increase in resistance with an increase in temperature whereas semiconductors, the components we use in electronics, will decrease with a corresponding temperature rise.

Factors which affect the resistance of a conductor

There are four factors affecting the resistance of a conductor:

- what the conductor is made from – copper, aluminium, silver, bronze, etc.,

- the cross sectional area of the conductor,
- the temperature of the conductor,
- the length of the conductor.

Material composition

The electrical conductivity of a conductor is very dependent on the size of its atoms and the number of free valance electrons available within their outer shells. Silver is one of the best conductors of electricity followed very closely by copper. These materials have large atoms and just one loosely held free electron. For practical purposes we only tend to use copper and aluminium as silver is far too expensive.

The cross sectional area

The resistance of a conductor is proportional to its length but inversely proportional to its cross sectional area. A 10 metre length of 16 mm^2 copper cable has far less resistance than an equivalent length of 1.0 mm^2 cable. To explain further, take a look at Figure 4.1 illustrating a water tower supplied with two outlets. One is far

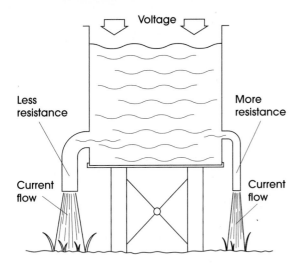

Figure 4.1 – *Current flow can be represented as water. The larger the outlet, as shown, the greater the potential current flow*

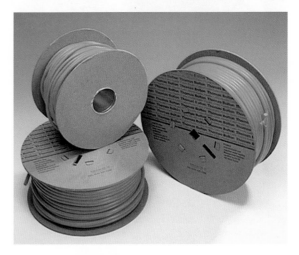

Figure 4.2 – *This is a reel of 2.5mm² blue 'arctic' cable*

bigger than the other. Let the water represent current flow and the pressure of the water, voltage. The larger outlet has the capacity to deliver a greater volume of water than its smaller companion. This is because the water experiences far less resistance – and so it is with electricity.

The temperature of a conductor

The temperature of a conductor must be taken into account when calculating the expected change in a conductor's resistance due to extreme temperature differences. Generally, a rise in temperature will produce a rise in resistance for most conductors of electricity. Nevertheless, there are a couple or so elements which have what is known as a *negative coefficient of resistance* and this we will talk about later. These materials, when exposed to heat, will decrease their resistance proportionally to a rise in ambient temperature. Silicon and carbon are just two examples of basic elements which show these characteristics.

The length of a conductor

The last factor affecting the resistance of a conductor is its length. The total resistance of a length of wire is always in direct proportion to its length – so if a very tiny conductor has a

measured resistance of 1 ohm per metre, then the total resistance of 50 metres will be 50 ohms – no more, no less! Figure 4.2 highlights a 100 metre roll of three-core, 2.5 mm² – 'Artic-cable'. As an in-house exercise, measure the electrical resistance of just one metre of this type of cable and then calculate the total phase conductor resistance for the whole roll.

What does the term resistivity mean?

The term *resistivity* describes the ability of a material to resist the flow of an electric current (in basic terms we are referring to a flow of electrons through a load-bearing conductor).

Different materials, equal in both physical dimensions and shape, show very different values when their electrical resistance is measured. In practice, the formal resistivity of a material is measured between the two opposite faces of a metre cube of the material under examination.

The unit of resistivity is the ohm/metre, and is identified by the Greek letter ρ (pronounced 'rho'). The resistivity for copper at 20°C is 1.8×10^{-8} Ω/m and for aluminium, 2.8×10^{-8} Ω/m.

To calculate the total resistance of a conductor please use the following expression:

$$R = \frac{\rho \times L}{a} \qquad (4.1)$$

where
  R is the total resistance of the cable in ohms,
  ρ is the resistivity of the conducting material in Ω/m,
  L is the total length of the conductor in metres,
  a is the cross sectional area in metres squared (m^2).

Please take into account that the natural resistance within a conductor will cause a voltage drop to occur. The total voltage drop is directly proportional to the current flowing in the circuit and is measured by multiplying the current flow by the voltage applied ($I \times R$).

The length, cross sectional area, resistivity and resistance of a conductor

This can be easily summed up by the use of Ohm's Law which states:
  'The current flowing in a circuit is directly proportional to the voltage applied to that circuit, but inversely proportional to the resistance at a constant temperature.'
  This is best remembered as:

$$I = \frac{V}{R} \tag{4.2}$$

where
  I is the current flow in amps,
  V is the voltage applied in volts (E is sometimes used and stands for 'electromotive force'),
  R is the resistance of the circuit in ohms (Ω).

Where resistivity can be applied

It is useful sometimes to be able to work out the voltage drop experienced on a long circuit whose current flow is constant. To do this, consider the following expression:

$$\text{Voltage drop} = I\left(\frac{\rho L}{a}\right) \tag{4.3}$$

where
  I is the current flowing in your circuit,
  ρ is the resistivity of the material the conductor is made from,
  L is the total length of your cable,
  a is the cross sectional area of the conductor.

Remember to bring all terms within your equation to standard units:

$$1 \ m^2 = 1\,000 \times 1\,000 \ mm$$
$$= 1\,000\,000 \ mm^2.$$

Series and parallel circuits

Series circuits

Figure 4.3 illustrates a typical series formation circuit of three 2 ohm and one 10 ohm resistor served by a 32 volt supply. All components are wired one after the other, indicating that the current is the same throughout the formation. Removing just one component will halt current flow throughout the entire circuit – very similar to our old fashioned Christmas tree lighting arrangements; remove one lamp, and they all go out!

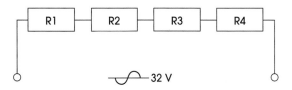

Figure 4.3 – *A typical series formation circuit*

The total resistance in ohms of any series formation circuit can be determined by the following expression:

$$R_t = R_1 + R_2 + R_3 \ldots \tag{4.4}$$

where R_t is the total resistance of the circuit and R_1, R_2 and R_3 are the individual resistance values.
  Substituting figures into (4.4) we have, as an example:

$$R_t = 2 + 2 + 2 + 10$$
$$\therefore R_t = 16 \text{ ohms (the total resistance of the series circuit).}$$

By applying Ohm's law, (4.2), the total current flow within the circuit can be found:

$$I = \frac{E}{R}$$

$$I = \frac{32}{16} = 2 \text{ amps.}$$

Confirming the voltage applied

The applied voltage can be determined by rearranging the terms of (4.2).

We do this by cross-multiplying the equation:

$I \times R = E \times 1$ or in familiar terms,

$$E = I \times R. \tag{4.5}$$

Replacing with known values:

$E = 2$ amps $\times$ 16 ohms.

The applied voltage, $E = 16$ volts.

Calculating the power dissipated

Power generated is measured in watts; a 3 kW electric fire (3000 watt) or a 2.2 kW (2200 watt) electric motor, for example. The SI symbol is W. Hence:

$$W = I \times E. \tag{4.6}$$

Replacing the terms with known values:

$W = 2$ amps $\times$ 32 volts

Total power dissipated = 64 watts.

Parallel circuits

A circuit is said to be connected in *parallel* formation when all the components are connected side by side as illustrated in Figure 4.4. To obtain the total resistance of a parallel circuit the following expression must be used:

$$\frac{1}{R_T} = \frac{1}{R_1} + \frac{1}{R_2} + \frac{1}{R_3} \ldots \tag{4.7}$$

where R_T is the total resistance of the parallel circuit and R_1, R_2 and R_3 are individual resistances.

As a practical problem, consider the following:

Calculate the total resistance of one 10, one 20 and one 30 ohm resistor wired in parallel formation.

Using expression (4.7) and substituting for figures:

$$\frac{1}{R_T} = \frac{1}{10} + \frac{1}{20} + \frac{1}{30}.$$

The next step is to find a common multiple that all three numbers will divide into exactly. This

number is 60. Divide the value of the resistor into 60 and then multiply by the number placed above it. For example, 20 ohms divided into 60 is 3. Three multiplied by 1 is 3 and so forth. Therefore:

$$\frac{1}{R_T} = \frac{6 + 3 + 2}{60} = \frac{11}{60}.$$

Now bring the equation into terms of R_T by cross multiplying then dividing each side of the equation by 11:

$$R_T = \frac{60}{11}.$$

Thus the total resistance in the parallel circuit = 5.45 ohms.

The answer will always be *smaller* than the smallest value of all those resistences in the parallel circuit – if larger, please take another look at your arithmetic.

The total resistance of a parallel circuit that has just *two* known values can be quickly calculated using the following expression:

$$R_T = \frac{R_1 \times R_2}{R_1 + R_2}. \tag{4.8}$$

A sideways look at Ohm's law

The most basic form of Ohm's law was published in 1827, but as Table 4.1 shows, there are many different variations of this original expression. It might be wise to learn these alternative formula as *Ohm's law* is far from satisfactory should you have a problem concerning *current* flow when your only available data is the power generated in *watts* and the resistance of the component in *ohms*.

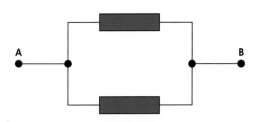

Figure 4.4 – *This is known as a parallel circuit*

Table 4.1 – *These are alternative ways of calculating current, power and resistance together with the applied voltage in DC and non-inductive AC circuits.*

Current (I)	$\frac{E}{R}$	$\frac{W}{E}$	$\sqrt{\frac{W}{R}}$
Power (W)	$E \times I$	$I^2 \times R$	$\frac{E^2}{R}$
Voltage (V)	$\sqrt{W \times R}$	$I \times R$	$\frac{W}{I}$
Resistance (R/Ω)	$\frac{E}{I}$	$\frac{W}{I^2}$	$\frac{E^2}{W}$

Magnetic circuits

Magnetic field

This can be defined as an invisible field of force which is present around any load bearing conductor or magnetic body such as a permanent magnet – see Figure 4.5. The strength and direction of the magnetic field strength is often referred to as *flux density,* symbol **B.** The SI unit of *field strength* is the *amp per metre*, written as $A\,m^{-1}$.

It has long been considered that the lines of magnetic force travelling from a North to a South seeking pole are virtually *parallel* and adopt an oval format with respect to each other. The lines of force are so positioned that they do *not* touch one another nor in any way cross over. This arrangement is pure magnetic energy and where the lines are crowded together it is at its strongest; where they are further apart the magnetic field is at its weakest. Figure 4.6 illustrates the lines of force surrounding a bar magnet

It has been shown that a load bearing conductor is completely surrounded by an invisible, circular, magnetic field positioned at right angles to the conductor in the form of ever-increasing circles. If this conductor is placed between a strong permanent magnet it will deflect its lines of force as shown in Figure 4.7.

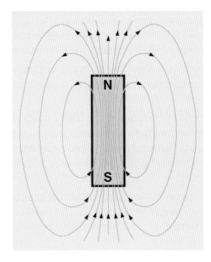

Figure 4.6 – *Magnetic lines of force surrounding a bar magnet*

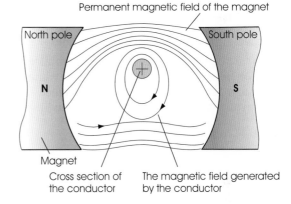

Figure 4.7 – *The effect of placing a load bearing conductor into a permanent magnetic field*

The magnetic effect of a load bearing conductor can be increased by winding the wire into a small coil of many turns.

First principles

- All magnets have a North and a South pole. Unlike poles will attract one another whilst like poles (a North and a North, for example) will repel one another.
- If a small compass is placed near the North pole of a bar magnet it will swing round, repelled by its North seeking pole.
- The strength of a magnet cannot be increased beyond a defined limited. Magnetic forces are energy.

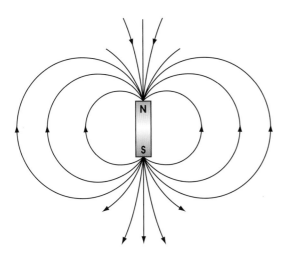

Figure 4.5 – *A magnetic field surrounding a load bearing conductor*

- Iron, cobalt, nickel and one or two alloys (mu-metal for example) are attracted to magnets. These are called ferromagnetic material.
- Material such as plastic, brass, copper and aluminium are not attracted to magnetic bodies.
- The ends of a magnet are called *poles*. It is here the magnetic attraction is at its greatest. The field around a single pole magnet has one small neutral point but two around a bar magnet.
- The space immediately around a magnet is known as a magnetic field. The field contains magnetic flux which has both magnitude and direction.
- Magnetic lines of force always present themselves as closed loops; no line ever begins or ends.
- The spacing between the lines of force is known as the field intensity (**B**).

Discovering the shape of a magnetic field

To plot a magnetic flux pattern, evenly sprinkle a few iron fillings onto a sheet of 160 g/m² white paper and place over the top of a bar magnet. Now gently tap the stiff paper with a screwdriver or pencil. The iron fillings will immediately adjust themselves into a traditional magnetic field pattern as shown in Figure 4.8. It will seem as though each tiny iron filling has suddenly become a bar magnet within its own right. You will be able to plot the course of these magnetic lines of force, seemingly leaving one end of the magnet and entering the other.

Additional data you will need

1 *Flux density* can be described as the number of magnetic lines per unit of cross sectional area.

2 *Magnetic flux*, symbol ϕ, is the product of the average magnetic field times the perpendicular area it penetrates.

3 A *magnetic field* can be loosely described as the area aaround a small stand-alone compass that can be affected by a nearby permanent bar magnet.

Magnetism is everywhere and it doesn't require ferromagnetic material to produce it – space, the sun and some of our planets are just a few examples.

Electrical inductance

A load bearing conductor served by an AC supply is always accompanied by a variable electromagnetic field. This is caused by the changeable nature of the voltage which produces a fifty cycle per second wave form. When this voltage is applied to an inductor, Figure 4.9, it has the effect of inducing (generating) a secondary

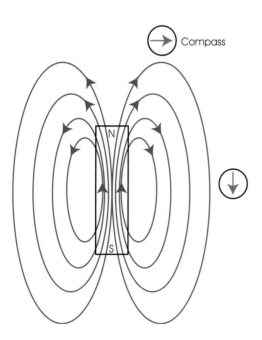

Figure 4.8 – Plotting the magnetic field with iron filings

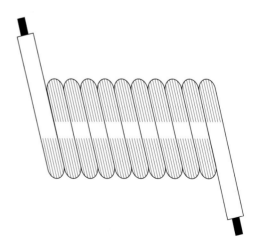

Figure 4.9 – A secondary voltage is induced into the windings of the inductor

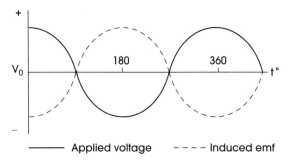

Figure 4.12 – *A BSEN symbol for an inductor*

Figure 4.10 – *The induced voltage is directly opposite to the principal supply*

voltage within the windings. This 'mirror' voltage is directly opposite the voltage of the principal supply as illustrated in Figure 4.10. Since this secondary induced voltage is 90 degrees out of phase with the applied voltage it will act as a current limiter. You might feel that this is just another type of resistor. This is not a resistor as current limiting is caused through inductance and is known as *inductive reactance*. The SI unit of inductance is called the henry (H), and is identified by the symbol X_L.

The value of an inductive circuit is calculated by use of the following expression:

$$X_L = 2\pi fL \qquad (4.9)$$

where f is the frequency of the voltage applied (50 or 60 hertz), L is the value of the inductor in henries and π is taken to be 3.1428.

In a pure inductive circuit the current lags behind the voltage by 90 degrees as Figure 4.11 will show. The stored energy is temporarily held within the electromagnetic field surrounding the inductor and this is returned to the circuit each time the magnetic field collapses. In the UK this will amount to fifty times every second!

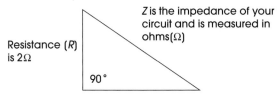

Z is the impedance of your circuit and is measured in ohms(Ω)

Resistance (R) is 2Ω

90°

The inductive reactance X_L is measured in ohms and equals $2\Omega_{fL}$ (3Ω)

So the impedance $Z = \sqrt{R^2 + X_L^2}$

Figure 4.11 – *In an inductive circuit the current lags behind the voltage by 90°*

Unfortunately this is not 100% efficient. Less energy is induced back into the inductor as some of it is scattered and wasted in the form of heat generated to overcome the natural resistance of the winding. Figure 4.12 illustrates how an inductor is drawn when designing a circuit.

Inductive components

Listed below are a few of these which are available today:

- Site and general purpose transformers.
- Inductors used within the electronics and engineering industries.
- Chokes for the control of inductive lighting arrangements.
- Coils used for relays and contactors, etc.
- Industrial magnets employed in industry.
- Magnetic heads used in audio and video recorders.
- The common door bell.
- Electrical locking arrangements.

A sideways look at generating electric current

Dynamic induction

Dynamic induction can be explained when a body which has electrical or magnetic properties can cause or induce a voltage into another body without direct 'hard-wire' contact. A well known example of this is your site transformer.

It can also be explained as when a variable (AC) current of electricity motivates a mirrored current in a nearby conductor which has been formed into a closed loop. This will happen if a seemingly 'dead' conductor is lying close to a conductor carrying a heavy load. The electromagnetic emissions produced by the heavy current transfer themselves across into the 'dead' conductor as an induced voltage – Figure 4.13 illustrates this. The induced voltage is small but it could mislead you if you were to foolishly test this conductor using a neon-type screwdriver.

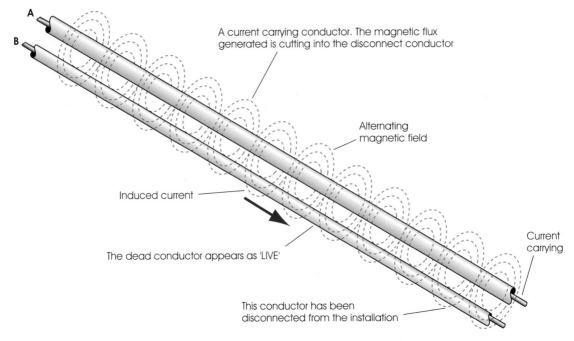

Figure 4.13 – *The load bearing conductor 'A' induces a measurable emf into the dead circuit 'B'*

Static induction

In the previous paragraphs we have seen how electromagnetism can induce a voltage in an adjacent coil without actually being connected to it. Similarly, an electric charge can materialise on an object without actually touching it. This process is known as electrostatic induction.

The formal definition of static induction is:

'The action by which a body possessing a charge of static electricity develops a charge, which is both equal and opposite in nature, within a neighbouring body without touching it.'

A practical example follows. Walk around a large nylon carpet wearing shoes which have man-made soles. Take with you an old fashioned neon-test screwdriver which will act as a crude 'instrument' for detecting static electricity. After a couple of minutes, place the blade of the screwdriver onto an electrically earthed item. You will immediately witness a bright luminous display from the body of the screwdriver without receiving a nasty shock.

This is caused by static electricity – triboelectricity. This type of electricity can also be generated by rubbing a hard rubber rod with clean dry fur. If the rod is directed towards a small pith ball suspended from a silk thread tied to an insulated scaffold, the pith ball will be attracted towards the charged rod. Likewise, if the pith ball is replaced by a small hard rubber ball which has been vigorously rubbed with fur, the charged rod will repulse the ball because both ball and rod have the same electrical charges. This is a fundamental rule; unlike poles attract; like poles repel – Figure 4.14 illustrates this.

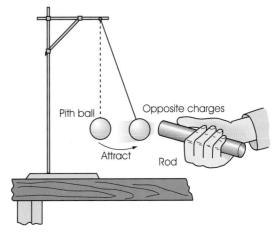

Figure 4.14 – *When a hard rubber rod has been rubbed with dry fur the pith ball will move towards the charged rod*

What is magnetic flux?

A force field always accompanies an active magnet – this you will experience for yourself when approaching a north seeking pole with another magnet of the same polarity.

A magnetic force field occupies an area directly surrounding the magnet and weakens with respect to the distance from the magnet. Within the force field lies the lines of magnetic flux, the strength of which will depend on its *flux density* or, to put it another way, the number of lines of magnetic flux there are within a given area. The more parallel lines there are, the stronger the magnetic effect will be.

Magnetic flux can be plotted using a tiny map compass or with iron filings sprinkled onto stiff paper, which is then laid on top of your target. The stiff paper must be gently tapped with a pencil and the iron filings will then position themselves to the pattern of the magnetic lines of flux as shown in Figures 4.15 and 4.16.

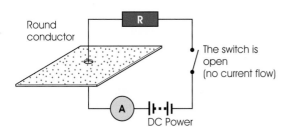

Figure 4.15 – Iron fillings sprinkled over a sheet of white paper with the circuit switched 'OFF'

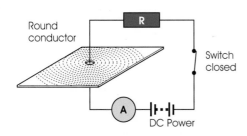

Figure 4.16 – When the circuit is switched 'ON' the iron fillings adopt a circular pattern around the conductor aligning themselves with the magnetic lines of force

Self and mutual induction explained

Self induction

- Remember that a load bearing conductor is always accompanied by a magnetic field (the shape of which mirrors the cross sectional profile of the conductor), see Figure 4.17. Electricity entering a coil of wire will immediately be changed into magnetic energy as shown in Figure 4.18.

- On breaking the circuit, the magnetic flux collapses and this energy is transferred back into the coil as energy

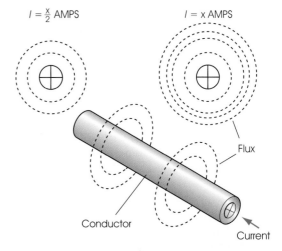

Figure 4.17 – The shape of the magnetic field always follows the profile of the cross sectional area of the conductor whilst on load

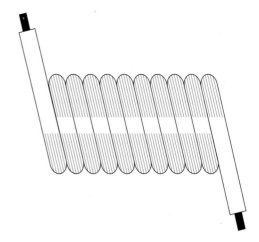

Figure 4.18 – Coiling a conductor will concentrate the magnetic fux

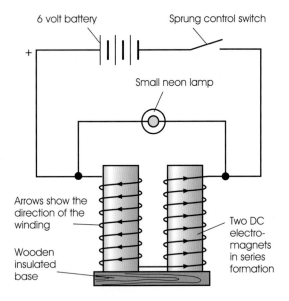

6 volt battery

Sprung control switch

Small neon lamp

+

Arrows show the direction of the winding

Wooden insulated base

Two DC electro-magnets in series formation

Figure 4.19 – Experimental self induction

- This can be demonstrated by connecting a 12 volt battery to a solenoid (this is just a tightly wound coil of wire) as described in Figure 4.19 and controlled by a flip-switch.
- Wire a neon lamp, with a minimum working voltage of 180 volts, across the terminals of the coil as shown in the illustration. The lamp will be seen to flash each time the circuit is broken. This flash is caused by a high self induced voltage entering back into the coil.
- This demonstration is best carried out in half-light.

Mutual induction

- If wire is wound in the form of a coil on a laminated iron former and is accompanied by a secondary coil wound on the opposite side of the former, an electromotive force will be induced into the secondary circuit if the primary circuit is connected to an AC supply.
- Energising the primary circuit will create a current flow within the coil.
- An alternating magnetic field is spontaneously created within the primary circuit windings.
- The magnetic flux generated is easily transferred to the secondary circuit through the medium of the iron core. This produces a

mutually induced voltage within the secondary coil.

- The value of the secondary voltage is dependent on the number of turns of wire within the secondary circuit and the value of the applied voltage to the primary circuit.
- This is a similar principle to the way your site transformer operates.

The force applied to a load bearing conductor suspended within a magnetic field

A force is brought to bear on any load bearing conductor placed within a magnetic field – take a glance back to Figure 4.8. The magnet fields illustrated show both the permanent magnet and the field around the conductor physically interacting with one another – they are distorted and the force applied to the conductor is both equal and opposite.

When the circuit is made and current flows, a magnetic force field is created around the entire length of the conductor. This magnetic field acts at right angles to the length of the wire and will physically move the conductor. Increasing the voltage will have the effect of moving the conductor faster. By manually adjusting the position of your magnets so they are not quite adjacent to one another will have the effect of slowing the conductor down when the circuit is energised. The force on the wire is proportional to the current flowing, the length of the conductor and the magnetic field strength.

Properties influencing mutual induction

- The current flowing within the circuit.
- The number of turns of wire you have within your coil.
- The rate of change of magnetic flux – see later in the chapter.
- The rate at which the magnetic flux is cut by a conductor – its velocity.
- The material the wire is made from (silver would be excellent but rather expensive, with copper a very good second choice).
- The temperature of the wire – the natural resistance of both silver and copper will increase with an increase in temperature.

Putting theory to practice

Figure 4.20 shows a simple arrangement of a typical bar magnet being moved in and out of a coil of wire attached to a galvanometer – this is an instrument for measuring very minute currents. When a strong permanent magnet is plunged rapidly into the centre of the coil the galvanometer will move briefly to one side. Removing the magnet will produce a similar but opposite effect – the needle will move in the reverse direction. This is electromagnetic induction at its most basic level.

Let your imagination take command. Let us suppose that the South pole of your bar magnet is moved from a point very close to the entrance of the coil where 100 magnet lines can cut through your coil to a position further within the coil where 1000 lines of force are present – and all within 0.25 of one second.

The rate of change of magnetic flux cutting through the coil will be:

$$\frac{1000 - 100 \text{ magnetic lines}}{0.25 \text{ second}} = \frac{3600 \text{ lines}}{\text{per second.}} \quad (4.9)$$

Now, swiftly remove your bar magnet from out of the coil in a rapid 0.005 of one second. This will reduce the flux lines from 1000 to just 100 very quickly.

The rate of change of magnetic flux cutting through the coil will now be:

$$\frac{1000 - 100}{0.005} = \frac{900}{0.005} = 180\,000 \text{ per second.}$$

The inducted emf is now a staggering 50 times greater than before. If we understand that the initial emf induced was, say, 5 millionth of one volt, the second movement produced 250 millionths of one volt!

Please remember: an induced emf is directly proportional to the rate of change of magnetic flux cutting through a coil together with the value of the magnetic flux density – the number of lines of flux linking the circuit.

Inductive components and equipment

There are many; here are just a few of them:

- The windings within a site generator.
- The field coils serving your site electric drill.
- Chokes used in fluorescent and other inductive lighting arrangements.
- A coil ignition serving a large site generator.

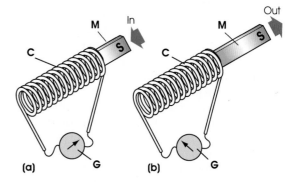

Figure 4.20 – *Michael Faraday's experiment with a coil of wire and a bar magnet – C, coil; G, galvanometer; M magnet*

- The windings within your site 10 kVA transformer.
- Cycle dynamo.
- Telephones and microphones.
- Loudspeaker.
- Lock serving a door entry system.
- The induction furnace – this is based on a high frequency alternating current producing eddy currents of which heat is produced as a by-product.
- Choke as used in electronics – where a high induction coil of low resistance allows direct current to flow but prohibits alternating current due to the phenomenon of self induction.

Capacitors

A few definitions

Capacitance – This can be described as a system of conductors or conductive material and insulators which has the ability of storing an electric charge when a potential difference exists between the conductors. The SI unit of capacitance is the farad (F).

In electrical engineering it is sub-divided into microfarads (μF) and picofarads (pF).

$$\text{Capacitance } (C) = \frac{\text{the stored charge in one conductor } (Q)}{\text{potential difference between the two conductors } (V)} \quad (4.10)$$

Electrical field – This is an area around which an electrical charge experiences a physical force often because of the presence of other charges. The electrical field strength (E), which may be selected anywhere within an electric field, is said to be the force per unit charge, sensed by a charge, when placed at that point.

Electric stress – This, within a capacitor, is known as *dielectric stress*. The electrical pressure between the two plates of the capacitor produces an elliptical deformity (oval warp) within the orbits of the electrons serving the dielectric (the insulated bit between the two plates). This is illustrated as Figure 4.21. Dielectric stress is often expressed in volts per millimetre and is directly proportional to the voltage applied and is said to be the maximum potential which can be placed on a material without it breaking down – in practice, burnt, holed or destroyed, etc. A capacitor will experience a physical breakdown if the voltage is increased beyond its dielectric strength. This physical stress within the dielectric is called electrical polarization (P) and is defined as follows:

$$P = D - E\varepsilon_0 \qquad (4.11)$$

where D is the displacement, E is the electric field strength and ε_0 is the electric constant

Listed in Table 4.2 is a selection of materials used for dielectrics. The right hand column provides for common values in kilovolt per millimetre of thickness.

Table 4.2 – *A selection of common dielectrics*

Dielectric material	Application	Maximum voltage – dielectric stress. (kilovolt per millimetre of thickness)
Air	Used for radio tuning	3–6
Paper	Used for low frequencies	4–10
Glass	Used for high frequencies	5–30
Waxed paper	Used for low frequencies	40–60
Mica*	Used for high frequencies	40–200

*Mica: a naturally occurring silicate mineral substance which has a layered structure. It is also used as an insulator in electrical engineering.

Dielectric – This is the non-conductive part of a capacitor; the component which separates and insulates the plates of the capacitor and where all the hard work is done. When both plates are fully charged it causes a displacement of charge within the dielectric but not a flow of charge.

Potential difference (pd) – This can be expressed as the difference between two points within an electric circuit, or the work done in moving a unit positive charge from one point to another. The practical unit for potential difference is the volt: 'The circuit has a potential difference of 12 volts', for example.

Charge and capacitance – The total capacitance of a capacitor (measured in farads/microfarads) is dependent on three factors which are built into the component at the design stage:

1   The total surface area of the plates.
2   The distance between the plates (this, in practice, will equal the thickness of the dielectric).
3   The material the dielectric is made from (mica, glass, paper, etc.).

When an electrical pressure is applied across the two plates of the capacitor, the positive plate will constantly attract a flow of electrons from the negative plate. This process will continue until the voltage across the capacitor is equal to that of the battery. Disconnecting will leave the capacitor in a fully charged state.

At this stage, something very interesting happens. All the negative electrons forming the

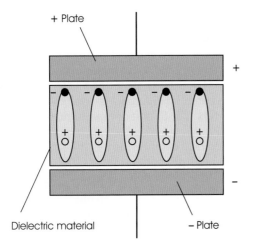

Figure 4.21 – *The orbits of the electrons become deformed into oval shapes*

+ Plate

+

Dielectric material

– Plate

dielectric are repelled from the fully charged negative plate and attracted to the positive plate.

Please remember, as with magnets, like poles repel; unlike poles attract. This is a basic law – keep it in mind.

The electrical pressure exerted between the two opposing plates generates an elliptical deformity within the orbits of the electrons serving the sandwiched dielectric. This deformity has been discussed in previous paragraphs.

A capacitor's energy or charge is stored within the dielectric material which is sandwiched between the two opposing plates. A simple way to demonstrate the release of this energy is to short circuit the two disconnected leads serving the capacitor – take care if you do this yourself. Once short circuited, the deformed orbits of the negatively charged dielectric cease to be under any stress and the electrons stored in the capacitor's negative plate are attracted towards the capacitor's positive plate at a colossal speed. The result is a familiar electrostatic crack accompanied by a brief visible display when the energy within is released.

Capacitors have a very broad working range and those serving a 'capacitor aided motor' are generally rated at a 400 volt working range. Machines that use capacitors to start them which are then automatically removed from circuit use the 230 volt working range variety. In practical terms, if the latter is connected into a circuit with the former, a breakdown can be expected!

Different types of capacitors

Each of the many varieties of capacitors has its own particular role to play in engineering. Here is a snap-shot view of some of them:

- *Air capacitor* – used in radio receivers for tuning.
- *Paper insulated* – used for many applications in electrical engineering. This type of capacitor can vary from a few picofarads to many microfarads. Widely employed in motor and power factor correction circuits.
- *AC electrolytic capacitor* – this type of capacitor is only designed to be in circuit for a very short time. It has the ability to store large amounts of capacitance in relation to its size and is very suited to start small induction motors. The downside to this type of capacitor is that it is best not to exceed any more than

between six to eight 10 second starts per hour if a breakdown is to be avoided. Figure 4.22 compares an electrolytic capacitor with a paper insulated type.

- *Trimmer capacitor* – used in electronic circuits on printed circuit boards. Often this type of capacitor is designed with a polypropylene dielectric for high resistance. Some manufacturers will colour code their products to indicate the range value. Yellow, for example, could be from 2 to 10 pF whilst green would indicate a value of between 2 and 22 pF. Adjustment is made with a small screwdriver. The working voltage is about 100 volts DC – see Figure 4.23.
- *Silver mica* – used in electronics where high stability is required. Ideal for radio frequency circuits. Values are in the picofarads range.

Figure 4.22 – *An electrolytic capacitor*

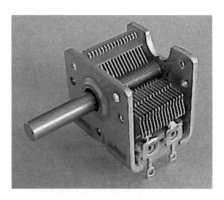

Figure 4.23 – *A trimmer capacitor*

Voltage ranges are from about 50 to 500 volts DC. Figure 4. 24 will illustrate these.

- *Multi-plate capacitor* – Figure 4.25 illustrates this type of capacitor. The useful plate area for this type of capacitor will be $A \times (n-1)$ where A represents the area of one plate and n the number of plates in the system.
- *Variable capacitor* – This is a very special type of capacitor which is used with high voltage systems. The range is wide, starting from about 50 to 1500 pF obtainable from different models. The dielectric is formed from a pressurised inert gas, factory set to 3 bar

Figure 4.26 – *A variable capacitor*

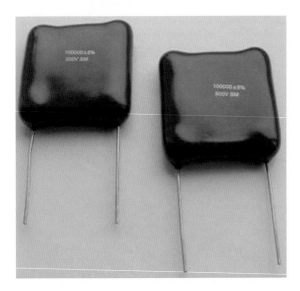

Figure 4.24 – *Silver mica capacitors*

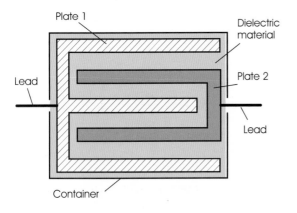

Figure 4.25 – *The basic essential of a multi-plate capacitor*

(approximately 44 pounds per square inch) and adjustment is made by a small shaft. Figure 4.26 illustrates this.

The dangers associated with capacitors

Please use this list to browse the hidden dangers linked with capacitors:

- Capacitors can keep their charge and if you unconsciously pick one up which is fully charged you are liable to receive a very unpleasant 'stabbing' electric shock.
- Violent, momentary capacitor shocks can cause fibrillations within your heart muscle. This could be very serious if you had a heart condition.
- Some older types of metal case capacitors contain oil which is extremely toxic and has, after 1975, proved to be cancerous in nature. The oil, a polychlorinated biphenyl (PCB) based substance, was designed and manufactured because of its high insulating properties. It was also found to be very stable under high temperatures and was not flammable – at the time it was considered to be an outstanding breakthrough. This type of oil is a *serious health hazard*. Once on your

hands it enters your body through your skin and will remain there permanently. Physical contact with this oil will make you sick and will leave a rather unpleasant smell throughout your body. You will be avoided!

■ Use rubber gloves when handling leaking metal-can capacitors – they may contain PCB cancer producing oil – *do not take this risk!* Obtain professional advice as to where to dispose of any such oil.

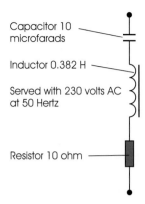

Figure 4.27 – *A capacitor, inductor and resistor in series formation*

The effects of resistance, inductance, capacitance and impedance within a circuit

The effect of resistance, inductance and capacitance will have on an AC circuit will be to oppose the flow of current within that circuit.

Defining reactance

The mathematical symbol for reactance is X and is measured in ohms. It is defined as the property of a circuit that contains capacitance and/or inductance, and jointly, with any resistive component incorporated, makes up the *impedance* of the circuit. You are then left with the following list:

■ capacitive reactance (in ohms),
■ inductive reactance (in ohms),
■ resistance (in ohms).

Loosely think of capacitive and inductive reactance as an AC electrical opposition to the flow of current within a circuit containing all three components. In this exercise we will see how to find the values of each of the listed items in terms of ohms. Once calculated, the total impedance, Z, for the series circuit will be established.

A theoretical exercise showing how this is achieved

Figure 4.27 illustrates a simple series formation circuit containing one 10 microfarad (µF) capacitor, a 0.382 Henry (H) solenoid and a 10 ohm (Ω) resistor. These three items are connected to a 230 volt, 50 hertz supply.

The total impedance, Z, of this series circuit is found by use of the following expression:

$$Z = \sqrt{[R^2 + (X_L - X_c)^2]} \quad (4.12)$$

where R is the total resistance of the resistor in ohms, X_L is the value of the inductive reactance in ohms and X_c is the capacitive reactance in ohms.

To calculate the impedance of this circuit we must first work out the value of the inductive and the capacitive reactance.

First, the inductive reactance, by using the following expression:

$$X_L = 2\pi f L \quad (4.13)$$

where L is the symbol we use to represent the value of the inductance in henrys, whilst f represents the frequency in hertz.

Substituting for known values:

$$X_L = 2 \times 3.142 \times 50 \; 0.382 = 120 \; 024 \; \Omega.$$

Next we will find the total capacitive reactance and this is done by using the following expression:

$$X_c = \frac{1}{2\pi f C} \quad (4.14)$$

where C is the symbol we use to represent capacitance in farads.

Substituting for known figures:

$$X_c = \frac{1}{2 \times 3.142 \times 50 \times 0.00001} = 318 \; 268 \; \Omega$$

We now have the values of all components within our circuit:

$R = 10$ ohms (taken directly from our example)
$X_L = 120.924$ ohms of inductive reactance
$X_c = 318.2$ ohms of capacitive reactance.

It would be just fine if we could add these three figures as though calculating a DC series resistive circuit – but this is impossible as there is no such thing as pure impedance (every coil has a small measure of resistance), nor pure capacitance as every capacitor has an element of resistance. We must take a different route to obtain the answer to our problem.

To resolve the total impedance *(Z)* of our circuit we must return to (4.12) and substitute figures for symbols:

$$Z = \sqrt{[10^2 + (120 - 318.2)^2]} = \\ \sqrt{[100 + (-198.2)^2]}.$$

Please remember in this calculation a minus multiplied by another minus quantity produces a positive result:

$$Z = \sqrt{39\,383.24} = 198.4 \ \Omega \ \text{impedance}.$$

The current drawn from this AC circuit can now be calculated from the following expression:

$$I = \frac{E}{Z} \qquad (4.15)$$

where *E* is the applied voltage and *I* is the current drawn from the circuit.

Applying figures to the expression:

$$I = \frac{230}{198.4} = 1.15 \ \text{amps}.$$

Calculating the power developed within an AC circuit

We can easily find the value of the power generated within a DC circuit by multiplying the applied voltage by the current flow (I × V = Watts). As an example, assume 10 amps are being drawn from a 230 volt resistive circuit. Then:

$$\text{Power in Watts} = 10 \times 230 = 2300 \ \text{watts}.$$

This, we are unable to do in an AC circuit because the current and voltage are rising and falling at different times. Multiplying these AC values together will only provide *apparent* power – not *actual* power. Apparent power is measured in kVA, such as a site transformer would give and can be gained by multiplying the applied voltage by the current and dividing your answer by 1000.

When an AC voltage is applied to a so-called pure inductive circuit, the current, with respect to the voltage, will be out of phase with each other by 90° or, to put it another way, when the voltage is at its highest, the current will be at its lowest.

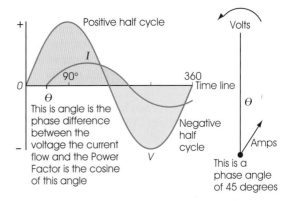

Figure 4.28 – *When the voltage is at its highest the current is at its lowest*

Figure 4.28 will provide additional clarification of this.

Looking at the illustration you will see that the phase difference between the voltage and current is out of step by 44°. The power factor of this circuit is the cosine of this angle – which when referring to the cosine table is found to be 0.7193.

The actual power in watts = current × applied voltage × power factor. (4.16)

Referring back to the figures serving our simple example:

The actual power in watts = 10 amps × 230 volts × 0.7193
Power = 1654.39 watts (or 1.65 kW).

To recap this important concept:

The formal definition of a power factor is the cosine of the angle of phase difference between the voltage and current. If the angle of difference is φ then the power factor is cos φ. This angle can vary from –90° to +90° for general inductive and capacitive loads. If the current *lags* behind the voltage, your load is inductive and is described as having a lagging power factor whose angle is *positive*. If the current within your circuit *leads* the voltage then your load is capacitive and is said to possess a leading power factor whose angle is *negative*.

The theoretical average power developed within a circuit:

- *Resistive circuit* – multiply the applied voltage by the current flowing within the circuit. This will give you the power in watts generated.

- *Pure inductance* – In an inductive circuit the power developed is zero as the current is out of step with the voltage by 90°. The power developed in this type of circuit is zero.
- As a footnote: there is no such thing as a pure inductive circuit as there is always a small element of resistance within the copper coil.
- *Pure capacitance* – In a pure capacitive circuit, the power developed is zero.

Semiconductor devices

You may be surprised to hear that the semiconductor has been around since the early days of wireless transmission when it was used within a 'cat's whisker' radio as a receiving diode.

A semiconductor is a device which is neither a conductor nor insulator but falls somewhere between these two categories. It has the unusual property of decreasing in electrical resistance with a rise of temperature – which is the opposite to electrical conductors such as silver, copper, aluminium and steel.

They are made from materials which are not really metals but in some way they bear a resemblance to them such as lead telluride, silicon, selenium and germanium. Impurities are added at the manufacturing/design stage of the component.

How are they made?

Semiconductors are usually prepared from crystals of germanium or silicon. They are melted at the processing stage, purified and then gas impregnated with a controlled impurity such as antimony, arsenic or phosphorus.

At room temperature a pure semiconductor has a naturally built-in electrical resistance somewhere between those of an insulator and conductor and will decrease in resistance with an increase in temperature. This is known as a negative coefficient of resistance.

The atoms of both germanium and silicon are so arranged they form a cubic crystal lattice – and that means that each atom is dependent on the other. Both elements have four free electrons within each of their outer shells, as illustrated in Figure 4.29. Because there are four and not just one free electron, as in an atom of silver (see Figure 4.30) the semiconductors are recognised

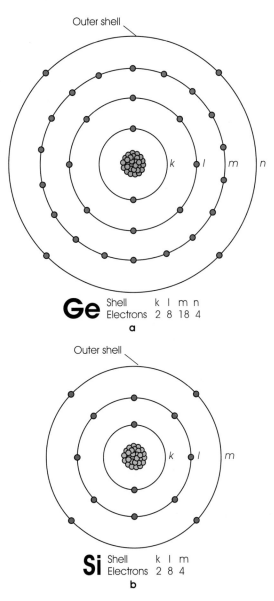

Figure 4.29 – *The atoms of geranium and silicon are so arranged that they form a cubic lattice*

as being neither a good conductor nor a bad insulator.

A brief summary of the crystal lattice

A semiconductor is made from a crystalline element, structured in the form of a cubic lattice. This means that pairs of free (valance) electrons are shared – one provided from each atom. Or, to put it another way, one free electron from each atom is coupled with an electron from a neighbouring atom. This is known as a covalent bond and could be loosely described as a common 'free electron linkage' system. (Please see Figure 4.31 for schematic details of this.) Each atom is inherently placed at a defined distance from its bordering atom and is bonded to four of its adjoining neighbours. The last illustration will make this point clearer.

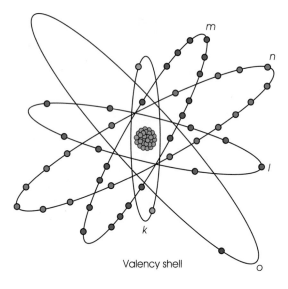

Figure 4.30 – An atom of silver has one free electron – this makes it a good conductor of electricity

At a temperature of 0 degrees Kelvin (absolute zero) all free electrons are held firmly in place, no current could possibly flow and our semiconductor would act as though it was an insulator. At room temperature the electrons become far livelier and are able to break free from their orbits and wander around the crystal lattice, as illustrated by Figure 4.32.

Once an electron is free, its parent atom will no longer be neutral (meaning that it is neither positive nor negative) but will now have a net *positive* charge, because the nucleus, the middle part of an atom, has a positive charge – the electrons orbiting around the nucleus are considered negative.

The space that is left by an electron leaving its parent atom is known as a *hole* and it is the place where an electron would need to be in order to complete the crystal lattice. This is schematically illustrated as Figure 4.33.

A balanced arrangement of holes and free electrons is a good means of electrical conductivity within a semiconductor. An incomplete positively charged atom will attract another free electron from a neighbouring atom, as Figure 4.34 shows. Once the defective atom admits its neighbour's electron the atom loses its

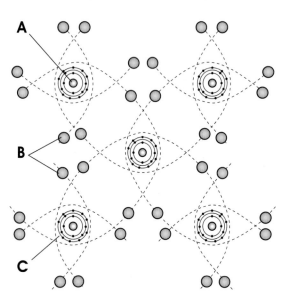

Figure 4.31 – One free electron from each atom is coupled with an electron from a neighbouring atom. This is known as a covalent bond

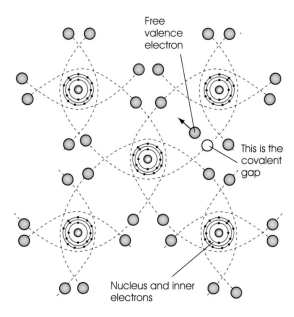

Figure 4.32 – At room temperature (18 to 21°C) the electrons become much livelier and can break free from their orbits and wander around the crystal lattice

'positively charged' status and is neutralised – it has once again become neither positive nor negative. This will leave a hole in the adjoining atom which will now have a clear *positive* charge. This charge will be sufficiently strong enough to attract another electron from a different atom and the whole process continues with no sign of stopping until the switch is thrown.

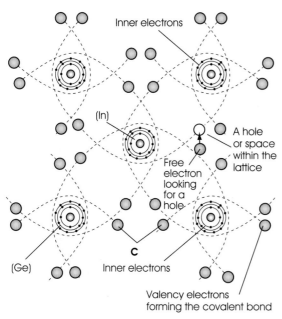

(Ge) Germanium atom. (In) Indium atom

Figure 4.33 – *The space that is made by an electron leaving its parent atom is known as a hole and is the place where an electron is needed to complete the crystal lattice*

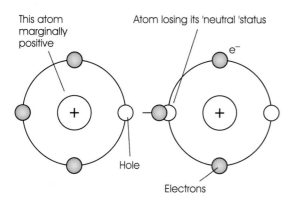

Figure 4.34 – *An atom which has a hole has a net positive charge and will attract another free electron from a neighbouring atom: A, atom (positively charged); B, a neutrally balanced atom (but not for long!); C, electrons*

Doping – this is the name we give when certain impurities are added to the semiconductor at the construction stage. These donor impurities can be one of the following to produce an *n-type* semiconductor:

- antimony (Sb),
- arsenic (As),
- phosphorus (P).

All three of these elements have a *valency* of five – this means that each one of them has *five free electrons* and therefore produces a surplus of one electron with respect to the semiconductor material. Figure 4.35 illustrates this point schematically.

Four of the five donor electrons form a covalent bond which is very necessary for the construction of the crystal lattice. The fifth donor electron is free to wander and is available for conduction. A semiconductor doped in this fashion is known as an *n-type semiconductor*. Figure 4.36 shows the semiconductor germanium doped with the donor impurity antimony and highlights the excess electron.

Any type of impurity will not necessarily provide the vital means required to do as intended. The impurity added *must* form part of the crystal lattice structure. It must occupy a place where an atom would normally have been;

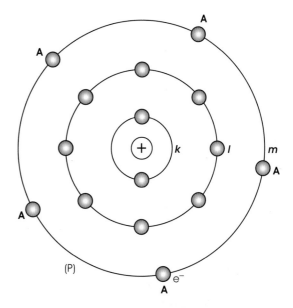

Figure 4.35 – *An atom of phosphorus whose atomic symbol is 'P'. A, free negative electrons*

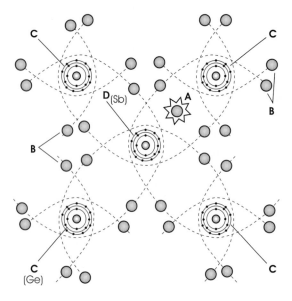

(Ge) Germanium atom. (Sb) Antimony atom

Figure 4.36 – *Germanium doped with the donor impurity antimony. This combination becomes an 'n'-type material. Only four of these free electrons will make the covalent bond for the lattice. The fifth electron can wander throughout the crystal. A, excess electron; B, free electrons forming the covalent bonds; C, an atom of germanium; D, an atom of the donor impurity antimony*

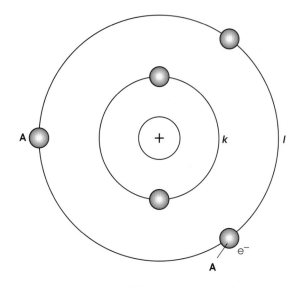

Figure 4.37 – *An atom of boron – atomic symbol 'B'. A, free electrons*

otherwise electrical conductivity will be considerably weakened.

Producing a 'p- type' semiconductor – doping a semiconductor with either gallium (Ga), boron (B) or indium (In) produces what is known as a 'p-type' semiconductor. Each donor impurity element has three free (valance) electrons in its outer shell. These are known as acceptor impurities – Figure 4.37 illustrates an atom of boron.

When the semiconductor silicon has been doped with an acceptor impurity such as boron (an element with a valence of three) the resultant compound is known as a *p-type* material. Figure 4.38 illustrates an atom of indium combined within the lattice work of the semiconductor germanium. The impurity atom has created holes within the lattice work – please recall that a hole is just a place where an electron should be to complete the crystal lattice.

An incomplete atom surrenders its neutral status as it has a net positive charge. This positive

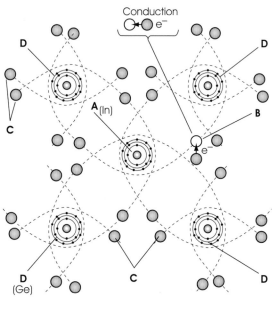

(Ge) Germanium atom. (In) Indium atom

Figure 4.38 – *When geranium is doped with the acceptor impurity indium it becomes a 'p' type material. Only three of the four covalent bonds may be made. This causes holes to be created and thus provide a route for the conduction of electricity. A, an atom of indium; B, hole; C, free electrons forming covalent bonds; D, an atom of germanium*

charge will attract an electron from a neighbouring atom in order to win back its neutral status. This will cause a hole to appear within an adjoining atom and so the process continues.

Types of semiconductor devices

Figure 4.39 illustrates a semiconductor device.

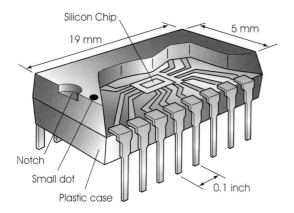

Figure 4.39 – *Semiconductor device*

Semiconductors within a rectifier circuit

Diode

The semiconductor diode is very similar in principle to a non-return valve serving your central heating system. The valve will allow water to flow in one direction but not to return. This is a similar rule which applies to the diode – in one direction it will allow the passage of current; the other way the resistance is so high that it stops current from passing through. The diode is made in many different physical sizes – each size is an indication of its current carrying capacity. Small and tiny diodes are used in radios, mobile phones and television receivers, whereas much larger types are designed to be used in power and rectification circuits.

About the diode

The anode component of the diode, the positive bit, is made from a shaped piece of *p-type* material and is factory fused to the cathode, a portion of n-type material, which is the same size and shape of its partner. These two components form the diode.

A diode will only permit electricity to flow in one direction. One way the opposition to current flow is very high whilst in the other direction the resistance is very low. The direction where the diode offers little resistance to current flow is indicated by the arrowhead on the schematic symbol, shown as Figure 4.40.

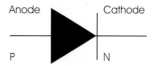

Figure 4.40 – *A schematic symbol of a diode*

How it works

The meeting point at which both p- and n-type materials get together is known as the *depletion layer* – see Figure 4.41. It is called this because

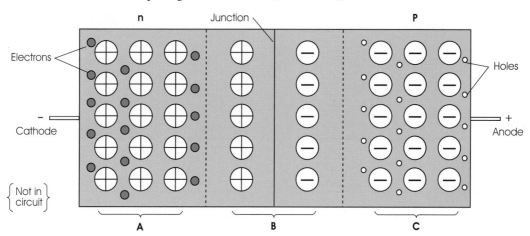

Figure 4.41 – *The depletion layer. A, 'n'-type material with free electrons; B, the depletion layer; C, 'p'-type material with holes*

the surrounding area each side of this junction has a shortage of electrons on the cathode side (negative) and a lack of holes within the anode (positive) side. This effectively stops n-type electrons moving across the junction to mix with p-type material. At this junction the cubic lattice arrangement is completely unaffected throughout the length of the bond – only the impurity valence electrons (the free electrons) vary on either side.

Applying a voltage

Forward bias – applying a small voltage to a semiconductor diode will allow free electrons (or carriers as they are sometimes known) to accelerate towards the p–n junction – see Figure 4.42. Some will have enough energy to bridge the coupling – increasing the voltage will allow current flow. It is worth pointing out that when the current is increased, the voltage drop across the diode remains steady. A forward bias germanium diode will drop about 0.2 volt whereas a silicon diode experiences a drop of around 0.6 volt. Semiconductors are not linear devices and therefore do not act in accordance with Ohm's law.

Reverse bias – reversing the voltage as shown in Figure 4.43 by connecting the anode to the negative terminal will attract electrons from the n-type material to the positive plate. Holes from imperfect atoms in the p-type material are drawn to the negative plate as unlike charges attract. This will widen the depletion layer until an electron balance is reached and electrical conduction will cease – see Figure 4.44.

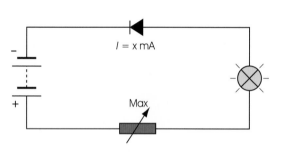

Figure 4.42 – *Forward biased circuit*

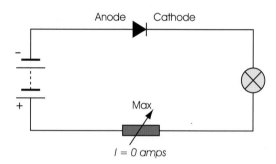

Figure 4.43 – *Reverse biased circuit*

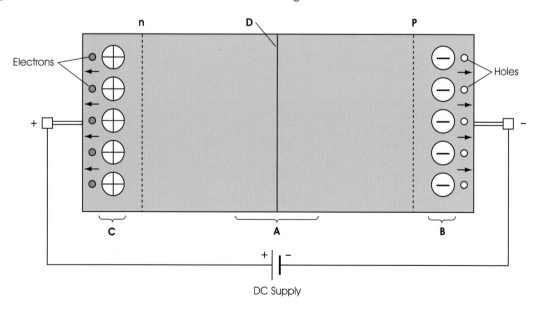

Figure 4.44 – *Widening the depletion layer will stop the flow of current. A, extended depletion layer; B, 'p'-type material pulled away from the junction; C, 'n'-type material pulled away from the junction 'D'*

At this point the internal resistance is very high and if the voltage applied is increased any further a total component breakdown can be expected.

Peak inverse voltage (PIV)

If a small reverse bias voltage is applied to a semiconductor diode, little will appear to happen. If you increase the voltage steadily the current will remain fairly stable until a voltage point is reached where a very large increase in current will suddenly take place. This huge saturation current will disperse power throughout the diode and effectively destroy the semiconductor device.

If this critical voltage, known as the *peak inverse voltage (PIV)*, is not exceeded the device will operate as intended.

Different diodes have different PIV values and information concerning this can be obtained from the manufacturer's information sheets available at your stockist.

The thyristor

This type of semiconductor is a silicon controlled rectifier (SCR). It has three terminals as illustrated in Figure 4.45:

- the anode [+] – the first terminal,
- the cathode [–] – the second terminal,
- the third terminal opens and closes the switching action.

The main termination points are the anode and cathode where current flow is found. The roll of the gate terminal (sometimes known as the base) is where a switching or control voltage is applied.

The device presents the same characteristics as the diode. It has a very high resistance to current flow in one direction. If a voltage is applied to the gate which is positive with respect to the cathode, the switching mechanism will trigger and current will flow. Once conduction has taken place, current flow will be maintained until such time the applied voltage is reversed or falls to zero. In practice it is always wise to place a resistor in series with the gate to protect the switching action from high currents. This will be seen by glancing back to the last illustration.

When a thyristor is placed within a circuit served with an alternating voltage it will only conduct on the *positive* half cycle and will repeatedly turn off during the *negative* half cycle. This is a very handy feature as the next paragraph will explain.

Thyristors are used to rectify AC circuits (Figure 4.46). They are also used as switching devices and a means to control heavier currents as in electric motor speed management applications and dimming circuits.

Waveforms

In order that our CD players and audio equipment can operate as intended we require a *direct current* source of power. We can either use batteries, which can be expensive and short lived, or turn to electronics for an answer.

There are two main principles of generating a DC voltage:

- by means of half-wave rectification,
- by means of full-wave rectification.

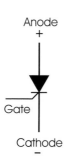

Figure 4.45 – *The thyristor semiconductor*

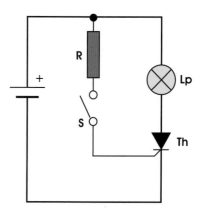

Figure 4.46 – *The thyristor in circuit*

In both cases the voltage has to be greatly reduced by means of a transformer before being applied to an electronic circuit.

Half-wave rectification

With this cheaper form of rectification only the *positive* cycle (the top bit of the wave form pattern) is permitted to pass through the series connected diode, as Figure 4.47 shows. This means that the diode only conducts electricity during *alternative* half cycles and so only produces a pulsating form of DC current. This is okay but this type of rectification must only be used for the simplest of electronic circuits such as a door bell or battery charger circuit.

One simple solution to this problem is to wire a *capacitor* in parallel across the load as shown in Figure 4.48. When the diode is conducting

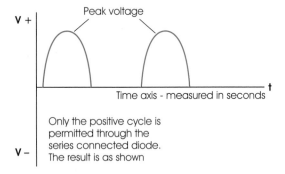

Figure 4.47 – *Half wave rectification – the circuit*

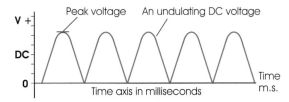

Figure 4.48 – *Placing a capacitor across the load releases a charge into the circuit*

(working as intended) it also charges the capacitor and the charge is released into the circuit when the diode is at rest by means of the load resistance. This system will work far more efficiently when the load is greater and less efficiently when the load is smaller.

When, at every half-cycle the diode is at rest, the peak inverse voltage (PIV) is equal to the peak voltage of the secondary output side of the transformer so it is important to check out the type of diode you are to use in case the voltage is too high. If you have the wrong type of diode in place or the voltage is too high, your diode will self destruct!

Full wave rectification

This is a far better way to obtain a DC voltage output as both the positive and the negative half-cycles are rectified. Although not 100% perfect it is noticeably better than half-wave rectification. With the negative half-cycle upturned it produces a variable uni-directional wave as shown in Figure 4.49. There are minute variations in voltage, resulting from the original alternating current's sine-wave pattern losing its negative half-cycle.

A simple full wave rectifier uses two diodes, a transformer with its secondary side centre-taped, to provide an accurate voltage value across the semiconductors and a capacitor connected across the load. The roll of the capacitor (sometimes known as a reservoir capacitor) is the same as the half-wave rectifier circuit – the only difference is that the capacitor is now charged twice per input cycle.

The disadvantage of this type of rectification is that the ripple content of the DC voltage output increases in proportion to the load current. Fortunately, this can be checked by the use of a suitable filter/smoothing circuit.

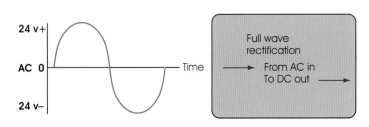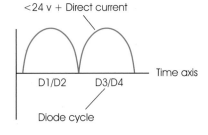

Figure 4.49 – *Full wave rectification*

The bridge rectifier

This type of rectifier requires four diodes but the advantage is that it doesn't need a centre tapped input transformer and can be used for higher DC output voltages – see Figure 4.50. As with the two other types of rectifier, a capacitor placed in parallel with the load will help to reduce the DC ripple effect, although there will be some.

Other means of rectification

There are quite a few, but the following represent the principal types:

- silicon rectification,
- copper oxide rectification,
- the thermionic or glass mantle diode valve,
- the mercury arc rectifier.

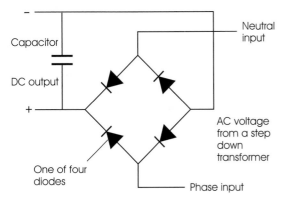

Figure 4.50 – *Bridge rectifier*

The transistor

Used as a switch

A transistor can be compared with a site socket outlet. A small flick of the switch will allow a large amount of energy to flow through to the connected appliance.

Take as an example a simple circuit designed to warn of the presence of moisture within a sensitive preparation. As soon as moisture is detected via the probes a very tiny current of a few milliamps trickles from one probe to the other. From this point the tiny signal current travels along the wire to the base or gate of the transistor. This small amount of current is able to trigger the gate's switching mechanism to allow some 20 to 100 times greater current to flow from the collector terminal (c) to the emitter terminal (e). The circuit is then made. Figure 4.51 illustrates this explanation and shows the gate in the open position after the detection of moisture.

Used as an amplifier – a practical approach

A transistor's emitter can losely be described as a common terminal, such as you would have within a two-way switching arrangement. It is common to both the base and collector circuits.

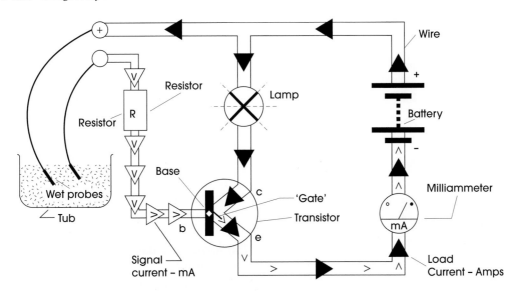

Figure 4.51 – *The transistor's gate in the open position*

Figure 4.52 illustrates a p–n–p transistor connected to a battery, a variable resistor, a monitoring microammeter and a milliammeter to form a simple demonstration amplifier.

By shifting the position of the variable resistor you will be able see a physical change in the collector current resulting directly from an altered base current. This can be further demonstrated by adding in a high resistance headphone to replace the mircoammeter and a low resistance headphone to use as intended but in place of the milliammeter. The high resistance headphone will now be your improvised microphone. Quietly utter a few well chosen words into the mic, cupping your hands above the microphone to avoid your voice carrying across to the other headphone. Ask a friend to listen in using the low resistance headphone. Although not perfect, it will demonstrate that amplification can occur with this wiring arrangement.

If you wish to carry out this demonstration you will require the following components:

- a p–n–p transistor Type OC 72,
- a microammeter from 0 to 1000 µA in range,
- a milliammeter from 0 to 20 mA in range,
- a 20 kΩ to 30 kΩ slider type resistor,

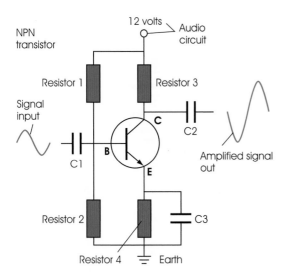

Figure 4.52 – *A transistor, (OC 72), is used in circuit as an amplifier*

- one 1.5 volt battery,
- insulated wire to suit your circuit.

Because the emitter is positive in potential, relative to the base, the emitter base is forward bias.

Electronic components and circuits

Table 4.3 lists seven common electronic components together with various relative data which will be of interest to you.

Identifying components and their values

Electronic components can be identified in a number of ways:

- *Colour coding* – resistors, capacitors, cables and wire, etc.
- *The shape of the component* – flat packs, metal-can relays, capacitors, semiconductors and integrated circuits.
- *Component leads* – the number, shape/profile as with resistors, triacs, metal-can relays and integrated circuits, etc.
- *The material from which the component's outer casement is made* – metal is used for variable voltage regulators and metal-can relays whereas plastic is used for transistor–transistor logic (TTL) devices and other integrated circuits. Light emitting diodes are made with coloured see-through plastic and many a fuse is designed with a glass transparent body.
- *By alphanumerical means* – Coded letters and numbers printed on the side of the component such as some types of resistors.
- *By simple observation* – by reading types, values and tolerances, etc., printed on the side of the component or package.

Exploring the use of common components

Table 4.4 lists and describes uses for common electronic components

Table 4.3 – *Common electronic components with relevant data*

Component	Symbol	Within a circuit	Comment	Illustration
Resistor		0 – 12 V d.c. Power source / Variable resistor / Resistor / Ammeter / V / A	Cylindrical in shape with coloured rings. A component which is placed within a circuit because of its electrical resistance. Can be wire wound or carbon based	
Capacitor	Standard / Electrolytic	12 volts / Resistor 1 / Resistor 3 / C2 / C1 / Resistor 2 / C3 / Resistor 4 / Earth	This is a device which is capable of storing an electric charge. In its most basic form a capacitor is formed from two metal plates sandwiched between an insulating medium such as mica or paper.	
Transistor	C Collector / Base B / Emitter E	LED / Resistor / Probe / 6 v DC / B / E	A semiconductor device which is able to amplify and rectify AC to DC. Often found in appliances and control circuits. The three connection points are the base (b), the collector (c) and the emitter (e).	
Integrated circuit	IC / RS V+ / DI / TH OP / TR / V– CV / NE 555		An integrated circuit is a microcircuit which has been built in a semiconductor chip. It is a complete system, miniaturised under one roof which includes universal counters, line drivers and clock timers. Use in the manufacture of computers.	Silicon Chip / Notch / Dot / PVC case / Pins
Thermistor		Resistor / Thermistor / Earth	This is a component which has a negative temperature coefficient and will swiftly decrease in resistance in proportion to an increase in temperature. Often used as a temperature measuring device.	

Table 4.3 – *continued*

Component	Symbol	Within a circuit	Comment	Illustration
Triac			A triac is a simple three-connection semiconductor which is used in incandescent dimmer circuits. It is capable of handling up to 350 watts of power.	
Diode	The arrow indicates the direction of current flow		This semiconductor has two electrodes known as the cathode (–) and the anode (+) and only permits current flow in one direction. This way it used in rectification circuits. It can also be employed in amplification circuits. The forward bias current is proportional to the applied voltage but the reverse bias current is exceedingly tiny.	Either a black band, a coloured spot or a coloured band identifies the cathode (the negative)

Table 4.4 – *Electronic components and their use*

Electronic component	Use or purpose	Your illustration
Photo cell	As its internal circuitry is electronically altered when light fall on the cell, it is used for outside lighting switching arrangements and the technology is included in photographers' exposure meters	Cadmium sulphide photocell
Photodiode	This is a semiconductor which is sensitive to the presence of light or can be so constructed as to measure its intensity – as in a light meter. The device is designed in reverse bias mode so that when in darkness the current is very small. When exposed to light the current flow is directly proportional to the amount of light falling on it.	

Table 4.4 – *continued*

Electronic component	Use or purpose	Your illustration
Phototransistor	Phototransistors are very much like photodiodes; the difference being is that the photoelectric current is *amplified* within the component and also tends to be far more sensitive to light than its diode cousin. Some types of phototransistors are designed to be used for switching purposes.	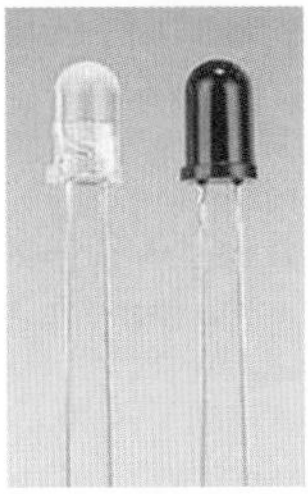
Optocoupler	This device comprises a gallium arsenide light emitting diode optically coupled to a silicon phototransistor. Used in power systems for workstations and in isolated AC to DC power supplies and converters.	
Infra-red source and sensor (a) The source (b) The sensor	(a) Illumination by infra-red means provides you with clear imagery when surveying in total or partial darkness. Used in combination with sensitive IR sensor equipment. (b) It is very easy to see imagery in areas of total darkness. The equipment illustrated may be either hand held, or head mounted. Used mainly for homeland security purposes by the police and private security agencies . . . Other uses are to be found in engineering surveillance, night-time rescue missions and military exercises.	(a)  (b) 
Solid state temperature device (a) Laser beam temperature measurer (b) Thermistor	(a) The illustration shows a solid state hand held, non-contact temperature measurer. It uses two laser beams projected at the target which forms a single point. Ranges are to choice – typically, between -32 to 535°C. (b) This solid state device is used for temperature control. It has a very low resistance at low temperatures and increases in resistance when the temperature is higher. The semi-conductor is very small in size and is fitted into a suitable enclosure for mounting. Sometimes referred to as a thermocouple.	(a) 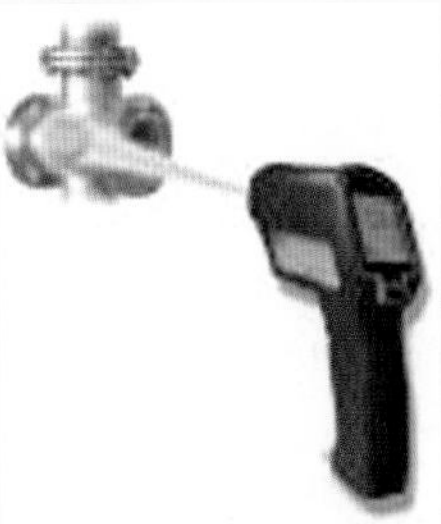 (b) 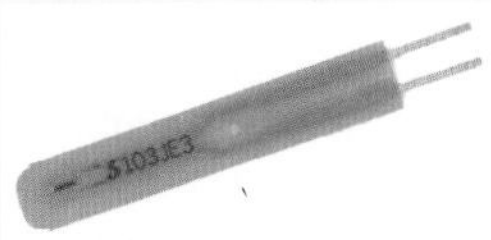

Table 4.4 – *continued*

Electronic component	Use or purpose	Your illustration
Fibre optic link	An optical fibre cable is used for data, telephone channels and television programmes. The 'cable' is very small, just 0.125 mm in diameter and is made from a thin glass core, which compared to the cladding around it, has a much higher refractive index. A low power laser or an infra-red light source is used as an optical transmitter. The advantages gained from using optical fibre systems are as follows: • Free from electrical interference. • Lighter and easier to install than copper cables • User security greatly improved. • Voltage drop not relevant. • A greater communications distance is able to be achieved. • Capable of carrying far more information than by conventional methods.	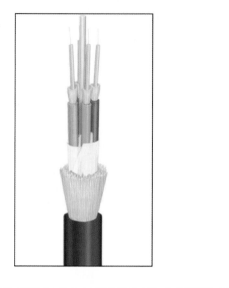

Handy hints

- It is very handy to keep an old stiff-wire type of coat hanger as part of your tool kit. When straightened out and formed with a small hook on one end it will provide handy access to where normally it would be difficult to get to.

- If your company has a global positioning device, then ask to use it when taking journeys which could end up with you getting lost. You will find it will be a reliable friend to have within your company vehicle.

- Most category 2 diffusers are fitted with a plastic membrane when purchased. This is to stop site contamination from interfering with the polished reflectors. To maintain the 'as-new' look, only remove the membrane before handing the installation over to your client.

- Avoid having *two* fused connection units serving one cooker hood circuit – especially if the one nearest to the hood is hidden behind a high level fitted facia board. Instead, position a fused connection unit on the kitchen work-top wall and a 20 amp flex outlet plate to serve the cooker hood.

- Vibration from site hand-held power equipment and tools can be responsible for permanent injury to the hands and fingers. This is known as hand, arm, vibration syndrome (HAVS) and can cause severe pain and lack of feeling at the tips of the fingers, a 'pins-and-needles' feeling at the tip of your fingers, a loss of grip in your hand and painful wrists.

- Avoid using aluminium steps to carry out electrical work – they conduct electricity. It is wiser and safer to use wooden or glass-fibre steps.

- A guide to the value of the circuit protective conductor, if measured end-to-end within a final ring circuit, is to multiply the value of the phase conductor, measured end-to-end, by a factor of 1.67.

Handy hints

- Lighting schemes for very large projects are very often computer aided and are considered highly specialised.

- Fluorescent lighting can be mounted at any angle and will work efficiently when supplied with a voltage from 230 to 276 volts AC. The lamp is supplied with a low pressure inert gas such as krypton or argon.

- The diode current is directly proportional to the voltage applied.

- A semiconductor zener diode is designed to work in a reverse bias mode and will trigger at a predetermined voltage. It is used in circuits as a means of regulating voltage.

- A colour coded resistor must be read from left to right. Band 1 represents the first number, Band 2 the second number, while Band 3 represents the multiplier or the number of zeros following the first two numbers. The last coloured band stands for the tolerance of the resistor as a percentage.

- Never flash-test portable appliances which contain electronic components – they will be destroyed!

- Never over-accommodate trunking with cables – it will warm up.

- Use only the correct equipment within high risk flammable areas. Standard 110 volt power tools will offer no protection against igniting flammable dust, vapour or gases.

Summary so far . . .

- *The symbols we use for resistance are R and Ω and can be described as the ratio of potential difference, in volts, placed across an electrical component to the current, in amps, which is flowing through it.*

- *The term resistivity describes the ability of a material to resist the flow of an electric current. The unit is the ohm/metre and is identified by the Greek letter ρ, pronounced 'rho'.*

- *A magnetic field can be described as an invisible force field, present around any load bearing conductor or surrounding any magnetic body. The strength of the field is its flux density and is identified by the Greek letter β, pronounced 'beta'.*

- *Dynamic induction is when a body with magnetic properties can cause or induce a voltage in another body without direct 'hard-wire' contact.*

- *A capacitor is a component which has the ability of storing an electric charge when there is a potential difference between the conductors. The unit of capacitance is the farad (F).*

- *The dielectric is the non-conductive part of a capacitor which is sandwiched between the plates.*

- *In a pure inductive circuit the current, with respect to the voltage, is out of phase by 90°.*

- *Reactance, X, measured in ohms is the property of a circuit which contains capacitance and/or inductance and, together with any resistance, makes up the impedance of a circuit.*

- *Sharp, violent capacitor shocks can cause heart fibrillations which could be serious if you had a heart condition.*

- *Capacitive and inductive reactance is an AC's opposition to current flow within a circuit. It is not a value of resistance.*

- *The power factor (pf) of a circuit is the cosine of the phase difference between the voltage and the current. It can vary from minus 90° to plus 90° for inductive and capacitive loads.*

- *The semiconductor diode is similar to a non-return valve serving your central heating system. It will permit water to flow only in one direction.*

Summary so far . . .

- Because a diode will only permit the flow of current in one direction but not the other, it provides a very good means of rectification – turning AC current into DC.

- The bridge form of rectification does not have to be centre tapped from the supply transformer. It is able to provide for a higher DC voltage output.

- An integrated circuit is a complete circuit which has been miniaturised on a semiconductor.

- A triac is a three connection point semiconductor used in incandescent dimmer circuits which is capable of handing up to 300 watts of power.

- Thermistors are used as a solid state temperature control device. Their internal resistance is low when cold but increases when the temperature is higher.

- Transistors can be used as a switching device or as an amplifier – they vary in physical size depending on duty.

- Other forms of rectification include; silicon rectification, copper oxide rectification, thermionic (glass mantle valve) rectification and mercury arc rectification, where a potentially high load is required to be served – often found in research centres.

- Optical fibre systems use an invisible infra-red or low power laser light source as an optical transmitter. The fibre cable is very small and often measures just 0.125 mm in diameter. The system is free from any type of electromagnetic interference.

Review questions

1 Describe what is meant by the resistivity of a material.

2 List four factors which affect the resistance of a conductor.

3 Calculate the total resistance of four 100 ohm resistors in parallel and also four 100 ohm resistors in series formation.

4 In your own words, describe a magnetic field and please state the first law of magnetism.

5 Explain the term dynamic induction.

6 Define electrical reactance within an AC circuit.

7 What is the name given to the process of adding certain impurities to a semiconductor at the construction stage?

8 Briefly describe in simple terms what a dielectric is, then list three common materials which dielectrics are made from.

9 Describe electrical stress within a capacitor.

10 List major dangers associated with capacitors.

11 Describe two major roles of the common transistor.

12 List the components and semiconductors used in a simple rectifying circuit.

13 Briefly describe the function of a semiconductor diode.

14 List three applications of optical fibre cable and its advantage over traditional copper cable.

15 When an AC voltage is applied to an inductive circuit, the current, with respect to the voltage, will be out of phase with each other by how many degrees?

Electrical supply systems, protection and earthing

Introduction

This chapter is all about the types of electrical supply systems that serve our industry and the ways of protecting ourselves and others from the dangers associated with electricity.

There are six underpinning knowledge requirements which are obligatory:

- Electricity supply system – from the power station.
- A review of industrial distribution systems – steel conduit busbar trunking, etc.
- All about transformers.
- A study of high and low voltage switch gear.
- Protection by means of earthing.
- Abnormal conditions for which we need to provide protection.

Practical activities

There are no formal activities accompanying this chapter but as a candidate you will have to provide evidence of the expertise learnt within the occupational unit evaluation.

Tests

As with other chapters, you will be given a multiple choice question paper. You will then be assessed in accordance with City and Guild's test requirements.

Electrical supply systems

In the UK, most electricity is generated between 11 000 and 33 000 volts AC by a high speed steam turbine alternator. Other prime movers are as follows:

- wind turbine – often rated about 400 kilowatt,
- wave motion – along the coast line,
- hydroelectric – waterfalls,
- oil – the use of diesel alternators,
- gas turbine,
- nuclear energy.

Once generated, the electricity is transformed in value to 132 000, 275 000 or 400 000 volts and delivered to the *National Grid* (conceived in the 1920s and established in 1936) which owns, maintains and controls all high voltage electricity lines within England and Wales.

Regional power stations scattered throughout England and Wales come under central control. You could, for example, be getting your electrical power from almost anywhere as Figure 5.1 shows. The interconnected power stations are placed, or taken off line at will, depending on the demand for power. If additional power is needed it is drawn from Scotland or from stations in mainland Europe. This effectively proves a very flexible way of power management.

The transformed voltage is fed into the National Super Grid from any one of the regional power stations. This nationwide grid supplies 400 000 volts to all four corners of the country.

Reducing the voltage

National Grid substations provide the means to reduce the power taken from the grid by transforming it back to 11 000 volts again; Figure 5.2 shows a typical substation situated in

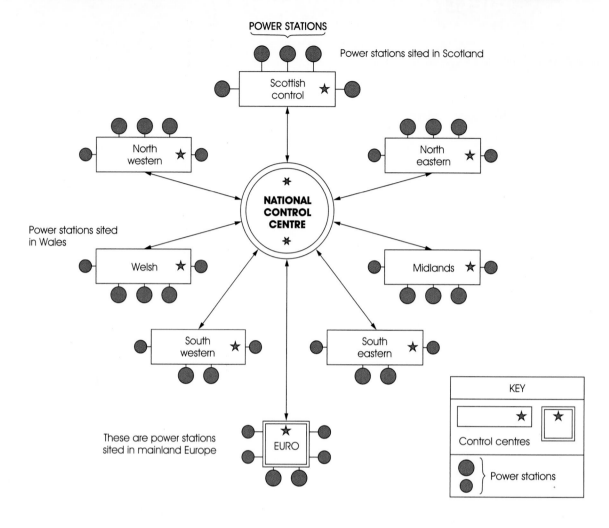

POWER STATIONS

Power stations sited in Scotland

Scottish control

North western

North eastern

Power stations sited in Wales

NATIONAL CONTROL CENTRE

Welsh

Midlands

South western

South eastern

These are power stations sited in mainland Europe

EURO

KEY

Control centres

Power stations

Figure 5.1 – *Regional, Scottish and European production of electricity*

southern England. The reduced voltage is then transmitted over short distances across country attached to insulators fixed to wooden 'T' poles. These are routed to community transformers where it is further reduced for industry and commerce to 400 volts and 230 volts for domestic use. Figure 5.3 shows an example of these poles serving rural areas of Britain whilst Figure 5.4 illustrates a typical community transformer.

Upon leaving the transformer by way of heavy service fuses a three phase and neutral supply distributes for the needs of the local rural community. Each conductor is attached to the top of tall wooden poles as shown in Figure 5.5.

Why the need for high voltage transmissions?

Cable, whether made from copper or aluminium, produces power losses caused by the natural resistance of the conductor. This power loss, which is throughout the cable, creates unwanted heat and a voltage drop. Therefore it stands to reason a very heavy cable would have to be used to transmit power over long distances and this is impractical.

Power lost within a conductor is equal to the sum of the current squared and the internal resistance of the conductor. Hence:

$$\text{Power loss} = I^2 \times R \qquad (5.1)$$

Figure 5.2 – *Typical local substation*

Figure 5.3 – *Wooden 'T' poles carry 11 000 volts to community transformers*

Figure 5.4 – *A typical community transformer*

Figure 5.5 – *The distribution of rural power to villages and farms*

where *I* is the current flow in the conductor in amps and *R* is the internal resistance of the conductor in ohms.

When electricity is transmitted at very high voltages the power loss is far smaller and this can be demonstrated by the following example:

Power totalling 20 000 watts (20 kW) is transmitted through a steel-cored aluminium conductor having a calculated internal resistance of 0.7 ohms.

Determine the internal power loss within the conductor when the voltage applied is (a) 230 volts and (b) 275 000 volts.

Here's how we do it.

(a) First, by applying a voltage of 230 volts:

$$\text{Internal resistance current consumer in amps} = \frac{\text{wattage (W)}}{\text{voltage (V)}} \quad (5.2)$$

Now substitute for known values:

$$\text{Current consumed in amps} = \frac{20\ 000}{230} = 86.9 \text{ amps.}$$

$$\text{Power loss within the conductor} = P \times R \quad (5.3)$$

$$\text{Power loss} = 86.9^2 \times 0.7 = 5286.12 \quad (5.28 \text{ kW}).$$

(b) Next by applying a voltage of 275 000 volts

$$\text{Current consumed in amps (from (5.2))} = \frac{20\ 000}{275\ 000} = 0.072 \text{ amps.}$$

The power loss within the conductor (using (5.3)) = $0.072^2 \times 0.7 = 0.0036$ watts.

Compared with the high voltage example the energy wasted at 230 volts is totally unacceptable and excessive and because of this, power is transmitted throughout the National Grid at a very high voltage.

How are three phase four wire supplies and a single phase supply produced from a three phase generator?

The most familiar method used in power distribution systems is the delta/star connection technique in which high voltage enters the community transformer and is applied to the primary delta windings. The secondary winding of the transformer is arranged in star formation, the mid-point of the star being where the neutral conductor originates from. The star neutral point is also grounded to the general mass of earth, often by the use of earth plates, making both the neutral and earth conductors the same electrical potential – zero volts. Figure 5.6 illustrates this point further in schematic detail.

Loads connected in both star and delta formation

An electric motor is a good example of how three phase equipment is either connected in star or delta formation.

Figure 5.7 illustrates the internal wiring arrangements serving a small squirrel cage induction motor wired in star formation. The three phase supply is connected to terminals L1, L2 and L3. The machine is started by means of a direct on-line electromechanical starter.

Star–delta machines are started in star formation by a sophisticated arrangement. After a short period of time (usually seconds) the starter

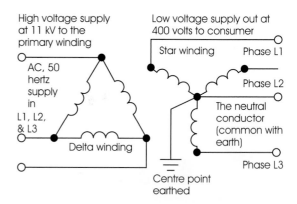

Figure 5.6 – *Delta/star transformer connections*

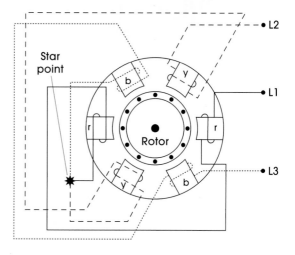

Figure 5.7 – *Star formation wound induction motor*

automatically places the motor windings into delta mode. This system is only appropriate to machines whose *stator windings* (the part which doesn't move) are designed for delta connection whilst running. Some machines are wired so either delta or star configuration may be used by the relocation of copper links.

Calculating phase values of voltages and current in a three phase balanced load

Star connected

Please refer to Figures 5.8 and 5.9 which will schematically show the electrical relationship between phase values and currents within a three phase balanced load. In a triple-phase motor circuit the voltage between each of the phases is at a fixed rotating angle of 120° – because of this the voltage, and the currents generated within

the windings, cannot be added directly together. Instead we have to use a special formula in which the number $\sqrt{3}$ (which works out to be virtually 1.73) is incorporated when calculating theoretical three phase problems.

When I_P represents the phase current, I_L the line current and V_P and V_L are the phase and line voltages respectively, the following is said to be correct:

$$V_P = \frac{V_L}{\sqrt{3}} \text{ where } \sqrt{3} = 1.73. \qquad (5.4)$$

Delta connected

$$I_P = \frac{I_l}{\sqrt{3}} \text{ where } \sqrt{3} = 1.73. \qquad (5.5)$$

A practical example

Calculate the line current drawn from a 400 volt, three phase, 7.5 kilowatt electric motor. The six-pole, 900 rpm machine has a power factor of 0.85 and is found to be 85% efficient.

The only data which is significant to this calculation is:

- the applied voltage, which is 400 volts,
- the kilowatt rating of the machine, expressed in watts (7 500),
- the power factor which is expressed as a number; 0.85,
- the efficiency of the motor, found to be 80%.

As our motor's efficiency is only 80%, it stands to reason that the input must be greater. To find this figure, the following expression is used:

$$\text{Input power in watts} = \text{rated power} \times \frac{100}{80}.$$

Substituting for known figures:

$$\text{Input power in watts} = \frac{7500 \times 100}{80}.$$

Therefore, power in watts with an efficiency of 80% = 9375.

To determine the power in a balanced three phase circuit, the following expression must be applied:

$$\text{Power in watts} = \sqrt{3} \times \text{current} \times \text{voltage} \times \text{power factor (pf)}. \qquad (5.7)$$

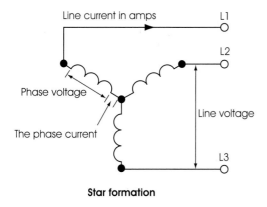

Star formation

Figure 5.8 – *Phase winding values in star formation*

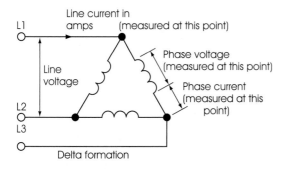

Delta formation

Figure 5.9 – *Phase winding values in delta formation*

But to calculate the current drawn in such a circuit (5.7) has to be transposed (reshuffled) in terms of I. By dividing throughout by $\sqrt{3} \times V \times$ pf, then substituting for known figures:

$$I = \frac{9375}{1.73 \times 400 \times 0.85} = \frac{9375}{588.2} = 15.94 \text{ amps.}$$

Determining values within a three phase system

Kilowatts (kW)

$$\text{Kilowatts} = \frac{\text{watts} \times \sqrt{3}}{1000}. \quad (5.8)$$

Kilovolt amps (kVA)

$$\text{kVA} = \frac{\text{voltage} \times \text{current} \times \sqrt{3}}{1000}. \quad (5.9)$$

And to find the current:

$$\text{Current in amps} = \frac{\text{kVA} \times 1000}{\text{volts} \times \sqrt{3}}. \quad (5.10)$$

kVAR (reactive power) and power factor

One of the most important considerations when calculating the power factor of an electric motor is the total kVAR value. The power factor of an electric motor, varying from as low as 0.5 to 0.95, is normally specified as a percentage or the cosine of an angle.

A practical example follows:

An electric motor draws 300 kVA of apparent power but just 200 kilowatts of real power from the mains. To find the amount of reactive power drawn from the supply we must first determine the power factor of the motor as this requires improving to 95%.

Given that

$$\text{Power factor} = \frac{\text{value of} \atop \text{real power}}{\text{value of apparent} \atop \text{power}}, \quad (5.11)$$

then substituting for known figures:

$$\text{Power factor} = \frac{200 \text{ kilowatts}}{300 \text{ kVA}} = 0.66$$

i.e. 66%.

To calculate the quantity of reactive power drawn from the mains, the following expression must be used (reactive power is used to energise the magnetic fields whereas active power produces the motive force):

$$\text{kVAR (reactive power)} = \sqrt{[(\text{apparent power})^2 - (\text{real power})^2]} = 223.6. \quad (5.12)$$

The way to advance the electrical power factor to 95%

To calculate the kVAR of the capacitor required to bring the power factor to 95%, the following expression must be used.

First we find the apparent power:

$$\text{Apparent power (kVAR)} = \frac{\text{value of} \atop \text{real power}}{\text{power factor}}. \quad (5.13)$$

Substituting for known values:

$$\text{Apparent power} = \frac{200}{0.95} = 210.5 \text{ kVA.}$$

Next we calculate the reactive power required:

$$\text{kVAR reactive power} = \sqrt{[(\text{apparent power})^2 - (\text{given real power value})^2]} \quad (5.14)$$

Substituting for known values:

$$\text{kVAR reactive power} = \sqrt{[(210.5)^2 - (200)^2]} = \sqrt{[44\,310.25 - 40\,000]} = 65.7 \text{ kVAR.}$$

To determine the required kVAR to improve the power factor to 95% we use the following expression:

$$\text{kVAR of capacitor} = \text{original kVAR (5.13)} - \text{kVAR reactive power (5.15).}$$

Substituting for known values:

$$\text{kVAR of capacitor} = 223.6 - 65.7 = 157.9 \text{ kVAR.}$$

Now select the nearest capacitor rated to this value from the manufacturer's tables.

Another way is to use the manufacturer's power factor correction tables which, in due course, is far quicker than the method discussed here.

Neutral current

If you have a perfectly balanced load such as a three phase induction motor, all three phases carry the same current so there is no requirement for a neutral conductor – all currents cancel out at the star point; zero volts.

You might have seen three phase and neutral installations where the neutral conductor is much smaller in cross sectional area compared to the phase conductors. This is fine if, for example, the installation is to serve only three phase motors and a couple or so lighting circuits. The system would be a virtually balanced load and any imbalance registered would be small compared to the actual line load. The greater the imbalance within your installation, the greater the current produced within the neutral conductor.

Problems which occur are caused by circuits which generate masses of third harmonic currents. These distort the current wave form. The guilty parties are:

- transformers carrying a small load,
- fluorescent lighting luminaries,
- circuits which have ballast units incorporated,
- sodium and mercury vapour lamps,
- any sort of discharge lighting circuit.

These and many others will distort the current wave form to three times the normal frequency (that is, 150 hertz) but unfortunately they do not cancel at the star point as normal frequency currents will. These rogue currents add up, resulting in the neutral carrying a very hefty third harmonic current – sometimes even greater than the line current. For this reason it is wise not to reduce the size of your supply neutral to a three phase installation unless you are sure your installation will be technically balanced throughout.

Please make sure that your neutral conductors are completely separate from each other. Never be tempted to steal a convenient neutral from an existing circuit to serve a newly installed single phase supply such as Figure 5.10 illustrates. Each individual single phase circuit must have its own separate neutral conductor. These conductors must be in the same order as your phase conductors leaving their individual protective devices.

Industrial distribution systems

Table 5.1 lists nine industrial type distribution or installation systems and suggests typical practical applications for them. Illustrations are provided for review.

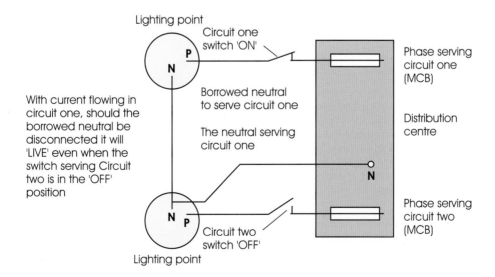

Lighting point
Circuit one switch 'ON'
P
N
Phase serving circuit one (MCB)

With current flowing in circuit one, should the borrowed neutral be disconnected it will 'LIVE' even when the switch serving Circuit two is in the 'OFF' position

Borrowed neutral to serve circuit one

The neutral serving circuit one

Distribution centre

N

N P
Circuit two switch 'OFF'
Lighting point

Phase serving circuit two (MCB)

Figure 5.10 – *Never be tempted to steal a neutral from a nearby circuit to serve a new circuit you have installed*

Table 5.1 – *Industrial distribution systems and their applications*

Method of distribution	Typical application	Illustration
Flexible conduit	Anywhere where ridge conduit would be inappropriate – supplying a feed to an electric motor.	
Solid conduit	Industrial and commercial use. In areas where protection is needed.	
Trailing cables and overhead collectors	In hazardous conditions. Electric portal cranes serving building sites. Industrial workshops. For power supplies to serve large movable hosting equipment. Transport and conveyor systems, etc.	
Busbar trunking	Often used for commercial purposes where the busbar trunking takes the form of an insulated multi-compartmental system used for local power distribution. Lighting busbar trunking is usually metal clad and used in industrial/commercial schemes where the call for extensive switching is unwarranted. Example: hypermarkets, factories and large workshops.	
Under-floor dualing	Dualing is the term used when wiring an installation for both voice and data. The cable is terminated at a succession of 'plug-in', take-off-points. The multi-stranded cabling used for dualing is varied and ranges from single to many cores. The 0.2 mm^2 cable is often lapped screened within a PVC sheath and routed within walls, under floors and within PVC-u trunking throughout the installation.	
Cable trunking	Used in commerce and industry as a bulk cable carrier. Examples: supermarkets, airports, factories. Smaller versions are used for domestic installation work.	CE Ⓕ Ⓝ

Table 5.1 – *continued*

Method of distribution	Typical application	Illustration
Rising mains	High rise flats and commercial buildings as a means of power distribution. Metal clad and well protected.	Principle supply conductors (three-phase and neutral) busbars. Insulated supports. Fire proofing between floors and bus-bar insulation. Steel trunking cover. Copper bonding link. Steel conduit cable carrier. Supply 'take-off' box served with fuses. *This system is ideally suited for office blocks and other high-rise buildings*
1. PILC/SWA, 2. PVC/SWA 3. MIMS [MI cable]	1. This cable is often used for underground installations where mechanical damage could occur. Used also for above ground power distribution within factories and heavy industry. PILC/SWA stands for *Paper Insulated Lead Covered/Steel Wire Armoured* cable. 2. PVC/SWA cable is a robust cable ideally suited for power transmission within industrial and commercial premises. It can be applied to control circuits and temporary wiring requirements to serve construction sites. Smaller cables range from 1.5 to 16 mm^2. Larger sizes are from 25 to 400 mm^2. 3. *MI cables* are mineral insulated (copper or aluminium sheathed) cables and were first produced in 1936. An ideal system of wiring for a wide range of commercial and industrial installations. Highly reliable and will not deteriorate with age. MI will operate in temperatures up to 250°C and will continue to function up to 1000°C. Ideal to install in boiler houses and other very hot environments.	COPPER CONDUCTORS, COPPER SHEATH, INORGANIC INSULATOR (MGO), LSF CORROSION RESISTANT OUTER SHEATH (OPTIONAL)
LSF cables	This stands for *Low Smoke and Fume* insulation and is applied to a range of different cables. It is used for panel work, conduit and general fixed wiring where there is a requirement for flexibility during the installation period and where there is a need to reduce the risk of toxic fumes and smoke in the event of a fire. Can be used in temperatures up to 70°C. The recommended bending radius is six times its diameter.	
FP 200® cables	This type of cable is ideal for a fire alarm system or an emergency lighting arrangement supplied from a central battery unit. The conductors are within an aluminium sheath, sleeved with a high grade PVC laminate. The outer sheath can be removed by scoring around the cable with a sharp knife then flexing until the work piece snaps – withdraw by twisting the cable whilst pulling. The insulation serving the conductors is made from silicone rubber which, under the conditions of extreme heat, is chemically converted into silicon dioxide. In this condition the burnt insulation becomes a very good insulator.	

All about transformers

How they work

The transformer is a simple device; it has no moving parts and can be made to any size or type required.

It consists of two lightly insulated copper coils of wire, called the *primary* and *secondary windings*. These windings are individually wound on a common laminated iron core and are electrically separated from each other.

Every load bearing conductor has an invisible electromagnetic force field around it as pictured in Figure 5.21. The heavier the load, the more intense the magnetic field will be. Just imagine the flux as ripples on a still mill pond after a rock has been thrown in. Let the rock you've thrown in represent the load bearing conductor.

The transformer works on a similar theory. It relies on the principle of the rise and fall of a magnetic force field, similar to the ripples on a pond, but produced by an alternating voltage applied to the primary winding. Because an alternating voltage is constantly rising to a peak and collapsing fifty times every second (50 hertz) it creates a fluctuating magnetic field in the primary coil. By mutual induction, an alternating current is induced in the secondary winding, as Figure 5.22 will illustrate. You can obtain this effect another way by rotating a simple loop of wire within a fixed North/South magnetic field. Please stretch your imagination a little further and see that as the loop cuts through the lines of force at right angles, an electromotive force (emf) is induced into the coil – Figure 5.23 will explain further.

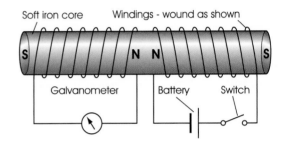

Figure 5.22 – *A current is induced into a secondary winding by mutual induction*

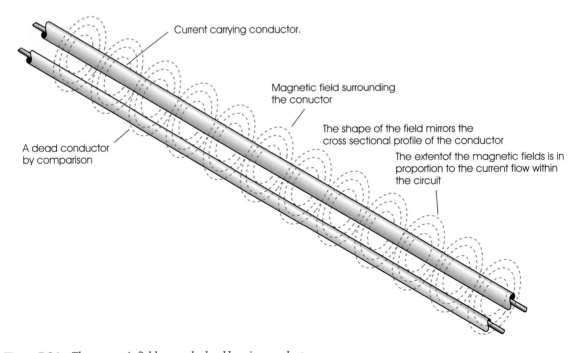

Figure 5.21 – *The magnetic field around a load bearing conductor*

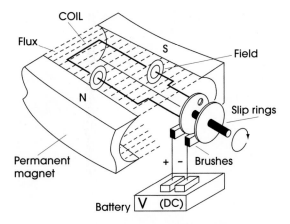

Figure 5.23 – *A loop of wire cutting through a North–South magnetic field will induce an emf into the wire. Coil references are annotated as c1, c2, c3 and c4. S, the slip rings and B, the brushes serving the galvanometer, G*

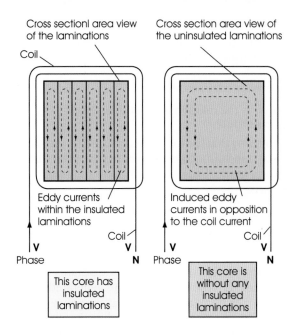

Figure 5.24 – *Eddy currents are reduced when the core of the transformer is laminated*

Input/output and losses

An isolating transformer is a type of transformer that will produce the same voltage output which is supplied to its input primary windings – or will it?

Unfortunately, no – even though each winding is identical and has exactly the same number of turns of wire, losses will incur. Although very efficient, the transformer has two reasons for losses:

- *Copper loss* –this is due to the heat generated within the copper due to the natural resistance of the copper. We call this the I^2R loss (power in watts).
- *Iron loss*, of which there are two types – eddy currents and hysteresis losses.

Eddy currents

Eddy currents occur within the steel core of a transformer. They are generated in the same way as the secondary winding receives energy from the primary winding – by mutual induction caused by a constantly changing magnetic field motivated by the frequency of the supply. Once the lines of magnetic flux cut across the metal core of the transformer, currents are induced into it. The currents are now on a tread-mill – they have nowhere to go other than to be driven round and round as though on the London Eye. This produces a loss of useful energy.

If the core of your transformer were to be made from solid iron or steel its electrical resistance would be very small and therefore it would produce high currents. To minimise this effect, the core is laminated – each slim lamination is insulated from each other. The resistance between these laminations greatly reduces the eddy currents generated within the core as the thin steel sheets do not provide an easy conductive path for the currents. Figure 5.24 will illustrates this concept clearly.

Eddy currents occur on other machines and equipment such as electric motors and ballast units serving lighting equipment. The use of low resistance wire also helps to minimise the effects of eddy currents.

Hysteresis losses

This is a magnetic phenomenon and is due to the energy used within the iron core during the constantly changing amplitude of the magnetic field. After the magnetic field has briefly collapsed a tiny amount of residual magnetism is left over. This left-over magnetism has to be neutralised and the energy used to do this corresponds to a theoretical loss.

To lessen this effect silicon steel is used as it experiences difficulty in holding onto the residual magnetism generated within the core. Silicon steel also provides a low magnetic resistance to the magnetic lines of force.

What is the relationship between input and output?

Providing that the losses incurred are tiny and your transformer has a power factor of one, the power input almost equals the power output; hence:

$$P_{\text{secondary}} = P_{\text{primary}} - P_{\text{losses}} \qquad (5.15)$$

Or

$$V_p \times I_p = V_s \times I_s \qquad (5.16)$$

where V_p is the voltage serving the primary winding, I_p is the current in amps generated within the primary winding, V_s is the voltage induced into the secondary winding, I_s is the current drawn from the secondary winding and P is the power in watts.

How do I calculate the efficiency of a transformer?

The efficiency of a transformer is always expressed as a percentage and can be determined by use of the following expression:

$$\text{Efficiency as a percentage} = \frac{\text{output of the transformer in kilowatts}}{\text{input (kW)} + \text{Cu loss (kW)} + \text{Fe losses (kW)}} \times \frac{100}{1} \qquad (5.16)$$

where Cu is copper and Fe is iron.

As an example consider the following:

A construction site transformer rated to deliver 10 000 watts of apparent power has an actual output power of 9800 kW. From information gathered from the manufacturer it was found that the copper losses were found to be 0.33 kW and the combined iron losses totalled 0.2 kW. Please calculate the efficiency of the transformer.

Substituting for known values:

$$\text{Efficiency of transformer} = \frac{9.8}{10 + 0.53 + 0.3} \times \frac{100}{1} = \frac{9.8}{10.83} \times \frac{100}{1}.$$

Figure 5.25 – A construction site transformer

Therefore, efficiency of the transformer = 90.4% (i.e. 9.6% of potentail energy wasted).

Commercial and site transformers are about 80 to 90% efficient whereas power line (National Grid) transformers are often more than 98% efficient. Figure 5.25 illustrates a typical construction site transformer.

The way to calculate the current drawn from a transformer

Figure 5.26 outlines in schematic form a basic 2:1 stepdown transformer. Please assume that no losses occur and that our theoretical transformer is 100% efficient. The primary winding has 400 turns of wire and is served by 200 volts, AC supply. The secondary winding produces 100 volts which supplies a load with a resistance of 10 ohms.

By applying Ohm's law the current drawn from the secondary winding may be calculated:

$$I = \frac{E}{R} \qquad (5.17)$$

where E is the secondary voltage, I is the secondary current in amps and R is the resistance of the secondary load in ohms.

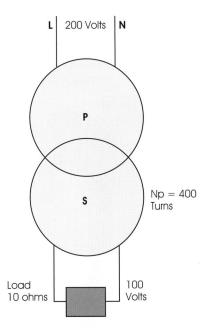

L | 200 Volts | N

P

S

Np = 400 Turns

Load 10 ohms

100 Volts

Figure 5.26 – *This is a basic two to one (2:1) step-down transformer*

Substituting for known values:

$$I = \frac{100}{10} = 10 \text{ amps.}$$

Once the secondary current has been calculated, use the following expression to determine the current drawn from the primary winding:

$$\frac{E_p}{E_s} = \frac{I_s}{I_p} \tag{5.18}$$

where E_p is the voltage applied to the primary winding, E_s is the voltage within the secondary winding, I_p is the current in amps within the primary winding and I_s is the current drawn from the secondary winding.

Now, substituting for known values:

$$\frac{100}{200} = \frac{I_p}{10}.$$

Now cross multiply and evaluate in terms of I_p:

$$I_p = \frac{1000}{200} = 5 \text{ amps (primary current).}$$

Finding the power in watts generated

Quickly glancing at these calculations you would be forgiven if you thought you were getting something for nothing, but by applying (5.19)

and (5.20) and finding the total power in both the primary and secondary windings you will see that this is not so.

$$\text{Primary power} = I_p \times E_p \tag{5.19}$$
$$\text{Primary power} = 5 \times 200$$
$$\text{Primary power} = 1000 \text{ watts}$$

$$\text{Secondary power} = I_s \times E_s \tag{5.20}$$
$$\text{Secondary power} = 10 \times 100$$
$$\text{Secondary power} = 1000 \text{ watts.}$$

Calculating the number of turns of wire

To determine the number of turns of wire within the secondary winding, we must use the following expression:

$$\frac{E_p}{E_s} = \frac{N_p}{N_s} \tag{5.21}$$

where N_p is the number of turns in the primary winding and N_s is the number of turns in the secondary winding.

For an example, please refer to Figure 5.26: this clearly shows that the primary winding has 400 turns. Then substituting for known values:

$$\frac{200}{100} = \frac{400}{N_s}.$$

Cross multiplying:

$$200N_s = 400 \times 100.$$

Dividing both sides of the equation by 200 to rearrange the expression in terms of N_s

$$N_s = \frac{40000}{200} = 200 \text{ turns.}$$

With 400 primary turns of wire in mind, it is clear that the transformer has a ratio of 4:2 or 2:1.

How may I check my calculations?

One way to check your calculation is to express the ratio in terms of the current drawn from both sets of windings – given that:

$$\frac{N_p}{N_s} = \frac{I_s}{I_p} \tag{5.22}$$

where N_p is the number of turns in the primary winding, N_s is the number of turns in the secondary winding, I_p is the current drawn from the primary winding and I_s is the current drawn from the secondary winding.

Substituting figures and presenting in terms of I_p:

$$\frac{400}{200} = \frac{10}{I_p}.$$

$$400I_p = 200 \times 10$$

$$I_p = \frac{2000}{400} = 5 \text{ amps.}$$

Now we must calculate the value of the current drawn in the secondary winding by referring back to (5.22) and substituting for known values:

$$\frac{400}{200} = \frac{I_s}{5}.$$

Cross multiplying and bringing the expression in terms of I_s:

$$200I_s = 400 \times 5$$

$$I_s = \frac{2000}{200} = 10 \text{ amps.}$$

By simple calculation (current × voltage) you will see that both sets of transformer windings generate the same amount of power.

Multivoltage transformers

You may come across a multivoltage transformer in an industrial setting where the transformer has been adapted to accept several different voltage ranges to serve the primary or secondary winding.

How we express the rating of a transformer and why it is in kVA

All transformers are rated in kilovolt-amps [*apparent power*], shortened to kVA. A kilovolt is 1 000 volts, so 230 volts would be correspond to 0.23 kVA. The actual *true power* output of a transformer is never used – it will vary depending on the total losses incurred. Large metal-clad construction site transformers are usually rated at 10 kVA. The much smaller, portable, glass fibre insulated type supporting just two or three outlets, shown in Figure 5.27, are rated at 5 kVA.

The term kVA is always applied – this is because the warmth created by the transformer is totally dependent on the current generated within the windings, so the term *true power* in kilowatts is never used. As the power factor decreases the current flowing in both sets of windings will increase but the kilowatt rating will remain unchanged.

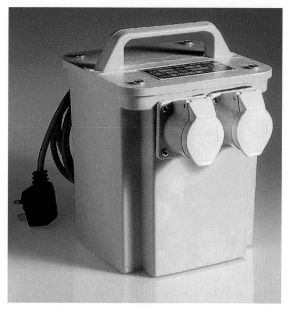

Figure 5.27 – *A portable 5 kVA transformer*

Finding the line current from a kVA rating

Let us assume that we wish to find the line current of a 400 volt, three phase 10 kVA construction site transformer having a rated power factor [pf] of 0.9 so as to select a suitable size cable to serve it. Please ignore the efficiency of this machine or volt drop due to length.

How it's done

First, we must apply the following expression:

$$\text{Line current} = \frac{\text{kVA rating} \times 1000}{\text{volts} \times \sqrt{3} \times \text{pf}}. \quad (5.23)$$

Now, substituting for known values:

$$\text{Line current} = \frac{10 \times 1000}{400 \times 1.73 \times 0.9} = \frac{10000}{622.8} = 16 \text{ amps.}$$

The line current evaluated is far smaller than if your 10 kVA (0.9 power factor) site transformer had been designed for single phase use. This can be verified using the following expression:

$$\text{Single phase line current} = \frac{\text{transformer's kVA rating}}{\text{supply voltage (kV)} \times \text{pf}}. \quad (5.24)$$

Now, substituting known values:

$$\text{Single phase line current} = \frac{10}{0.23 \times 0.9} = \frac{10}{0.207} = 48.3 \text{ amps.}$$

Please remember to install a Type 'D' or 'C' miniature circuit breaker to serve your 10 kVA site transformer as Type 'B' regularly switch themselves 'OFF' during the start-up process.

The auto transformer

An auto transformer is a special type of transformer made from a single length of lightly insulated wire wound many times around a thinly laminated steel or iron core. It works on the principle of self induction. The primary and secondary windings are wired in series with each other and the magnetic lines of force created by the primary slices through the secondary winding and induces a voltage into it – hence the term 'self induction'. The transformer is so designed that the output voltage can easily be varied and is ideal for manually starting an electric induction motor – Figure 5.28 illustrates this in schematic form.

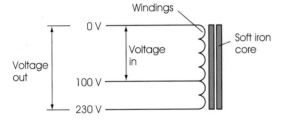

Figure 5.28 – *A step-up auto transformer – schematic profile*

Throughout the length of the secondary winding, output tapings provide steps to either increase or decrease the voltage supplied to the motor. These 'steps' are hard wired to an adjustable voltage selector mechanism which in turn provides a variable voltage to the motor.

The step-up version is often employed for use in medical equipment such as X-ray machines.

The toroidal (round, doughnut shaped) auto transformer is another version and has many advantages to offer over the more traditional transformer:

- It is compact and smaller in size compared to a traditional transformer providing an economical answer to many a problem.
- Easily installed in existing equipment.
- Its weight is about 30 to 70% lighter.
- It has very low stray magnetic fields.

- Can be mounted horizontally.
- Obtainable from 250 VA to 1000 VA.

Associated dangers

As there is a direct electrical connection between the input and output supplies, the auto transformer's useful function is significantly limited. Care must be taken to avoid direct electrical contact with live parts.

Low and high voltage switch gear

Low voltage

Low voltage is any voltage which is above 50 volts, AC but not greater than 1000 volts AC between conductors or 600 volts AC between the phase conductor and earth.

Switch gear is designed in all different shapes and sizes:

- Isolators – triple and single phase quick make/break for AC/DC use.
- Switch-fuses – Figure 5.29.
- Splash proof switch disconnectors to IP 54.
- Safety switches – single and double pole.
- Low voltage switch panels.
- Isolators – single and three phase AC.

Figure 5.29 – *Common switch fuse*

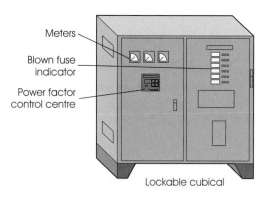

Meters

Blown fuse indicator

Power factor control centre

Lockable cubical

Figure 5.30 – *A lockable modular panel board*

Figure 5.31 – *This overland means of power distribution carries 11 000 volts*

- Time switches – usually single phase.
- Changeover switches – single or three phase.
- Fireman's switch – usually rated at 20 or 32 amp.
- Lockable modular panel board system – Figure 5.30. This type of panel can accommodate miniature circuit breakers, fused switches, contractors distribution centres and other types of control and monitoring equipment. Often both the back and front panels are designed to be removed for ease of access. The bus bars are inclined to be very large and typically rated from 1000 to 10 000 amps. The short circuit capacities for units of this type are usually between 2000 and 1000 amps for duration of 1 second in time.

There are many different types of low voltage switch gear; this is just a snap-shot of some of them.

Medium and high voltage switch gear

Medium voltage

Medium voltage will normally exceed 1000 volts, AC or 1500 volts DC between conductors but is no greater than 24 000 volts. An example can be taken from the 11 000 volt cross-country distribution systems which serve community transformers from National Grid sub-stations as shown in Figure 5.31.

Medium voltage switch gear is used exclusively for switching and power distribution management within small sub-stations.

Figure 5.32 – *Medium voltage switch gear*

Typically, panels similar those illustrated as Figures 5.32 and 5.33 are assembled from modular components such as circuit breakers, load break switches, overcurrent equipment, metering apparatus, bus bars and cable termination points. Special vacuum contactors and switches are used as means of control as the nominal voltage can be as high as 24 kV in value.

Figure 5.33 – *Medium voltage control panel*

High voltage switch gear

High voltage starts from 24 000 volts – examples follow:

- The grid distribution system providing 132 000 and 275 000 volts nationwide. Figure 5.34 illustrates a classic 33 000 volt distribution network.

- The super grid, which delivers 400 000 volts across country – Figure 5.35 illustrates a distinctive pylon sited in southern England.

High voltage switch gear is only found within National Grid power stations where energy is produced at about 11 000 volts and transformed to an extremely high voltage for delivery into the grid.

Figures 5.36 and 5.37 provide a brief insight into the type of switch gear used for high voltage installations. Switch gear installed within a power station's central control room is modular in design and is either air or gas insulated. The voltage rating on all electrical equipment is 420 kV but the current rating is far lower than you might expect as this is because the working voltage is very high.

The function of low voltage switch gear

We shall be reviewing three typical switching devices we are regularly associated with during our daily routine.

Figure 5.34 – *This type of pylon carries 33 000 volts across the countryside*

Figure 5.35 – *This is a super-grid pylon and carries up to 400 000 volts*

Figure 5.36 – *High voltage switching panel*

Figure 5.37 – *A high voltage panel*

1   *Circuit breakers* – for manual or automatic disconnection of the supply, resulting from an over-current or a short circuit condition from the phase conductor to earth. They can be divided as follows:

   –   *Miniature circuit breakers (MCB's)*. There are three types: thermal and magnetic, magnetic/hydraulic and assisted bi-metal. These are sub-divided into various categories (Type 'B' for lighting circuits, 'C' for medium size motor circuits and 'D' for heavy site transformers and motor circuits). Different short circuit capacities are available (4.5, 6 and 9 kA, for instance). MCB current ratings are surprisingly between 0.5 amp and 125 amps whilst one to four pole varieties are freely available. They are also designed to accommodate either AC or DC supplies and most will accept voltages of up to 600 volts.

   –   *Moulded case circuit breakers (MCCB's)*. These devices carry out the same task as the smaller circuit breakers. The difference is they are much larger in physical appearance, able to cope with far bigger loads and their short circuit capacity is greatly strengthened. The duty rating is an 'M' number, for example M1 or M3 up to M9 and is printed on the side of the MCCB. M1 would represent 1000 amps short circuit capacity whilst M3 would signify a 3000 amp short circuit capacity. These are non-adjustable factory sealed time/current quantities – these, you cannot change. Figure 5.38 illustrates a typical triple phase MCCB

2   *Switches* – These are mechanical manual/auto devices which, when activated, are capable of disrupting the electrical supply to a circuit whilst under load or any abnormal condition such as an overload. There are numerous, of which you must be familiar with many. Circuit disconnectors are designed for many different purposes:

   –   splash proof safety switches to IP 54,
   –   three phase distribution centre main switches to BS 5419: 1977,
   –   toggle switches – see Figure 5.39,
   –   domestic and industrial switches,
   –   cupboard switches,

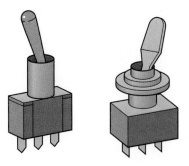

Figure 5.39 – *Toggle switches*

Figure 5.38 – *A three phase moulded case circuit breaker*

Figure 5.40 – *This is a standalone passive infra-red detector*

- infra-red switches (PIR type) – see Figure 5.40,
- light sensitive switches (photo-cell type).

These are all switching devices, not to mixed up with isolators and are all capable of switching a circuit either 'ON' or 'OFF' under load conditions.

3 *Isolators* – These are devices used to isolate all live conductors (and this includes the neutral) from the supply to the load and must be so fitted as to avoid accidental reclosure brought about by mechanical jolt or vibration. This usually means the use of a lock and key. All isolators must indicate whether the contacts are *open* or *closed* – either visibly or clearly indicated. Never use a sensi-touch switch or a semiconductor device as a means of isolation – they are just not reliable enough.

There are many forms of isolation you can choose from but what you decide will be ruled by the type and setting of your installation. Here are just a few for you to consider:

- typical metal clad isolators,
- copper links,
- plugs and socket outlets which serve your equipment,
- switch disconnectors,
- fuses – semi-enclosed, cartridge, HRC and HBC, etc.,
- any type of circuit breaker which act in accordance with the contact separation rules. BS 5419 defines these.

Figure 5.41 – *A lockable industrial isolation switch*

Figure 5.41 illustrates a lockable isolator, an ideal solution for when planned maintenance has to be carried out on machinery.

Earthing systems

There are several different types of supply/earthing arrangements. They are known by initials; BS7671 (312–03–01) covers the following:

- TT,
- TN-S,
- TN-C-S,
- TN-C,
- IT.

What do these letters stand for?

The first letter
T – The live components have one or more direct connections with earth.
I – The live components are connected through a high impedance.

The second letter
T – All exposed conductive parts within the installation are grounded to earth via an electrode buried in the ground.
N – All exposed conductive parts within an installation are connected by way of an earth connection point to the earth provided by the supplier.

The remaining letters
C – A combined protective conductor and neutral within the same cable.
S – A separate neutral and protective conductor.

TT systems

You will find this system within the countryside, especially in older properties where you see overhead cables (phase and neutral) entering the property. It is your responsibility to the consumer to supply a suitable earthing arrangement for his or her installation. This might be in the form of a series of earthing plates or steel, copper plated electrodes buried deep into the ground. This form of earthing is very expensive and it is usually left to the installer to provide a suitably rated residual current device (RCD) and an earth rod connected to the earthing bar of the installer's electrical distribution centre.

It is usual for you to fit a 30 milliamp rated RCD. This is to protect any socket outlets or other sensitive circuits you have installed within your installation from earth leakage problems. A 100 milliamp device will protect lighting arrangements, etc. By this I mean stray electrical 'fault voltages' which are exposed to conductive and extraneous material. Providing the outside soil resistance is no more than 200 ohms in value when measuring from your earth electrode to the start point of the community transformer, which is suitably earthed, your RCD will faithfully 'auto-trip' within 0.2 second or less if a problem of this nature arises. It will not protect against over-current or short circuit conditions – this will be left to your in-circuit MCB or RCBO device.

TN-S systems

This system is found within towns and cities – particularly on older properties. The lead sheath of the supplier's underground service cable is earthed at the community transformer and connection is also made to the star point of the HV transformer's *secondary* winding.

The consumer's installation is then earthed by connecting a suitable size earth wire from the lead sheath of the service cable, via a heavy clamp, to the earthing/bonding bar within the distribution centre.

As the star point of the HV transformer is neutral, any fault voltage to earth would be the same as connecting a phase conductor directly to neutral. The protective over-current device would be brought into play – the circuit will be isolated.

TN-C-S systems (protective multiple earthing – PME)

This is a system found both in towns and villages where the service cable to the property is a combined copper stranded neutral and earth wrapped around an insulated phase conductor. The complete cable is sheathed in PVC to insulate the neutral/earth conductor from touch. This type of cable is referred to as a PEN cable.

The supplier's PEN cable is connected to their service fuse assemblage and a suitable size earth wire (16 mm^2 is used when the double-insulated service tails are 25 mm in cross sectional area) is routed to the earth/bonding bar within the customer's distribution board.

As the neutral conductor is always grounded to earth at the star point of the high voltage community transformer's secondary winding, any stray voltage from the consumer's installation would be the same as applying a direct short circuit between both phase and neutral. The overcurrent device will respond accordingly.

To recap: the supply is a combination neutral and earth – but these are separated once within the installation.

TNC systems

TNC earthing arrangements stem from a privately owned isolating transformer, served from the public supply. All installation work is carried out using mineral insulated cable (MI), the copper sheath of the cable acting as both neutral and protective conductor. A residual current device cannot be used with this type of earthing arrangement as both neutral and the protective conductor are the same. This system is in decline.

To recap: no separate earth conductor. The neutral conductor is used as an earth throughout the supply and installation.

IT systems

This is when your supply originates from a portable site generator (Figure 5.42 illustrates this) or from a highly specialised private power source. The principal earth wire is connected from the terminal point at the generator to a site installed earth electrode, sited as near to the generator as practical. Never used for public supplies.

Figure 5.42 – A typical site generator

Why do we provide protective earthing arrangements to an electrical installation?

Low voltage systems

Many electrical appliances are enclosed within a protective metal enclosure such as fridges, cookers and dish washers. These and many others could become 'live' in the event of a phase to earth fault occurring. Anyone touching the appliance could receive a severe electric shock or even suffer death, but this would depend on the fault current flowing through the victim's body.

To stop this occurring, an earth connection is made to the exposed metalwork of the appliance by means of a plug and socket outlet serving the equipment via the general bonding system. This earth connection will limit the voltage on the exposed metal work to a reasonably safe level until the over-current protection device is brought into play and, with a miniature circuit breaker, this is occur within 0.4 or less of one second.

All extraneous conductive parts such as copper pipes and steel frames, etc., must also be connected to a common earth terminal. This will ensure that dangerous electrical potentials will never exist between any metal extraneous part – they would all be at the same electrical potential; zero volts. The main electrical earth connection will then provide an uninterrupted path for any stray or dangerous fault current to dissipate, and to electrically disrupt any of the following protective devices:

- all types of fuses from semi-enclosed to cartridge,
- residual current devices,
- miniature circuit breaker,
- RCBO,
- moulded case circuit breaker,
- earth fault (voltage type) breakers – these are not used so much these days.

Figure 5.43 shows the fault path to earth from a defective appliance whose protective conductor has been unintentionally disconnected from the 13 amp plug.

A final thought

Please remember, the electrical earth within a TN-S system (the lead sheath) is connected to

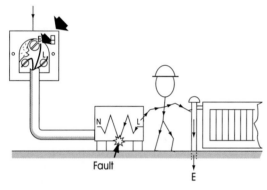

Fault

E

Figure 5.43 – *This illustration follows the fault voltage from a defected electric fire*

Figure 5.44 – *Semi-enclosed fuses (BS 3036)*

Figure 5.45 – *Cartridge fuse*

the star point of the local community transformer. A fault situation to earth is really a fault condition between both the phase and neutral conductors.

High voltage systems

This is a brief insight into high voltage earthing systems as space within this book is limited.

Good high voltage sub-station earthing is essential if the system is to be both safe and reliable. High voltage systems are protected in the following way:

- Earthing by way of copper earth matting, grid systems and plates buried deep in the ground.
- By the use of earthing switches within sub-stations.
- By earthing each steel pylon within the system.

Each high voltage system must be protected by a direct connection to earth or through a suitable resistor to earth. When a high voltage fault occurs it travels deep into the ground and produces a graduated surface potential at ground level. High voltage earthing techniques are a science in itself.

Protection – dealing with abnormal low voltage situations

Table 5.2 reviews ways of dealing with abnormal situations.

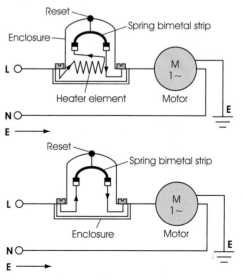

Figure 5.46 – *Miniature button-type motor overload device*

Table 5.2 – *Abnormal conditions and methods of protection*

Abnormal condition	Method of protection
Dangerous short circuit currents Example: exposed trapped conductors	1. Miniature circuit breaker 2. RCBO (this is a combined RCD and MCB used on single phase only) 3. Semi-enclosed fuse – Figure 5.44 4. HRC and HBC fuses 5. Other types of cartridge fuse – Figure 5.45
Earth leakage currents Example: Faulty immersion heater element	1. Residual current device (30 mA rated if serving sockets on a ground floor) 2. RCBO device (do not insulation test from phase to neutral – electronic) 3. Older installations might be protected by a voltage operated earth leakage circuit breaker – not in general use.
Overload currents Example: A small electric motor working far harder than intended	1. Miniature bimetal overload devices, with or without integral heater elements as illustrated as Figure 5.46 (single phase only) 2. Overload relays used as part of a starter unit to serve an electric motor. 3. MCB, RCBO and all types of fuses.
Protection by functional extra low voltage [FELV]. For example, an extra low voltage lighting arrangement. For technical detail, please refer to Regulation 471-14 [BS 7671]	When conditions cannot be met for a separated extra low voltage system, FELV can be used. To protect against direct contact, protection is to be provided by one or more of the following: • Enclosures with protection to IP2X • Insulation able to withstand a voltage of 500 volts (rms) AC for one minute One point of the FELV installation must be connected to an electrical earth.

Handy hints

- When tightening screws using a battery operated drill, to avoid damage to your hand or finger tips, keep the hand you are not using well away from the screwdriver bit.

- Keep an up-to-date electrical installations textbook in your company vehicle – you never know when you might need to refer to it.

- Never accept word of mouth variation orders or requests for additional work from your builder. It could be difficult to claim for payment for the work you have carried out without any paperwork.

- Each year throughout the UK there are about five fatal accidents involving portable appliances. For your own, and for the sake of others, insist that site portable appliance testing be carried out regularly – not just at the beginning of the job!

- You may not be insured if you leave your personal tools within your company vehicle overnight – this is worth checking out.

- Professionally prepared electrical drawings can occasionally contain errors – take a little time out to check them before work commences.

- Refractory bricks, used inside night store heaters, will give you unnecessary problems if installed wet. Once switched 'ON' the water will vaporise and cause damage to the wall decorations and possible start to corrode the internal working components and make them very wet.

- Nylon carpets and furnishings have a low melting point. Take care to prevent your hot masonry drill (especially if blunt) coming into contact with low melting-point fixtures. It might cost you an 'arm and a leg'!

Handy hints

- Never rely on the steel wire armouring around a SWA cable as your only means of earthing, especially under site conditions. When installed in the open there is always a risk of corrosion setting in or the cable end pulling away from the brass gland. This often happens if your SWA cable is connected to a site transformer which is frequently moved or re-sited occasionally. Install three core SWA cable for a single phase load and four core when wiring a temporary three phase installation such as a site cement silo.

- To convert imperial horsepower values into metric kilowatts, divide your horse power by 1.341. Alternatively, multiply your total horse power by 0.746.

- Changing a star–delta contactor coil can be a very scary task if you have never experienced this before. It is made easier by labelling all associated conductors and internally fitted wires as to where they go. This is the way to minimise mistakes. Be careful of internal pressure springs – you could lose them!

- If you are a job holder, adopt a common cable marking policy when working with two or more electricians. This will avoid uncertainty during the second fix stage of your installation. Keep all the destinations from your distribution centres in the same order when working on a block of flats.

- Tungsten halogen lamps will, in due course, blacken when used in conjunction with a dimmer switch, which has been set to a reduced voltage. You can return the lamps back to their former glory by just setting the dimmer control to maximum voltage. Leave 'ON' until the blackness has cleared.

- Equipment containing electronic components should not be high-voltage tested – it will destroy delicate circuitry. Remember this when carrying out Portable Appliance Testing work on site microwave cookers and extra low voltage battery charges, etc.

- If at all practical, balance your load in a three phase and neutral installation. This will help to prevent unwanted neutral currents, caused by the third harmonic phenomena from entering the neutral supply conductor.

Summary so far ...

- *Electricity is generated at 11 and 33 kV from hydro-electric, oil fuelled, gas turbine and nuclear power stations. Other means of generation are wind and wave motion.*

- *Electricity is transmitted across country using high voltage to avoid power loss. Voltages of between 132 and 400 kV are used.*

- *The greater the current imbalance within a three phase installation, the bigger the current is within the neutral conductor.*

- *Third harmonic currents are unwanted currents and are produced by transformers, fluorescent lighting and any type of discharge lighting circuits.*

- *A transformer is a simple device which has no moving parts. Two coils of wire which are electrically separated from each other, known as windings, are wound onto an iron or steel former. The supply voltage is connected to the primary winding whilst the load is taken from the secondary winding.*

- *All transformers are rated in kilovolt amps, shortened to kVA. Large metal clad site transformers are usually 10 kVA in capacity, the smaller insulated type with four outlets are 5 kVA.*

- *When calculating the efficiency of a transformer, remember to take into*

Summary so far . . .

consideration the iron and copper losses which are incurred.

- Use a Type 'D' miniature circuit breaker to serve your site transformer if tripping is to be avoided when switching 'ON'.

- There is direct electrical contact between the input and output sides of an auto transformer. Please beware!

- Low voltage is any voltage above 50 volts, AC but no greater than 1000 volts. Medium voltage starts above 1000 volts but is no larger than 24 000 volts, whereas high voltage starts from where medium voltage leaves off.

- Type 'B' miniature circuit breakers are ideal for domestic use, whereas Type 'C' and 'D' are preferably suited for motor circuits.

- Never use a sensi-touch switch or a semiconductor device as a means of isolation. They are not reliable enough.

- A moulded case circuit breaker can carry out similar tasks as a miniature circuit breaker. The difference between them is their greater current carrying ability and that MCCB's have a larger short circuit capacity.

- An 'IT' earthing arrangement stems from a site portable generator. An earth, attached to the main frame of the generator, is connected to an earth electrode, staked near by.

- The three most commonly used earthing arrangements are TT, TN-S and TN-CS.

- An RCBO must be taken out of circuit before a phase to neutral insulation test is carried out. The device contains an electronic circuit.

- The star point of a community transformer is always judged to be neutral and is connected physically to the general mass of earth by way of copper earthing plates.

- An RCBO is an ideal device to protect against over-current and earth leakage voltages.

- High voltage earthing is provided by copper plates, matting and grids buried deep into the ground.

- Miniature bi-metal overload devices are ideal for over-current protection on small single phase motors.

Review questions

1. List five methods of generating electricity.

2. Why do we need to generate high voltages in order to transmit power across the country?

3. How do we obtain a three phase, four wire supply from the secondary winding of triple phase community transformer?

4. The power factor of an electric motor is normally expressed as one of the following:
   (a) as an even percentage,
   (b) as the tangent of an angle,
   (c) as an even number,
   (d) as a cosine of an angle.

5. List four ways in which third harmonic currents can be generated.

6. Describe briefly where bus bar trunking can be applied.

7. Explain how eddy currents are generated within the core of a transformer.

8. In your own words, describe how a transformer works.

9. What is the danger associated with an auto transformer?

10. True or false: Transformers are always referred to in terms of 'true power'

11. Where would you most likely find the following supply/earthing arrangements?
    (a) an IT system,
    (b) a TN-S system,
    (c) a TN-C arrangement,
    (d) a TT system.

Review questions

12 Describe the best use for the following types of miniature circuit breaker:

 (a) type 'B',

 (b) type 'C',

 (c) type 'D'.

13 Describe a typical TN-CS earthing arrangement.

 (a) What other name is it known by?

 (b) Where would you most likely find this arrangement?

14 Describe the function of a RCBO device.

 (a) Where best is it served?

 (b) Why is it considered best not to insulation test this type of device?

15 State two devices you should never use as a means of isolation.

Introduction

Derived from Outcome 6 of the City and Guilds 2330 scheme syllabus, this is the smallest chapter serving this title and forms the last part of your Unit 1 element (Application Health, Safety, and Electrical Principles). For further details, please refer to the *CG 2330 Scheme Hand Book*.

How electric motors and generators function

A simple AC dynamo

Let us get down to basics. A simple AC generator, known as a dynamo, is a device designed for the continuous production of electricity. Dynamos, fitted to older bicycles, provided the means of electrical power to serve after dark cycling arrangements.

The theory is straightforward. Take two strong permanent magnets of opposite poles (one North, the other South) and place them facing opposite each other. Next, construct a rectangular shaped copper coil of wire, which is able to rotate freely between these magnets. Connect the ends of your coil to two slip rings, and attach to an imaginary coil support spindle as illustrated in Figure 6.1. The slip rings will now act as a voltage take-off point. Mechanically rotating the coil of wire will enable it to slice through the invisible lines of magnetic flux, which span the two magnets in parallel lines from one magnetic pole to the other. When the copper coil interfaces with the magnetic flux at right angles, a natural phenomenon occurs within the copper coil. Free, outer positioned electrons, which serve a multitude of invisible atoms are magnetically shifted off course and collide with one another and other neighbouring loosely held electrons, repeatedly. This movement of electrons induces a voltage into the coil. It is current flow in its most basic form; a flow of negatively charged electrons – just think of how the domino effect works! You then obtain your current from the ends of the coil via two carbon brushes held snugly against the slip rings.

The voltage obtained will depend on the following factors:

- the *magnetic flux density* of the magnets (whether they are strong or weak),
- the rotational speed of the coil,
- by increasing the number of turns of wire within the coil,
- by winding the coil of wire onto a soft iron-laminated armature – this will have the effect of increasing the magnetic flux throughout the coil.

Remember it is only *current* which flows throughout the length of a conductor – for current is the movement of free electrons. Voltage, on the other hand *does not flow*; it is simply the pressure, which allows electron flow to happen.

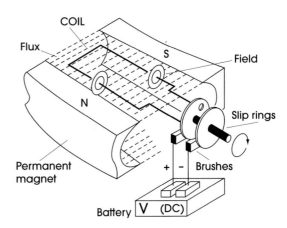

Figure 6.1 – *The free ends of the coil are connected to slip rings. (S) and carbon brushes (B)*

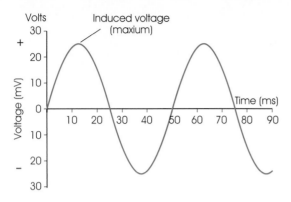

Figure 6.2 – *EMF generated from a simple alternator*

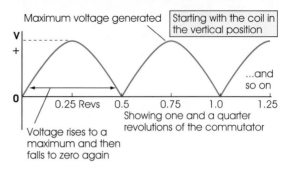

Figure 6.4 – *The emf produced from a simple dynamo*

Figure 6.2 illustrates graphically varying induced voltages within the coil as it rotates through 360°.

The function of a simple commutator

To produce a *direct* current voltage, you will need a *commutator*. Replace the two slip rings of your simple alternator with a single split ring, served with two separate carbon brushes. Illustrated in Figure 6.3, the two carbon brushes act as your voltage take-off point. The brushes neatly hug the split ring in such a way that when the single copper coil of wire is in the vertical position, that is, not cutting through any lines of magnetic flux, the two halves of the split ring are just about to change contact from one split ring to the other.

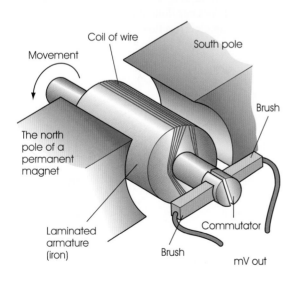

Figure 6.3 – *The principles of a simple direct current generator*

This is a *commutator* in its most basic form and a way in which to produce DC current.

Figure 6.4 shows a graph of voltage generated through one and a quarter complete cycles using this method.

Electric motors

Electric motors depend on the following factors:

■ the frequency of the supply,
■ to establish its speed, the number of pairs of poles,
■ the value of the applied voltage,
■ the magnetic field strength generated.

The synchronous motor

The synchronous motor is a good example to examine regarding constant speed and dependency of the frequency of the supply. It relies entirely on a rotating magnetic field, generated within its *stator windings*. (Remember; the *stator* is the part of the motor that houses the windings; these have no physical movement.) Speed is dependent on the number of pairs of poles and the frequency of the supply voltage.

Figure 6.5 illustrates how the rotor takes on the role of a permanent magnet, following a rotating magnetic field generated within the stator windings. Simple synchronous motors are not self-starting – but manually rotated to the speed of the applied electromagnetic field. Once done, the rotor will 'lock-in' and faithfully pace the field, and in doing so, will maintain a constant speed. If the applied load is too great, the rotor will find it hard to keep pace with the spinning magnetic field and will stall, eventually coming to a standstill.

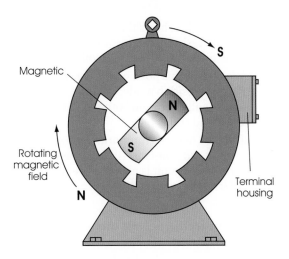

Magnetic

Rotating
magnetic
field

S

N

S

N

Terminal
housing

Figure 6.5 – *A very basic synchronous motor*

Designed for use on either single or three phase supplies, this type of motor will faithfully run air conditioning plants and timing devices.

The three phase synchronous rotor has recessed squirrel cage bars, all of which join to form a circuit. Two slip rings mounted on the shaft of the rotor are the source of DC energy for high resistance coils of wire placed in series formation beneath the squirrel cage bars.

A three phase supply, connected to the stator windings, creates a rotating magnetic field allowing induced current into the squirrel cage each time the stator's magnetic field cuts through the bars. As the cage is short-circuited, current will flow continuously, creating its own independent magnetic field and interacting with that of the stators. This has the effect of dragging the rotor around trying to pace the spinning magnetic field.

When the rotor reaches a speed, which is just below that of the synchronous speed, a DC supply, provided by a small generator mounted on the shaft of the rotor, automatically supplies current to the series wound rotor by way of the two slip rings. In practice, as the rotor moves towards the synchronous speed of the supply, the magnetic cutting action through the bars of the squirrel cage decreases and the current reduces in value until finally diminishing when reaching the synchronous speed.

How it works

The energised series wound rotor produces an electromagnetic field, which locks on to the stator's rotating field. At this point, the rotor will spin in unity with the rotating stator field.

An interesting, but rather expensive, machine that is best left to experts to diagnose when problems arise.

Direct current machines

There are not many DC motors on hand for inspection these days, unless, of course, you are lucky enough to visit a research establishment somewhere in the heart of Salisbury Plain. These motors are still available, but few and far between.

The main types in this class of machine are:

- *The 12-volt car starter motor* - this has a multi-segment commutator, heavily wound armature and field coils formed from insulated copper strip. They generate a huge starting torque with an equally large current flow.

- *The DC series wound motor* – served with a multi-segment commutator and a wound armature, placed in series formation with a set of field coils, as pictured in Figure 6.6. Current flow through the armature is the same as current flowing through the field coil. This type of machine will work equally as well on AC or DC supplies. They have a good starting torque but there is a notable drop in armature speed when load is applied. Removing your load will lead to the motor racing to extremely high speeds. Never use this class of motor where constant speed is required. Series wound motors are usually rated below 0.2 kW, and are principally used in portable hand tools and small electrical appliances found throughout our homes.

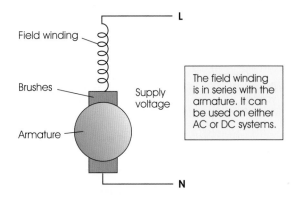

Field winding

Brushes

Armature

L

Supply
voltage

The field winding
is in series with the
armature. It can
be used on either
AC or DC systems.

N

Figure 6.6 – *A schematic layout of a serious motor*

- *The DC shunt motor* – this type of machine has a wound field coil in parallel with the armature as shown in Figure 6.8. Starting a motor of this type can lead to problems coupled to current control. In practice, placing a current limiting device into the circuit as pictured in Figure 6.9 is a solution to this problem. This device could loosely be described as a DC starter, having a built-in overload-sensing factor controlling the function of a 'no-volt-coil'. The illustration describes the role of a DC starter when serving a typical shunt motor. Increasing the resistance will decrease the magnetic flux and so increase the speed of the armature.
- *The DC compound motor* – as the name suggests, it is a combination of both series and shunt wound, as illustrated in Figure 6.7.

How it works

To control a direct current machine you have to provide an inverse variable resistance to both the armature and field coil at the same time. Please refer back to Figure 6.9:

- *Position 1* – this shows the starter with maximum resistance in the armature but with no resistance within the field coil circuit.
- *Position 4* – at this position the control circuit is in a mid-position mode; the total resistance shared equally between both the armature and the field coil.
- *Position 7* – here the rotor arm is magnetically held to the 'no-volt-coil'. All resistance from the armature circuit is now removed, transferring maximum resistance to the field coil.

Means of protection against over-current

Looking at the illustration you will see that both the armature and overload coil are wired in series formation, so an over-current condition would create a proportional increase in current all the way through the circuit. An increase in current would generate enough magnetic flux within the overload coil to attract the 'no-volt' short-circuiting terminal bar – labelled NV 1/2. Once shorted, the magnetic field would crumple whilst, at the same time, release the sprung-loaded rotor arm, effectively switching off the electrical supply to the motor.

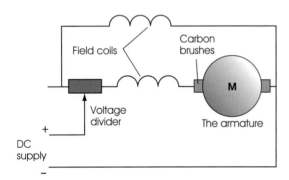

Figure 6.7 – *Compound motor*

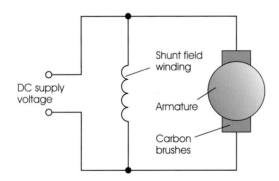

Figure 6.8 – *A schematic arrangement for a shunt wound DC motor*

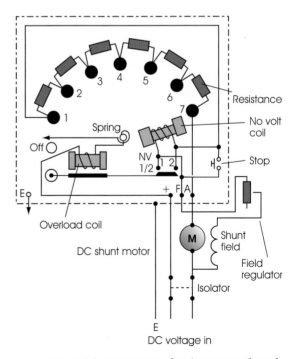

Figure 6.9 – *A DC motor starter showing means of speed control by means of a field regulator*

How to produce a rotating magnet field within a three phase stator

Operating on the principle of a rotating magnetic field, to obtain momentum, each voltage must be constantly changing and out of phase by 120°. The physical location of the windings around the stator adds an influencing factor, allowing the magnetic field to gyrate. The speed of the rotating magnetic field, called the synchronous speed, is dependent on the following:

- the frequency of the supply in hertz (pronounced,' hurts'),
- the number of pairs of poles and how they are arranged within the stator,
- the voltages serving the stator must be out of phase with each other by 120°.

As an expression: the synchronous speed together with the rotor speed in revolutions per minute (rpm)

$$= \frac{f \times 60}{P} \qquad (6.1)$$

where f is the frequency of the supply in Hertz per second and P is the number of pairs of poles.

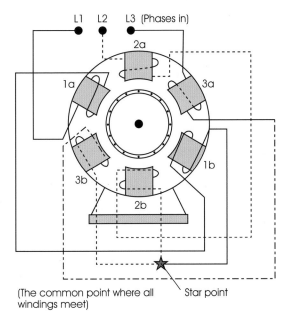

(The common point where all windings meet)   Star point

Figure 6.10 – *A two pole three phase machine. (1a and 1b are the two poles served by the phase, L1 of a 400 volt three phase supply. Poles 2a and 2b are served by the second phase and poles 3a and 3b are supplied by the third phase.)*

Unlike DC machines, AC motors are not equipped with conventional 'pole-pieces' but are loosely referred to as this. This means that a two-pole machine has in fact two poles serving *each phase* or six poles serving the complete motor, as Figure 6.10 will explain. Poles are wound so that

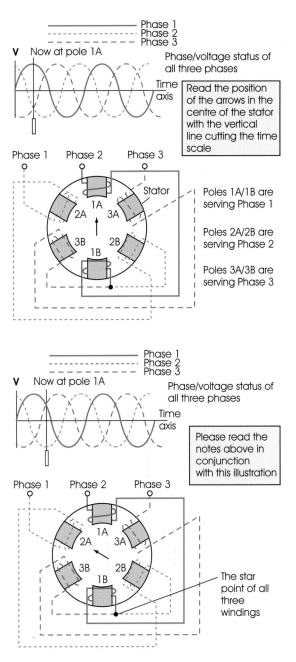

Figure 6.11 – *How a three phase magnetic field rotates around the fields of a stator*

the pole directly opposite will always have an opposite magnetic polarity. Take a closer look at the illustration and you will see that pole number 1a (L1) is a North-seeking pole whereas pole 1b (also L1) has a South magnetic polarity.

To grasp the theory of how a magnetic field rotates around the stator's field coils, refer to Figures 6.11 and 6.12. First, look at the three phase sinical waveform, serving our illustrative induction motor; show as Figure 6.11. Purposely sliced through the phases, supplying our motor, are six vertical numbered lines in succession. Line numbering starts at one, and ends at six, providing a means for reference. The vertical lines are to draw attention to the status of all three phases at any (numbered) point in time.

Placed within the centre of the stator, represented by Figure 6.12, are six numbered

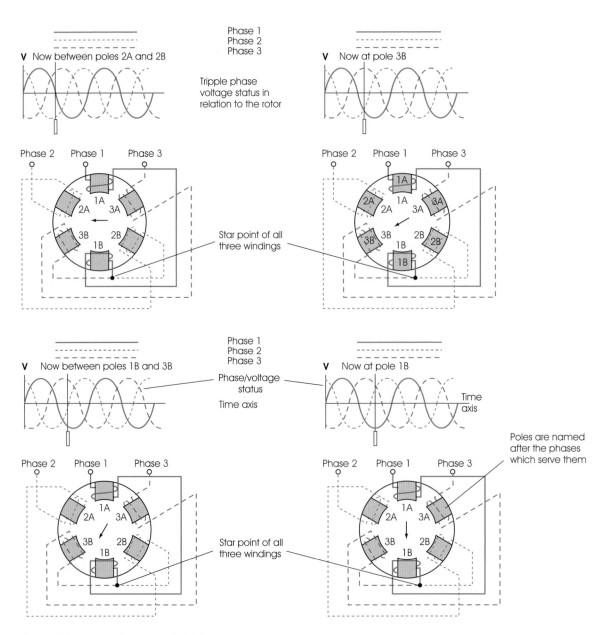

Figure 6.12 – *A rotating magnetic field*

arrows. Each arrow points to a peak, North magnetic, field strength and shares the same time reference as the vertical lines within Figure 6.11. For example, arrow number one relates to voltage position one on the sinical triple phase graph; arrow two connects to voltage position two, and so on. You will also see that L1 is at its highest possible peak at this moment in time but L2 and L3 are less so.

The three phase induction motor

Squirrel cage

There are many different types of electric motors available today. Arranged into different groups, the most common are as follows:

- synchronous motors,
- induction motor,
- commutator motor.

Figure 6.13 illustrates how to divide your motor into three main specific parts. These consist of the following:

- *The stator* – the stationary part that includes the field winding and cable terminal housing.
- *The rotor* – a cylindrical core (imagine a tin of baked beans) served with copper bars which are made common with each other. A shaft placed through the centre of the rotor allows the rotor to spin freely within the stator.
- *End shields* – cast aluminium circular discs or rounded-cornered squares precisely positioned on the ends of the stator to carry the rotor by way of suitable bearings.

Squirrel cage types

Induction motors are the most common of all types of motors used today and are subdivided into two classes known as *squirrel cage* and *double cage*.

As with the synchronous machine, examined at the beginning of this chapter, select squirrel cage machines for use on either single or three phase supplies. Figure 6.14 describes where the windings are within the stator and their connection to the supply voltage. Soft iron slots accommodate the windings, housed within the rolled-steel shell of the stator.

Connecting a three phase supply to the stator windings will produce an instant rotating magnetic field. Remember, at this stage, that there is no voltage connection to the rotor – it is just a cylindrical soft iron component supported with short-circuited copper bars mounted on a shaft, carried by bearings within the end plates.

The rotor

Copper bars, drawn into hollowed slots formed within the rotor, run at defined distances throughout its length. A common circuit, formed by means of a brass (sometimes copper) ring brazed to each end of the copper bars, shown as Figure 6.15, forms the squirrel cage component of the motor.

You have read, at the start of this chapter, how a coil of wire, cutting through magnetic flux at right angles induces an electromotive force (emf)

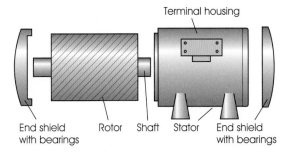

Figure 6.13 – *The motor is divided into three main parts*

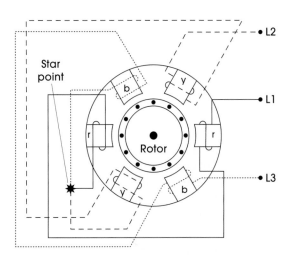

Figure 6.14 – *A typical wiring arrangement of a three phase squirrel-cage motor*

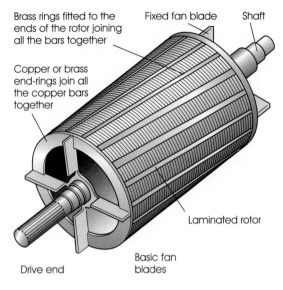

Brass rings fitted to the ends of the rotor joining all the bars together

Fixed fan blade

Shaft

Copper or brass end-rings join all the copper bars together

Laminated rotor

Drive end

Basic fan blades

Figure 6.15 – *A rotor which serves a typical squirrel-cage motor*

into the coil. It is this same electrical principle that enables a squirrel cage motor to function. The rotating magnetic field within the stator induces an emf into the copper rotor windings – often referred to as mutual induction. The short-circuited copper bars (windings) will allow current to flow throughout the rotor circuit. They, too, will set up their own electromagnetic fields, and in doing so will interact with the fields formed within the stator windings. The stator's rotating magnetic field, which we know as the *synchronous speed,* has the effect of dragging the rotor around with it.

The current produced within the squirrel cage windings is in opposition to the very current which created it. The only way respond to this change is for the closed loop, the rotor, to turn in the same direction as the rotating magnetic field generated within the stator windings.

In practice, the rotor will follow at a little less than the synchronous speed. If both speeds were in step with one another, the rotating magnetic flux would not be able to cut through the rotor windings. Mutual induction to produce an electromagnetic field then would never happen – our induction motor would never function.

Slip

As a footnote, the difference between the synchronous and rotor speed is the 'slip' and can vary from about 2.5 to 5.5% at normal loads. If

the load is very much increased, the difference between the two speeds will also increase. Place too much load on the machine on the machine, and it will stop.

Slip is a percentage of the synchronous speed and calculated by use of the following expression:

$$\text{Slip \%} = \frac{N - \text{rotor speed}}{N} \times \frac{100}{1} \qquad (6.2)$$

where N is the synchronous speed of the motor in r.p.s.

Speed

The range of speeds to serve induction motors is surprisingly limited. Take as an example, a machine with a *single* pair of poles. By applying the following expression:

$$N = \frac{f}{P} \qquad (6.3)$$

where N is the synchronous speed of the motor in revolutions per second, f is the frequency of the supply in hertz per second and P is the number of pairs of poles, the machine will deliver a theoretical speed of 50 r.p.s or 3000 revs per minute. Two pairs of poles will reduce the machine's speed to 25 r.p.s. or 1500 revs per minute. Three pairs of poles would allow the synchronous speed to fall to16.6 r.p.s. and so on.

A higher supply frequency will provide a higher synchronous speed as (6.3) and Figure 6.16 will show.

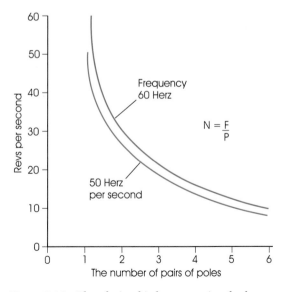

Figure 6.16 – *The relationship between pairs of poles, frequency and speed*

Advantages and disadvantages

Disadvantages are that the starting torque is low and the current can be up to three or four times greater than the running current. Advantages are that it is hardwearing and long lasting, reasonably priced and very efficient. The machine is easy to maintain and control/starting equipment is readily obtainable.

Where they are used

You can find the induction motor in a variety of applications. Many are in industry and serve agricultural installations such as extraction/ventilation plants, conveyors, feed augers and vacuum pumps where a reasonably constant speed is essential.

Double cage induction motor

Designed in a similar fashion and a close relative to the induction motor, is the double cage induction motor. Instead of just one set of 'caged' rotor bars, there are two as Figure 6.17 shows. One set has a higher resistance than the other set. Both are within the body of the rotor core.

When current flows through both sets of rotor bars, it produces a far better starting torque, in fact over *twice as much* as an ordinary squirrel cage motor. This provides a much lower starting current. When switching the machine to 'run'

mode, current is mainly restricted to the outer cage of the rotor. The lower resistance bars work in a similar way to the standard squirrel cage machine, producing a high efficiency performance on load. Rotor current at normal running speed will split between the two cages in reverse ratio to their respective cage resistances. This means that the inner winding now bears most of the current.

Disadvantage

The disadvantage of this type of machine is the cost, being up to 50% dearer than its regular cousin.

Applications

Used where a high starting torque is needed:

- industrial shearing machines,
- water pumps,
- large refrigeration compressors, etc.

Induction motor – wound rotor type

The triple pole commutator motor is expensive and complicated in construction. The stator is no different from the ordinary run of the mill induction motor but the rotor contains wound coils of lightly insulated wire and has as many poles as there are stator poles. As an example, a two pole wound induction motor would actually have six poles, two poles to serve each phase.

Rotor winding

The start of each winding to serve the pole pieces connects to one of three slip rings mounted on the rotor shaft and routed to form two pole pieces positioned directly opposite each other with opposed magnetic polarity. This then repeats for the second and third phases. The three loose ends of wires serving each group of windings connect to form a star point as shown in Figure 6.18.

The slip rings

The slip rings will allow you to connect an external control, via carbon brushes. This takes the form of three variable resistances mechanically interlocked to serve the stator windings as described in Figure 6.19. All three resistances will work in complete harmony when adjusted. Adjusting the external resistance will allow current flow management within the rotor windings, which in turn will also control the current flowing within the fixed stator windings.

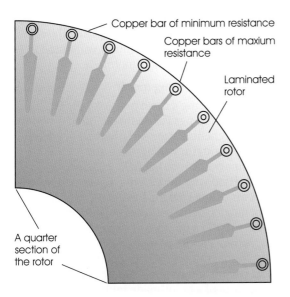

Copper bar of minimum resistance

Copper bars of maxium resistance

Laminated rotor

A quarter section of the rotor

Figure 6.17– *This is the end view of a double cage induction motor*

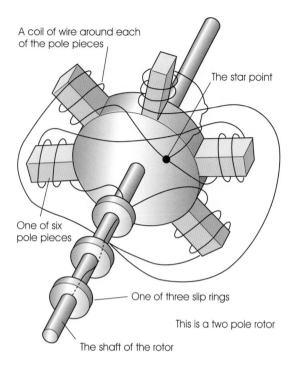

A coil of wire around each of the pole pieces

The star point

One of six pole pieces

One of three slip rings

This is a two pole rotor

The shaft of the rotor

Figure 6.18 – *Induction motor – wound rotor type*

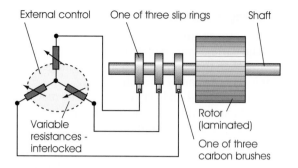

External control

One of three slip rings

Shaft

Variable resistances - interlocked

Rotor (laminated)

One of three carbon brushes

Figure 6.19 – *The windings of a wound rotor induction motor are connected to three brass slip rings mounted, but insulated, on the steel shaft forming the axis of the rotor*

The advantages of this type of motor are:
- high starting torque,
- limits the initial starting current,

while the disadvantages are:
- very expensive,
- can be complicated if fault conditions arise.

How it works

Connecting a three phase supply to the stator windings will instantly generate a rotating magnetic field within the windings, as described in previous paragraphs. As the magnetic field is rotating, it will unconsciously induce a voltage into the rotor windings, which also creates a magnetic field. The current flow will be proportional to the induced voltage within the rotor and the value of the external resistance connected. The magnetic field produced by the rotor attracts the rotating field generated within the stator. The two magnetic fields work in complete harmony with each other to produce rotary movement.

When the motor increases in speed the induced voltage will reduce. This is because the magnetic

cutting action between the rotor and the fixed stator windings lessens. To make clearer; if the rotor *could* spin at the synchronous speed of the rotating magnetic field, there would be no cutting action whatsoever. An increased rotor speed produce less torque as a reduced amount of current is flowing within the rotor windings.

Reducing the external resistance will have the effect of allowing additional current to flow within the rotor windings. This, in turn, will provide more torque within the rotor and the motor will increase in speed. Carry on doing this until the motor gets to its best possible speed. When you have removed all the external resistance from the rotor windings and the rotor circuit is shorted out, your motor will function like a typical squirrel cage motor.

Starting your slip-ring motor

Start by providing a three phase supply to the stator windings via a suitable rated electromagnetic starter with thermal overloads. Exciting the stator will produce an electromagnetic rotating field as described before. Hard wire the external resistance serving the rotor to a purpose made device, which automatically removes (cuts out) the resistance to the rotor in a number of steps. This progression continues up to the removal of *all* the external rotor resistance and effectively short circuits the system. Your device will be in the form of a tailor-made multi-contact switching mechanism suitably rated to serve the rotor current.

Single phase AC motors rated below 1 kW

Series wound

You will find this type of motor fitted to domestic appliances where the load is virtually stable, for example in vacuum cleaners, mains operated electric drills, hover mowers and some types of hair driers. Known also as a *universal motor,* ratings vary from a few watts to just under 0.2 kW (200 watts) – used for both AC and DC supplies although you might come across some with slightly greater power ratings.

Physical description

1 A wound armature served with a brass segmented commutator.

2 The current is supplied to the armature by means of two sprung loaded carbon brushes positioned horizontally opposite each other on the brass commutator.

3 The stationary field coils (the windings) are wound with lightly insulated thin copper wire.

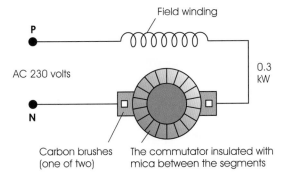

Figure 6.20 – *This is a schematic wiring arrangement of a direct current series wound motor*

4 Both the field coils and the wound armature are in series formation with each other as shown in Figure 6. 20. Current flow within the armature is the same as current flow within the stator.

Running characteristics

This type of motor has a good starting torque but there is a notable drop in speed when load is applied. When removing your load the motor can race to higher speeds. It is wise not to use this class of motor where, for example, constant speed is required.

How to reverse rotation

All you do is to interchange the internal leads serving the two carbon brushes as illustrated in Figure 6.21.

Speed control

Series motors are ideally suited for speed control. You can draw everyday examples from your Dad's old record deck, your Mum's kitchen appliances, to your own hand held electric drill.

You gain speed control in the following ways:

- series impedance control – linear resistance,
- solid state linear control – electronics,
- auto transformer control – electromechanical.

A transformer made from a single length of copper wire that is wound around a laminated iron core many times has, throughout its length, tapings V1 to V4. These are to provide steps to either *increase* or *decrease* the voltage supplied to the series wound motor. These connect mechanically into an adjustable voltage selector device, which is adjustable to your required speed.

Fine-tune your tapings to a higher value if your motor fails to respond to your minimum voltage input level. You will find the higher the voltage applied, the greater the starting torque.

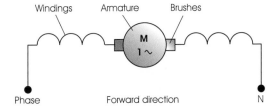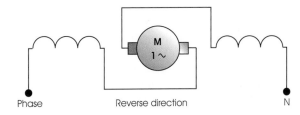

Figure 6.21 – *A series motor – forward and reverse rotation*

Advantages gained from using this method of speed control include:

■ reduced starting current,
■ torque directly proportional to the applied voltage,
■ smoothing starting sequence.

The disadvantages are:

■ high installation costs,
■ regular maintenance required,
■ only certain types of motors are suitable.

Please note that BS 7671 Regulation 555-01-02 forbids the use of a step-up auto transformer in an IT earthing arrangement supply system. This system, not in use for public supplies, has no direct connection between any live part (phase and neutral) and earth.

Split phase AC motors

Single phase induction motors are naturally self-starting but operate subject to a pulsating magnetic field. Therefore, the torque produced is also pulsating and is not constant but, by twirling the rotor, will gather momentum until reaching a pace which is just below that of synchronous speed. This is much the same way as a triple pole induction motor operates.

Altering the design of the machine by adding another winding displaced exactly 90° (electrically) to the stator and connected in parallel with the original wiring will make it self-starting.

Known as a split phase induction motor or resistance start motor, the two separate windings wired in tandem produce a two phase rotating magnetic field. This is due to a difference of impedance. In practice, the 'start' winding has a higher resistance than the 'run' winding and after a few seconds of rotation, a centrifugal switch or timed relay automatically removes the start winding from circuit. See Figure 6.22 for schematic detail. As the start winding is not pure inductors but has a built-in resistance due to the natural impedance of the coils, it causes the current flow to be *less* than 90° out of phase with the voltage. In practice, this type of motor will develop a phase differential of about 40° between the current and start windings. This, unfortunately, results in a poor starting torque together with a slight drop in speed when load is applied. We will have to put up with this difficulty

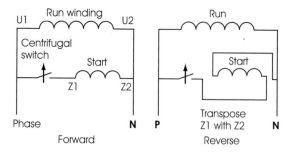

Figure 6.22 – *Induction split phase motor – forward/reverse rotation*

until room temperature superconductivity technology is with our grasp.

Applications

When the physical starting load is small or there is next to no load at all, such as in an industrial suction cleaner or workshop pedestal drill, the split phase motor has a role to play.

Physical description

1 Use a timed relay or centrifugal switch to isolate the start winding once reaching approximately 70% of the maximum speed.
2 The start and run windings connect to each another in parallel but 90° (electrically) out of phase with each other.
3 Typical squirrel cage rotor.
4 The windings are within the stator.
5 The electrical connection with your circuit is within the stator's terminal housing chamber.

How to reverse rotation

Interchange the leads serving the start windings. Details are in your last illustration, Figure 6.22, and shown as terminals Z1 and Z2.

Maximum rating

Generally limited to motors over 0.2 kW (200 watts).

Speed control

This is not practical due to the switching device within the start winding.

Additional information

■ The start winding has a higher impedance than the run winding.
■ There is no capacitor within the circuit.

AC capacitor aided motor

This is also known sometimes as an induction split phase permanent capacitor. A close relation to the capacitor start (where a capacitor connected to the start winding can be isolated from a circuit by a centrifugal switch) – this type of machine is known as a permanent-split capacitor motor. As its description suggests, the capacitor is in circuit all the time and helps in correcting the power factor as Figure 6.23 shows. This does away with any need to use a centrifugal switch or timed relay. Generally, oil-filled 400 volt working voltage capacitors serve this type of motor. A second capacitor, placed in parallel with the permanent 'run capacitor', can be added as an aid to starting – please look at the last illustration. A centrifugal switch or timed relay will remove the 230 volt auxiliary capacitor from circuit. After obtaining about 70% of the maximum speed, the switching device will come into play and remove the smaller capacitor from circuit.

The rotor used is of typical squirrel cage construction as described in previous paragraphs.

Applications

It has good starting and running torque and is often found within air conditioning units used within refrigeration plants or serving water filled unit heaters. You will experience a slight drop in speed under load conditions.

Physical description

1 Capacitor permanently in series formation with the auxiliary winding.
2 Run and auxiliary winding connected together in parallel formation.

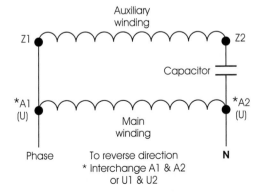

Figure 6.23 – *A capacitor aided motor*

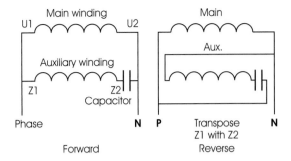

Figure 6.24 – *Induction split phase induction motor (permanent capacitor) – forward/reverse rotation*

3 Squirrel cage rotor.
4 Windings within the stator.
5 The electrical connection for your circuit is within the stator terminal housing.

How to reverse rotation

Interchange both internal wires, which lead to the auxiliary winding marked Z1 and Z2 shown in Figure 6.24.

Additional information

The auxiliary winding has higher impedance than the run winding and always remains within the circuit.

Testing for continuity

You can test continuity in a reasonably straightforward way.

First, disconnect the two sets of windings leading to terminal housing connections. At this point, make a written note as to what goes where – do not trust your memory! Recognize the 'tails' leading to your start winding as Z1 and Z2; the run windings are marked as U1 and U2. If you come across a very old machine, and there are a few of them about still, you will find the run winding coded A1 and A2 – a yesteryear equivalent to U1 and U2.

Ideally, use a low-scaled digital ohmmeter to measure the value of the impedance offered across the individual sets of field windings – not an analogue scale meter as shown in Figure 6.25.

The start winding will always record a higher resistance value than the run and, depending on the design of he motor, could be up to ten times greater in value.

This is an ideal test to show that you have at lest continuity within the field coils. It will *not*

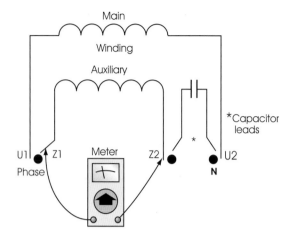

Figure 6.25 – *Continuity testing a split phase machine*

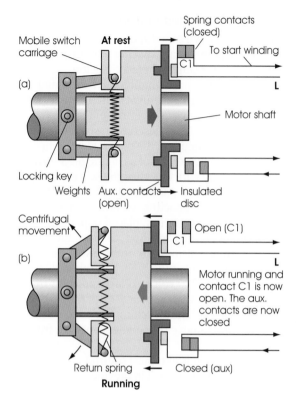

Figure 6.26 – *The centrifugal switch. (a) At rest; (b), position when the motor is running*

point towards whether any part of your winding has been short-circuited by mechanical damage or heat generated by over-current. To discover this you must first know the *total* impedance of both sets of windings – and that could be difficult!

If your machine has a centrifugal switch

The role of a centrifugal switch is either to disconnect the start winding or auxiliary capacitor from circuit once the rotor speed has reached about 70% of its potential speed. Many of these switches are disc-like, together with a couple of counterweights and control springs.

Overcome by centrifugal movement (spinning around fast) of the counterweights, the control springs give way, allowing the switching disc to do its job. Expect a successful disconnection of either the capacitor or the winding from the supply once the centrifugal switch has swung into open mode. Figure 6.26 illustrates the mechanical principle of this operation in diagrammatical form.

Switching off the machine causes the reverse to happen and the switching disc snaps back into its original position once the motor slows down – the centrifugal switch contacts will close. The switch is now ready for the next starting sequence. Please look back and refer to the top illustration in Figure 6.26.

Problems with centrifugal switches

Common conditions are:

- *Switching action 'ON' all the time* – This often results from broken springs or arc-welded

switching contacts. To make good, change the switching assembly for a correct replacement. Check the capacitor; change if necessary. Test both sets of windings in case of damage caused by over-current.

- *Centrifugal switch permanently open* – this happens due to mechanical failure within the switch. Machines falling casualty to this problem will develop a '50 cycle hum' that will continue until the motor is closed down by the local over-current protection. It is best to check out the run winding in case of damage caused through the long-lasting over-current sequence.

- *Switching assembly burnt out* – excess current developing within the start windings or resulting from a faulty capacitor will cause this type of problem. It would be wise to check both.

Machines served with current relays

Use a current relay as a substitute for the centrifugal switch where such a device might prove impractical when, for example, serving an underwater slurry pump.

The current relay coil, comprising a few turns of heavy gauge, but lightly insulated wire, connects in series formation with the run winding as shown in Figure 6.27. On start up both the run winding and the relay coil carry a very high current. This produces a strong magnetic flux throughout the coil. When suitably magnetised the relay switching assembly triggers and provides energy to the start winding. On reaching between 70 and 80% of full speed, the current flowing through the start winding lessens. This proportionately has an effect on current flowing through the relay coil. The magnetic flux then collapses sufficiently to allow the start winding contacts to reopen under gravity.

Testing your current relay

To test, physically check the mechanical action of the relay's switching assembly. To do this you might have to remove the relay from a protective container. Check that the contact cradle opens and closes smoothly, and does not catch on anything. Inspect the switching contacts and clean if required.

It is best to carry out a low voltage continuity test across the coil and the switching contacts of the relay. Aim for about 0.05 ohms or less across the contacts if possible. Please check for loose terminals within the relay as these can often contribute to future fault conditions if left unattended.

Always fit a replacement relay in the correct position so that the de-energised relay can open under gravity. Some units are completely factory sealed, making it necessary to provide a complete replacement should a fault condition occur.

AC capacitor-start motor

Place an electrolytic capacitor in series with the start winding, to obtain a greater phase angle and starting torque. The start winding, now capacitative, causes the current to lead the voltage to produce a better phase difference between the two sets of windings (i.e. the start and the run); Figure 6.28 illustrates this.

By adding the correct value capacitor, a phase angle of 90° will build up between the two sets of currents. This will obtain the maximum starting torque – a necessary part of any motor design.

A centrifugal switch or timed relay removes the start winding from circuit in a similar way to a split-phase induction motor when reaching about 75% of the full rotor speed.

Some manufacturers do not recommend starting this type of motor any more than seven or eight times per hour for fear of damaging the capacitor. You will find that capacitor-start machines are fitted with capacitors within the 230 volt working range, whereas capacitor-start and run motors operate using a much higher working voltage.

The design of this type of machine will produce a high starting torque but expect a slight drop in speed when on load. Used to serve compressors and pumps where a high initial starting torque is necessary, this is an ideal single phase machine for industrial use.

Physical description

1   Relay or centrifugal switch used to take out the start winding from circuit.
2   The stator provides anchorage for the capacitor; which sometimes can be remotely sited from the machine.

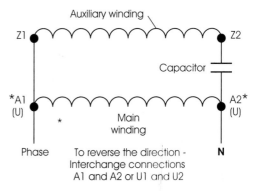

Figure 6.28 – *An induction split phase (capacitor-start) motor*

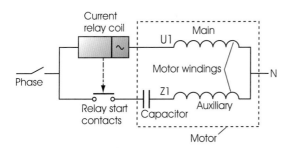

Figure 6.27 – *Schematic layout of a current relay*

3   Wound stator where your electrical connections are made within the terminal housing.

4   Typical squirrel caged rotor.

How to reverse rotation

Swap over the leads serving the start windings as shown in Figure 6.29. Recognise these as Z1 and Z2.

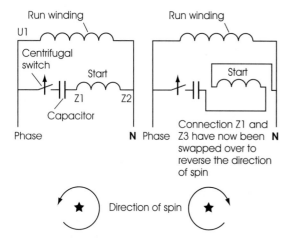

Figure 6.29 – Capacitor-start induction motor – forward/reverse rotation

Applications

Oil, water pumps, and refrigeration compressors are examples, but consider any other application where a good starting torque is required.

Additional information

1   Capacitors are 230 volt working range and electrolytic since the capacitor is only in circuit long enough to start the machine.

2   Machines are usually rated above 0.2 kW.

3   The motor has a high impedance start winding.

4   Reach your maximum starting torque by making certain the correct amount of capacitance is used. If you get it wrong, you will not produce a phase angle of 90° between the two sets of windings and your starting torque will be poor. Too much capacitance will decrease the starting torque within your motor. Get it right!

5   You will run into murky waters if you replace a capacitor with one of a lower voltage rating.

Standard motor enclosures

Table 6.1 lists seven motor enclosures you will most likely to come across within the course of your career.

Table 6.1 – *Standard motor enclosures*

Enclosure	Method of construction	Where they are used
Weather proof	Totally enclosed infrastructure – *not* watertight	External situations
Watertight	Completely watertight	Submersible pumps and marine installations
Protected	Live and moving parts protected by end shields bolted to the stator yoke. Ventilation ports often protected by wire mesh attached to the stator.	Machine tools and *dust-free* average temperature environments
Flameproof	A totally enclosed infrastructure designed to safeguard any spark or arcing within the motor.	Oil refineries, petrol pumps and coal mines. Any flammable or explosive environment
Pipe ventilated	A pipe fitted to the air intake point of the cooling system enables clean cool air to circulate within the motor. Once used the ventilated air travels via a pipe to the outside atmosphere.	Cement works, large boiler houses and situations where there is a dirty, dusty environment coupled with a high temperature.
Drip proof	Ventilated end shields provided with hooded louvers	Agricultural situations, laundries, and installations subjected to dripping moisture.
Totally enclosed	Mechanical protection for both the rotor and stator is within a ribbed frame; this also helps in cooling. An internal plastic or pressed steel fan provides for ventilation	In an industrial situation where the surrounding temperature is not too high.

Direct-on-line starters for electric motors

Figure 6.30 provides a simple block diagram of a typical wiring arrangement to serve a single phase electric motor. BS 7671 requires a means of protection to stop an electric motor from restarting again after stopping due to voltage failure or over-current. To satisfy Regulation 522, an automatically controlled electromagnetic starting device is used; but this rule does not apply to motors under 0.37 kW or if serving a pump controlled by a float switch, as Figure 6.31 shows. Regulation 552-01-02 confirms this.

How it works

The starter operates on the principle of converting electrical energy into linear motion by means of an electromagnetic coil, linked mechanically, with a switching assembly. The coil is energised either manually or automatically,

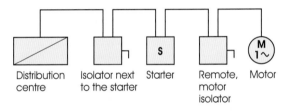

| Distribution centre | Isolator next to the starter | Starter | Remote, motor isolator | Motor |

Figure 6.30 – *Block diagram arrangement serving a single phase electric motor*

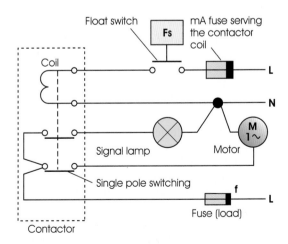

Figure 6.31 – *Wiring diagram of a pump motor controlled by means of a float switch*

such as by a float or pressure switch. Figure 6.32 illustrates, schematically, detail of the internal wiring arrangements for a typical three phase direct on-line (DOL) manually operated starter. Once pressing the green 'START' button, the switching sequence begins by bridging terminals 7 and 8 on the starter switch. This provides an uninterrupted supply to the coil. This action closes the switch – the supply is now in union with the load. Look at the illustration and you will see that coil terminal 'a' is served directly from the BROWN PHASE (1) whereas coil terminal 'b' is fed from the GREY PHASE (5) passing through the thermal overload switch (terminals 95 and 96) and auxiliary switching terminals 8 and 7.

When the starter coil receives its supply of electricity, it generates sufficient electromagnetic flux for the switching mechanism to overcome the strength of the service springs, and pull, magnetically, into the iron core. Once energised, the switching assembly automatically maintains a supply to the coil through the control circuit 'hold-on' terminals 7 and 8. You interrupt the control circuit, manually, by pressing the RED 'stop button'. Your motor circuit will also automatically close down when the thermal overload relay activates, but whichever method is used, the coil is electrically isolated from the electrical supply, and the load mechanically

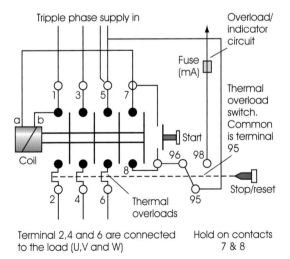

Terminal 2,4 and 6 are connected to the load (U,V and W)   Hold on contacts 7 & 8

Figure 6.32 – *The internal wiring arrangements of a typical three phase, direct-on-line, manually operated starter*

disconnected. Regulation 537-04-04 refers to this.

Wiring your starter

A suitable rated three phase supply is connected to terminals 1, 3 and 5 (Brown, Black and Grey conductors respectively). Take the load from terminals 2, 4 and 6 in the same phase/colour sequence. When energising the starter coil either manually or remotely (automatically), the switching mechanism activates, bridging terminals 1 and 2, 3 and 4 and 5 and 6 as shown in Figure 6.33. This effectively connects the load with the supply.

Replacement starter coils are available for a wide range of voltages and frequencies. Do remember to watch out for the small service springs, which play an important role in the mechanical operation of the switching cradle. It is very easy to lose them! Remember to place a 400 volt warning label on the front cover of your starter!

A quick word concerning eddy currents

The alternating magnetic field generated within the coil induces an electromotive force (emf) into the iron core and switching armature.

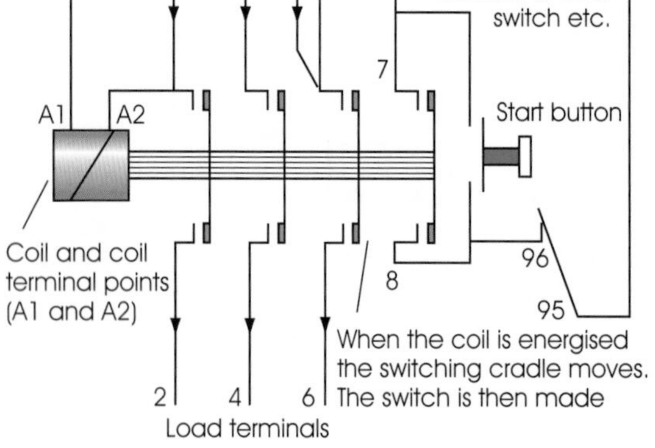

The three phase supply in Brown terminal 1; Black terminal 3 and Grey terminal 5

Coil circuit (through the overload terminals 95 and 96

Remote auto start terminals. Pressure or float switch etc.

7

A1 A2

Start button

Coil and coil terminal points (A1 and A2)

96

8

95

When the coil is energised the switching cradle moves. The switch is then made

2 4 6

Load terminals

Figure 6.33 – *The switching assembly is brought into play when the coil is energised*

These residual (lingering) currents, known as *eddy currents,* are generally unwanted as they can be to blame for heating problems within the iron core, leading to theoretical power loss.

To overcome this problem, the construction of the core and armature is from thin laminated sheet iron. Although not a perfect answer, it helps.

Motor over-current protection within your starter

The most widely excepted method of over-current protection is the *thermal overload relay* which, when coupled to the starter, forms an integral part of this unit.

Construction is from three self-regulating heavy-duty heater elements attached to three independent bimetal strips. The elements connect in series formation with the supply phases serving the connected load. When the load is experiencing over-current, the three integral thermal elements respond by bending in unison. This in effect mechanically isolates the control circuit and kills power to the connected load.

How it works

Each heating element twists around an insulated and anchored bimetal strip, allowing freedom of movement at one end. A large increase in current through the heating elements (remembering that they are in series with the load) will cause the bi-metal strip to bend a little. A series of small levers built into the system boosts this displacement and trips the spring loaded relay. Figure 6.34 will help to provide additional information and make this principle clearer. Please refer to Regulation 433-01-01.

With the overload relay unit firmly attached to the starter, connect your load to terminals 2, 4 and 6. Adjust the device to the rated current demand of your motor and select the 'manual/auto' switch to the desired mode. The manually operated mode will allow setting of the thermal overloads by means of the 'STOP/RESET BUTTON'.

Figure 6.35 explains how terminals 95 and 96 are internally used. If required, use terminal 98 to serve a remote warning lamp providing instant advice of an overload condition. Just wire the

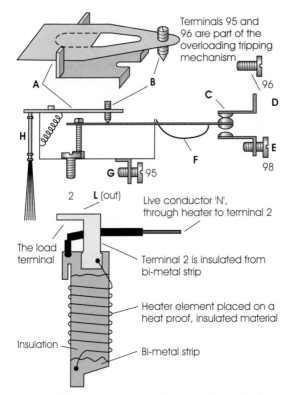

Terminals 95 and 96 are part of the overloading tripping mechanism

Terminal 98 is used as a source for a remote overload indicator lamp. When the tripping mechanism is activated, terminal 98 is energised.
(Remember, single phase indicators will require a neutral)

Figure 6.34 – *The mechanical principles of a thermal overload relay. A, tripping plate; B, adjustment screw; C, relay contacts; D, coil circuit terminal (96); E, coil circuit terminal (98) for 'remote overload tripped' indicator; F, sprung copper strip; G, coil circuit (95); H, pivoted lever attached to a linear slide integrally linked to the bimetal overload heaters*

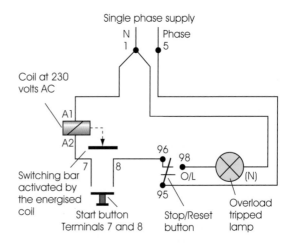

Figure 6.35 – *Direct-on-line starter. Details of the integral control circuit wiring for a remotely sited 'overload tripped' indicator lamp*

brown phase conductor from terminal 98 to the 'live' side of your lamp whilst providing a blue neutral supply to the other side.

Problems involving thermal overloads

Over-current devices operate between 110 and 120% of full load. You must set them to the maximum rated current demand of the motor. Setting it too high will cause damage to the motor and starter alike. In severe case, heavy currents can develop when the overload is unable to respond to fault conditions within the motor or circuit. This mechanical problem often results in the motor overheating or burning out. Overloads unable to act in response to excess current could become very hot or experience powerful terminal contact arcing – but whatever the reason, there is clearly a very high fire risk!

Unlike the rewireable fuse element, which ruptures when reaching about 190%, the thermal overload operates to provide improved protection. By their very nature they respond to variations in temperature. Because of this, they tend to be oversensitive during the summer months but slow to react when the temperature is low. In countries where there is a noticeable climatic difference between summer and winter, engineers readjust these to accommodate seasonal changes. Please keep this in mind, as our weather in the UK/EU is very unpredictable at times.

Trouble-shooting

Generally, thermal overloads present little trouble. If a problem exists, an open circuit in one of the heating elements or a faulty spring relay mechanism can often cause it. Check this out using a good digital continuity tester with the supply isolated.

Miniature bimetallic devices for small motors

Miniature bimetallic-type overload devices, screwed to the stator, provide thermal and over-current protection to machines rated under 0.37 kW (370 watts). Not all small motors have these – but many do.

Two varieties are available:

■ one designed with a small heating element, wired in series with a phase conductor – see Figure 6.36,

■ the other, a far more basic device, comes without a built-in heating element – see Figure 6.37.

Both of these types use bowl-shaped bimetallic discs forged from sprung metal. When in contact with heat, the disc, which is virtually a switch, provides a thermal snapping action and isolates current to the motor. When cool, the bimetal disc will snap back to its original position to allow for current flow. Some models are designed to be manually reset with a small insulated button fitted to the top of the over-current device.

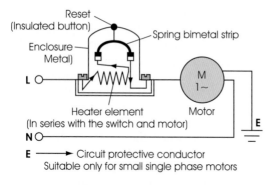

Figure 6.36 – Miniature bimetal overload device with a built-in heater element

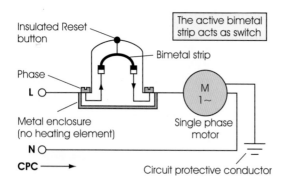

Figure 6.37 – Miniature bimetal overload device without an integral heater element

Testing

Testing is by means of a digital continuity/ohmmeter, targeted across the switching terminals of the over-current device. A value of 0.05 ohms would be ideal. Generally, these simple devices are reasonably trouble-free but occasionally the switching contacts may require cleaning or an open circuit condition can occur when the switching arrangement becomes mechanically unreliable. If in doubt – fit a new one.

Starters to serve single phase machines

The starter you choose will have a 230 volt AC coil, but select your overload unit with care taking into account the design current of the circuit.

Connect your conductors to the assembled starter are follows:

■ The neutral conductor to terminal one.

■ The phase conductor to terminal five.

■ Now, link a conductor of the same cross sectional area you are using for your circuit, from terminal two to terminal three. This will ensure all three overloads are in circuit.

■ Connect the neutral conductor of the load to terminal four and the load phase conductor to terminal six.

Remote warning of overload

When required, wire a phase conductor from terminal 98 to your remote warning indicator connected with a suitably sized over-current protection device – see Regulation 435-01-01. Take your neutral conductor to serve the indicator from terminal one.

Mechanical starters

A much older and far less favoured direct-online starter is the mechanically activated starter. Yes, you can still see them around where equipment upgrading takes low priority.

This type of starter, designed to switch loads of up to 2.5 kW (2500 watts) single phase, is purely *mechanical* in construction. It does not have an integral control circuit or a coil which, if energised, would bring a switching cradle into play. Over-current protection is very basic, taking the form of bimetal strips attached to mechanical tripping levers. You are able to select additional current settings by extending the tripping lever a few millimetres outwards.

Seen mainly in agricultural and light engineering situations, the mechanical starter has lost recognition in recent years.

Disadvantages

It requires occasional maintenance and protection against over-current is not as accurate as its modern counterpart.

Advantages

It is very cheap, easy to wire and install. Problems are usually mechanical and can be overcome fairly easily. It can be selected for use on either single or three phase supplies.

Starting controls serving DC machines

Starting a large DC motor can lead to problems associated with current control, so in practice a current-limiting device is designed within the circuit. Figure 6.38 shows such a device in its

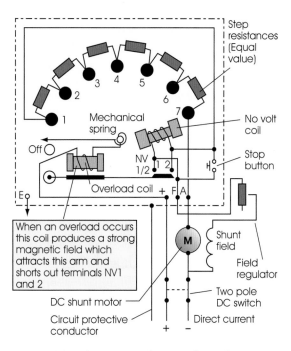

Figure 6.38 – *Direct current electromechanical starter controlling a shunt motor*

most basic form. This starting device, designed with an overload-sensing feature, controls the functions of a 'no-volt coil'.

The illustration, together with the following text, will take you through the initial starting procedure for a large DC shunt motor. Increasing the resistance (numbers 1 to 7) will decrease the magnetic flux and so increase the speed of the motor. These mechanically started machines provide an inverse variable resistance to both the armature and field coils in agreement with each other.

Position 1 (resistance 1)

This shows the starter with maximum resistance in the armature and no resistance within the field coils.

Position 4 (resistance 4)

The control circuit is now in a mid-position mode; the total resistance is shared evenly between the armature and the field coil alike.

Position 7 (resistance 7)

The rotor arm, magnetically held to the 'no-volt coil', maintains this position. The armature circuit has had all its resistance removed at this position, transferring the maximum resistance possible to the field coils.

The magnetic 'poles' serving a shunt wound DC machine are known differently to their AC opposite number. They are call field coils and wired in parallel with the armature.

Protecting a DC machine against overload

A proportional increase in current in both the armature and field coils would generate sufficient magnetic flux to attract the associated 'no-volt' short circuiting terminal bar (drawn below the footnote NV _). Once magnetically drawn to the coil and mechanically bridged across 'no-volt' coil terminals 1 and 2, the accompanying magnetic field generated by the NV coil would collapse and simultaneously release the spring loaded control arm from its magnetic hold. This effectively isolates the DC supply from the motor.

AC star–delta starters

This type of starter is complex and, by its composite nature, makes it often difficult to detect stubborn problems that arise from time to time.

Figure 6.39 illustrates the basic wiring and switching arrangements serving the typical star–delta starter, following through to the windings of the motor, whilst Figure 6.40 provides you with a systematic breakdown of the control circuit switching cycle. View this arrangement as though the motor and starter were at rest.

First, press the 'START' button to trigger the automatic switching sequence and expect the following:

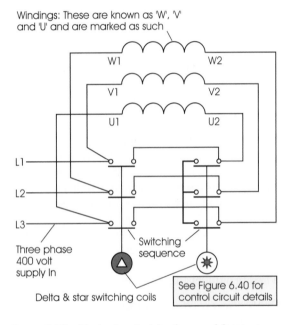

Figure 6.39 – *The basic principle of a star–delta starter – autoswitching*

1 With all the additional switches resting in the 'ON' position (for example, the 'STOP' button terminals, OVERLOAD SWITCHING and TIMING UNIT TERMINALS, etc.), coil **Y** is energised. This electromechanically brings in switching contact Y1.

2 The motor windings are now in star formation.

3 The two-way switching contacts also provide a path to energise coil L. For practical purposes, imagine an instantaneous switching sequence for coils Y and L.

4 With L now energised, contact L1 closes.

5 Timer unit T energises – this opens and closes terminal L2.

6 At the end of the timing cycle (this can be several seconds) terminal L2 then opens.

7 With terminal L2 open, current can no longer flow to coil Y. In this de-energised state, the switching contacts Y1 serving coil Y opens.

8 Switching contact Y1 now provides a supply, by means of closed contact L1 to the Δ coil. This places the motor windings in delta (Δ) formation.

9 Energising the Δ coil opens switching contact Δ1 providing interlocking facilities. This stops coil Y from activating.

10 Manually pressing the 'STOP' button will de-energise coils Δ and L. Your machine will stop.

The best way to master this starting cycle is under controlled conditions within the classroom. Your tutor will provide you with practical information and demonstrate this sequence of events in detail.

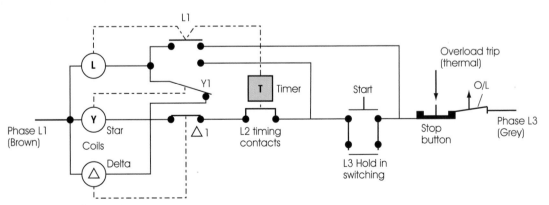

Figure 6.40 – *The basic control circuit and switching sequence serving a star–delta starter*

Speed control

Single phase machines

Single phase speed control is limited to split phase capacitor aided motors where automatic mechanical switching to isolate the start winding is not required. Machines using speed control have high impedance windings to limit the current they develop at lower speeds.

Both series and shaded pole motors are ideally suited for speed control. Please draw examples from old fashioned vinyl-record decks and hand held electric drills. Speed control is not suitable for all types of motor but for motors that are able to adapt, the following types of control are used:

- Series impedance control (linear resistance). This is just a linear resistance connected in series formation with the phase conductor.
- Solid-state linear control technology, meaning speed control using electronic circuitry.
- The use of an auto transformer where the output voltage output is electromechanically controlled.

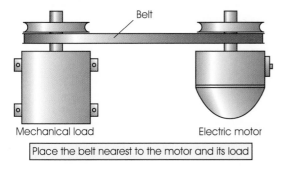

Figure 6.41 – *Schematic diagram of an auto transformer*

We can obtain speed control with an auto transformer. The construction of the transformer takes the form of a single length of thin, lightly insulated copper wire wound many times around a laminated iron core. Throughout its coiled length, output tapings V1 (the lowest voltage output) to V4 (the highest) provide output steps to either increase or decrease the voltage supplied to the motor. These in turn are connected, mechanically, to an adjustable voltage selector (this is a movable switch), the common of which supplies the machine with power. Figure 6.41 illustrates an auto transformer wired in series formation with the electric motor.

Table 6.2 describes the relationship between the supply voltage and torque when placing a capacitor-aided induction motor in circuit with an auto transformer speed controller.

At 35% of full voltage, the starting torque is just 11%. This would be sufficient when the load is light but would prove more of a problem if the load were greater. If the motor fails to respond to the minimum voltage range, adjust the output tapings on the transformer to a higher value.

The advantages gained from using this type of speed control are:

1 reduced starting current,
2 torque proportional to the voltage applied,
3 smoother starting sequence.

The disadvantages are:

1 higher installation costs,
2 regular maintenance required,
3 only certain types of motor are suitable.

BS 7671 Regulation 555-01-02 bars the use of a step-up auto transformer in an IT earthing arrangement, as there is no direct connection between any live part and earth.

Table 6.2 – *The relationship between voltage, current and torque using an auto transformer as a means of speed control to serve a capacitor aided induction motor. (Data kindly supplied by Salisbury College.)*

Percentage of declared voltage (U_0)	Starting current as a percentage of full load	Percentage of the maximum starting torque
35	76	11
40	110	24
62	260	57
80	384	96
100	>450	100

Three phase machines – speed control

First, ask yourself the following questions:

- Just how much speed reduction do I require?
- Is the speed reduction required in steps or has it to be smoothly changeable?
- What levels of torque do I require?
- Will the speed I obtained need to be constant or does it really matter?
- Will my selected speed need to be constant irrespective of load variations?

The type of speed control you choose will depend on the answers to your questions. Some forms of control will be very unsuitable whilst others will not. Think long and hard but do not dive into muddy waters!

- *Three phase commutator motors* have their speed controlled by varying the brush positions. The efficiency of this technique is far better than that of a slip ring motor with rotor-resistance speed control as there are no losses acquired within any external banks.
- *Squirrel-cage induction motors* are speed controlled by varying the supply voltage to the stator windings through the means of a voltage regulator or by selecting a multiple winding machine (a motor with many sets of 'poles') with the correct type of control gear. Please remember that the torque delivered is always proportional to the applied voltage.
- *Slip ring machines* are controlled by rotor resistance control or have numerous windings, served by special control equipment or a mixture of both. Alter the torque of a slip ring motor by placing a constantly rated variable resistance in series with the rotor circuit. This will provide a good range of torque values but always keep in mind the load placed on your machine as, when reduced, the speed of the machine increases.

Difficulties can be experienced maintaining a constant speed below 40% of full r.p.m. as slight variations in the load and within the external resistances due to temperature variations will cause the rotor speed to be unreliable. The efficiency of the machine tends to fall off at lower reduced speeds.

DC machines

Speed control serving DC machines is described in previous paragraphs, under the sub-heading 'Direct current machines'.

Handy hints

- Regular planned maintenance can often prevent expensive and inconvenient motor and control equipment breakdowns.
- Unblock all ventilation grills to allow free circulation of air.
- Examine carbon brushes and renew if past their useful life.
- Inspect the physical condition of the commutator. If required, please clean using a fine grade abrasive paper. Freshen up the mica insulation between the brass commutator segments with a very thin fine toothed saw.
- Examine bearings for wear. Re-grease after every 10 000 hours of running time. Never over-grease the bearings; this will cause premature wear through overheating.
- It is best to remove the bearings from a motor, which has been running continuously for two years, and inspect for serious wear. If satisfactory, wash thoroughly with a degreasing agent and pack with a suitable grease – if not, renew.
- Housing caps (containers in which grease is stored) must be filled with just two thirds of grease. This will ensure that a supply will always be in contact with the bearing between maintenance periods.
- Please clear the breather holes at the base of the motor from contamination.

Handy hints

- Check the mechanical action of the centrifugal switch by removing the cowl and fixing bolts at the end opposite to the motor drive. Clean and remove any impurities. Rub the switching terminals with a fine abrasive paper to remove any 'pitting', which might have occurred. Always check the electrical continuity of the switch using a digital ohmmeter before you reconstruct.

- When using one belt, in a two-pulley system, place the working belt nearest to the bearing. Closing your eyes to this rule can cause unnecessary belt-leverage, enough to cause premature wear to the bearings.

- Ensure your conductors serving the motor terminal housing are secure. Conductors crammed against the wall of the terminal housing could wear through their insulation and cause a fault condition. Vibration and the general movement of the machine over a period can cause this problem.

- An odd growling noise coming from a three phase motor will indicate the loss of one supply phase to the machine. Check the over-current protection device.

- To reverse the direction of rotation of a three phase motor, swap over the two outer phases.

- From time to time, check the physical condition of the capacitor. Make certain it is free from mechanical damage and swelling.

Summary so far ...

- *An electric motor depends on the following – frequency of the supply; for speed – the number of pairs of poles, the value of the applied voltage and the strength of the magnetic field.*

- *When connecting a three phase supply to a stator, it creates a rotating magnetic field.*

- *A series wound DC motor will work equally well on an AC supply. This type of motor has a multi-segment commutator and wired in series with a set of field coils.*

- *A DC shunt machine has a set of field coils in parallel with the armature.*

- *To produce a rotating magnetic field each independent supply has to be out of phase with each other by 120°.*

- *There are three types of AC induction motors – synchronous, induction and commutator motor.*

- *The 'slip' factor of an electric motor is a percentage of its maximum speed compared to the synchronous speed of the rotating magnetic field – see (6.2).*

- *Double cage induction rotors have two sets of rotor bars called the upper and lower; they produce a higher torque factor.*

- *The slip ring motor allows for an external resistance as a means of speed control.*

- *Achieve speed control in single phase AC motors by the use of a series impedance, an electronic management system or control by auto transformer.*

- *A capacitor-aided motor has a capacitor permanently connected in series formation in the auxiliary windings.*

- *A relay or centrifugal switch automatically disconnects the start winding from a circuit within a capacitor-start motor after reaching about 70% of the full speed.*

- *BS 7671 requires a way of protection to stop an electric motor from restarting after stopping. This does not apply to motors under 0.37 kW or if serving a pump controlled by a float switch.*

Summary so far . . .

- Miniature bimetal type overload devices attached to the stators of small motors under 0.37 kW provide thermal and over-current protection.

- Figure 6.40 provides a schematic wiring control circuit serving a typical star–delta starter.

- The advantages gained from using an auto transformer as a means of speed control are as follows: reduced stating current, the torque is proportional to the voltage applied and the system is able to offer smoother starting.

- Three phase commutator machines are speed controlled by varying their brush positions.

- A way in which speed control is applied to a squirrel cage motor is by varying the supply voltage to the stator windings by the use of a linear resistance device.

- Gain control of a slip ring motor by using a special peripheral rotor resistance device. Motors designed with many sets of windings, wired to suitable automatic control equipment, may also be used.

- Difficulty in maintaining a constant speed when the rotor falls below 40% of its maximum pace is commonplace with slip ring motors.

Review questions

1 An electric motor depends on four factors – briefly list them.

2 Describe the construction of a squirrel cage rotor.

3 What aspects govern the synchronous speed of a motor?

4 List three types of three phase induction motors.

5 TRUE or FALSE – The current produced within a squirrel cage winding is in opposition to the very current which produced it.

6 List three ways in which to use a double cage motor.

7 What would you consider the main disadvantage of choosing a double cage motor?

8 Provide three ways in which to speed control a series wound AC/DC motor.

9 Briefly describe why it is not possible to speed control a capacitor-start motor

10 How would you reverse the rotation of a capacitor-start motor?

11 List two types of overload protection you might use for an electric motor rated under 0.37 kW (370 watts), which are not miniature circuit breakers.

12 State both the advantages and disadvantages of choosing an auto transformer as a means of speed control.

13 What method would you use to speed control a typical AC slip ring motor?

14 Briefly state how a DC series motor is constructed.

15 Why should you never over-grease bearings, which serve an electric motor?

Introduction

Derived from Outcome 1, Unit 2, this is probably the longest chapter in this division and forms an essential part of your Unit 2 element (Installations (building and structures) inspection, testing and commissioning). The contents of this chapter will give you an idea about time-honoured procedures for commissioning, inspecting and testing your finished electrical installation.

Your formal assessment will take the form of a short written test of keystone knowledge which you will have obtained over your course of study within your learning centre. This will be coupled with practical assignments prepared by your college or training centre.

Practical activities

The practical activities linked to this chapter include the following topics:

◘ You must know the types of tools and equipment to use and the procedures to carry out to commission an electrical installation, taking into account appropriate codes of practice.

◘ You will have to describe the required clearances needed before any site work is carried out.

◘ You will have to demonstrate you are able to connect and complete an electrical installation in accordance with BS 7671 (our electrical regulations) and other suitable codes of practice. This assessment will be made within your training centre.

Expect a formal recorded judgment on all your practical activities carried out within your training centre. The marks you get will go towards your final award, graded as either Pass, Credit, or Distinction.

Written tests/assignments

City and Guilds will provide you with an assignment guide and an assortment of written questions.

Making an area safe before starting work

Normally it is possible to arrive at your place of work, make a quick check to ensure that no problem areas exist and start your first fix installation work with no further worries. This is fine when rewiring an occupied house or installing new work in a builder's 'one-off' development.

Where it is different is when the general pubic are involved or are within close proximity of your workplace, such as cable laying for decorative bollard lighting arrangements within a pedestrian precinct or maybe just carrying out electrical refurbishment work in an area where members of the public are free to roam such as a large hypermarket or superstore.

Where the public are concerned your attitude towards health and safety at work must be faultless and polished – you really must appear to be squeaky-clean and comply with the safety regulations. The very last thing you want is for someone to take legal action against you after tripping over an item of your equipment and breaking an ankle!

Listed in random order are measures and procedures you can make to ensure that your work area is safe before any work is carried out.

1 The use of barriers or plastic chains as shown in Figure 7.1.
2 Parking cones supporting black/yellow PVC non-adhesive barrier tape (Euro Tape®).
3 Placing warning signs in appropriate positions as illustrated in Figure 7.2 – with adhesive ground marking tape as shown in Figure 7.3.

4 Letting people know who will be affected by the nature of your work.
5 Isolating selective circuits as a protective measure.
6 Obtaining a *Permit to Work Order* before any work is started – see page 38.

Types of work that need to be included within an electrical installation

Expect a wide variety of different types of work before your installation is completed. To be a

Figure 7.1 – *Coloured chain and posts can be used for marking out hazards, etc.*

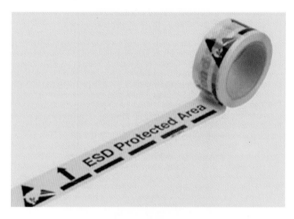

Figure 7.3 – *Adhesive ground marking tape can be used to cordon off hazardous areas*

Figure 7.2 – *Place your warning signs in places which will draw immediate attention to a potential hazard*

good electrician you will need to put your hands to many assorted skills – our profession is a great deal more than just laying cables and installing accessories, once the first-fix part of an installation has been completed. There is more to it than that!

Listed is a snap-shot of what to expect once you are qualified and is based on an electrical installation for a typical new three bedroom house:

- The ability to read a scaled electrical location drawing.
- Marking out where all your accessories and equipment are to be positioned. This is done by scaling off information from your electrical drawing and then reading data from your job specification. Greater detail will result from this exercise.
- *'Boxing out'*. This is the term used for cutting out block-work to accommodate your steel accessory boxes. For this you will require a chasing machine and a sharp mechanical chisel. Alternatively, a hammer and bolster will remove block-work or brick, but this old-fashioned approach is both hard work and time consuming.
- Fixing (by use of wood screws) your accessory boxes to the block/brick work and fitting open rubber grommets to protect the future cable installation.
- Sometimes you will have to remove floor boards – drilling and notching for cable runs. Figure 7.4 illustrates how to drill and notch out correctly.

- Laying your first-fix cables, marking the cables and clipping where required. At this stage, cables must be safely accommodated within the steel accessory boxes.
- Measuring and cutting to length PVC-u or steel channelling. This is placed over your cable as a means of mechanical protection before plastering is carried out and then as a way of drawing cables in and out if required. Figure 7.5 illustrates cable channelling. This is used solely for PVC sheathed cables.

After plastering has been carried out

- Cable termination and connection to your electrical accessory. Once the sheath of the cable has been split and trimmed, cut your conductors to length. Measure from the

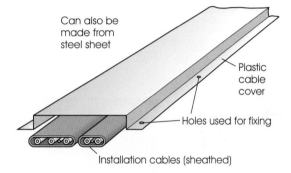

Figure 7.5 – *Cable channelling*

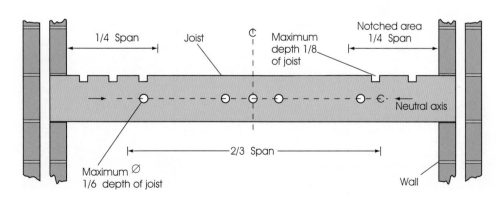

Figure 7.4 – *Drilling and notching out correctly*

trimmed sheath to the termination point of your accessory. Please remember to install a suitable length of green/yellow sleeving over the bare circuit protective conductor (cpc).

■ Square and straighten all accessories together with your distribution centre by using a small spirit level.

■ Choose the correct fixing devices to serve ceiling roses and heaver luminaries.

■ Keep your cable installation well away from hot sources and heating pipes.

Other types of work include the following

■ Metal work serving steel conduits, trunking and cable tray installations.

■ Installation of electrical equipment such as fire and intruder alarms, door entry systems, central heating and CCTV operational units.

■ The replacement of electrical equipment or extending circuits in an up and running installation.

Issues which need considering when planning an installation

Site accommodation requirements

Before placing your temporary accommodation on site, clear the surrounding area of rubble and rubbish. This will provide good bedding for your short-term buildings and also make available a safer and cleaner workplace for everyone concerned.

A clean site is a good start for all and offers safe access for delivery vehicles without fear of damage caused by sharp objects and strewn waste particles.

Your company is responsible for the following:

■ An office, where site business can be carried out. Your best option is a portable purpose-made rented unit. Figure 7.6 illustrates a typical Portakabin® office unit for building site use. It's clean, comfortable and secure.

Figure 7.6 – *A Portakabin® or a Rollalong® module will make an excellent site office*

- A heated covered area where meals can be taken with table, chairs and a means to boil water. Again, your best option is to rent a purpose-made canteen with a sink and running water.
- A steel, windowless materials store where cable, conduit and all your electrical accessories can be safely kept.

Male and female toilet facilities, first aid requirements and drying rooms are the responsibility of the main building contractor. If your company is the main contractor (and it does happen at times) you are then responsible for the health, safety and welfare of the workforce.

Your employer is obliged to provide a well stocked first aid box. Keep the contents in a wall mounted cabinet in your site office. This must be clearly marked 'FIRST AID' as illustrated in Figure 7.7.

Building fabric and external influences

It is far better to choose a strong purpose designed portable unit than a second-hand garden shed full of holes to carry out your site administration work and store your materials. Wooden structures are inexpensive, light and easy to move around but they are not secure, even with the best lock in the world, and can also be easily blown over in high winds. Older wooden constructions are often plagued with problem areas such as holes, badly fitted timbers and rot. This could let in the rain or snow during the colder months and dust and drafts in milder times. Be professional – rent a purpose designed portable unit for your requirements.

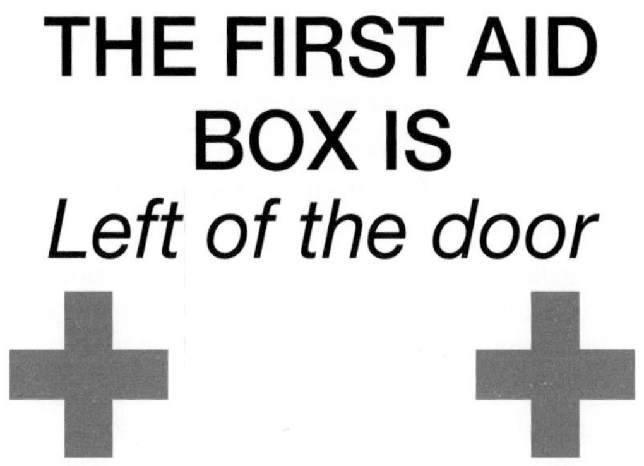

THE FIRST AID BOX IS
Left of the door

Figure 7.7 – *Your site first aid box must be clearly marked*

Storage of parts and materials, power tools and equipment

Select a windowless steel container with a good reliable lock to store your plant and material items you will be using for your electrical installation. Choosing a timbered structure is as bad as leaving the store door wide open during the hours of darkness. They are so easy to break into. Different sizes are available to suit your needs and requirements.

Remember to secure the door at all times if opportunist pilfering is to be avoided. Do not take chances with your own or your company's property.

The need to minimise disruption to neighbouring working areas

Disrupting the work of others can lead to clashes, conflict and problems. Imagine, for example, if you need to carry out work on a distribution centre sited within a small cupboard which is also home to the central heating boiler and the plumber also requires to make adjustments to his system. In cases such as these, negotiation is the best step forward. Cooperation is better than conflict.

If work has to be carried out in other areas such as a small bathroom, the last thing you want is to be working with other trades, and everyone getting in each other's way. In cases such as these it is best to have a gentle word with the site manager and let him/her deal with the situation.

Estimating the length of time to do your installation

This is difficult if you are the job holder, accountable for several operatives who work alongside of you.

As a rule of thumb, work on your own personal experience, rather than those of others who may be either faster or slower than you are. When estimating the length of time to complete a small installation, consider one part or part-area at a time, log your individual estimations and then find the grand total. With this provisional figure in mind, add to it a quarter of the grand total. This then will be a rough approximation of how long your job will take including any remedial works which might have to be done after your job has finished. Remember to include time for testing your installation.

If you have access to a site computer and appropriate software you will be able to obtain additional guidance. On-line guidance is also available but you may have to pay for the information extracted from the web site.

Your installation and the relationship of parties concerned

A successful electrical installation requires the participation of many people. Each has his or her role to play as members of an extended back-up team whose job is to ensure that every stage of your installation runs as smoothly as possible.

Table 7.1 identifies the parties concerned who support your installation and their relationship with one another. Please assume for this that your company is the nominated main contractor.

Regulatory requirements within electrical installation engineering

Table 7.2 outlines the role played by individuals in respect to regulatory and legal requirements within the workplace.

Understanding drawings and specifications for your site materials list

Working drawings (wiring diagrams or sometimes known as site plans)

Some of you will have to extract and list your material requirements from your site installation wiring diagrams, whilst others, who will be far more fortunate, will have others to carry out this task for you. Wiring diagrams use BSEN (British Standards European Norm) graphical symbols to show accessories, luminaries and point to point interconnections within an installation. These symbols are made clearer by studying Figure 7.8 whilst Figure 7.9 provides an example of this type of working drawing.

Making your list

Using a clean (readable) scale rule measure the quantity of steel wire armour cable, trunking, conduit and other cable carriers you require for your job.

Table 7.1 – *Identifying the role people play in support to your electrical installation.*

Person concerned	The role they play	Their responsibilities
Client	Your customer	To provide data and relative information to produce a fair and reasonable price accompanied by an accurate working drawing
Main contractor	To undertake the electrical installation and be responsible for their personnel	To ensure that the installation complies with the working drawing and current BS 7671 standards
Sub-contractor	Often involved with specialised installation work such as fire, intruder alarms and door entry systems, etc.	Directly to the main electrical contractor. To ensure their codes of practice are adhered to. To commission and issue relevant certificates
Suppliers (wholesalers)	To respond to material requests and technical product information required by the job holder or supervisor	Firstly to their customer (the company who you work for). Secondly, to the site ordering the material items for use in their installation.
Consultants	Provide expert and technical advice with problem areas associated with your installation.	They are responsible to your management to provide up-to-date information and expert opinion on all problem areas of your installation
Supervisor	Supervises site progress. Liaison between management and the site workforce.	Regular site visits to oversee progress of work. To discipline. To advise and inform and to cater for site material requirements. Responsible to the contracts manager
Company store man (large construction sites only)	Issuing materials such as cable, plant and accessories on production of a stores requisition order	Supplying material items for the installation. Administrative duties and ordering. Stock taking. Responsible to the contracts manager.

Table 7.2 – *The role people play in respect to regulatory requirements*

Regulatory requirement	The person who will bear responsibility	The role they play
The Electrical Regulations – now BS 7671: 2001	Overall, your contracts manger but at site level it will be the job holder – you will be responsible to your contracts manager	To ensure that the installation complies with BS 7671: 2001 and published amendments
The Health and Safety at Work Regulations (1974)	In general, the site manager has this responsibility but if you are the job holder you are also liable for the health and safety of your workforce	To ensure, as far as reasonably possible, the health and safety of the site workforce and visitors are protected.
The Building Regulations (1991)	Overall, the site manager but the job holder and your company's contract manager are also responsible as far as Part 'P' of the Building Regulations is concerned. This Part brings within its scope all domestic electrical installation work within England and Wales	They must comply with the spirit of BS 7671:2001. Part 'P' of the Building Regulations 2002 and 2004 (amended) places a legal requirement upon operatives carrying out electrical work in 'single' dwellings and blocks of flats which have communal areas attached to them. It also includes shared amenities such as a visitor's toilet, laundry room, common room and common room kitchens
The Environmental (Information)Regulations (1992) This regulation has been drafted to prevent/ regulate the discharge of pollution into the air. Control of airborne contamination such as toxic waste, greenhouse gases, industrial bonfire smoke and asbestos dust,etc. is by way of your local authority. Heavier pollution is the responsibility of HM Inspector of Pollution	This is the local responsibility of your site manager who will have overall control of his construction site. However, you as a jobholder must also ensure that these regulations are not broken by the operatives you oversee. These regulations are enforced by your local authority and HM Inspector of Pollution	To understand the spirit and meaning of these regulations and to ensure that dangerous emissions are not leaked into the atmosphere during the course of the installation work. To make certain that equipment used is suitable for its working environment – for example, it would be very unwise to run a small petrol driven generator in a tiny confined windowless room or work in a battery charging room without adequate ventilation

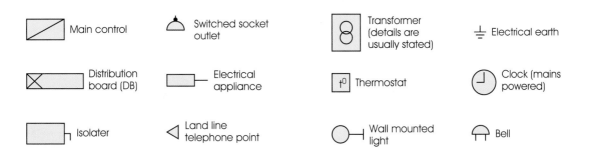

Figure 7.8 – *Use only BSEN symbols to build up your wiring diagram – avoid creating your own*

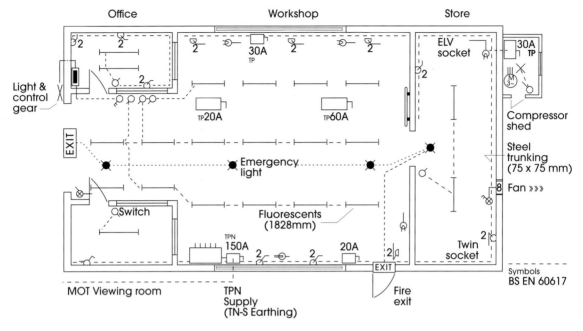

Figure 7.9 – *A point-to-point wiring diagram shows wiring and interconnections within an installation*

Please remember an electrical location drawing is only two dimensional and that depth will need to be taken into consideration. If you are not sure, take time to check out the proposed installation and adjust your measurement as necessary. After all, it is better to have three metres too long than three metres too short!

Accessories and other items can be quickly extracted from the drawing and listed. The secret is to familiarise yourself with the location symbols so you know exactly what you are looking at when reading your drawing.

A provisional list such as this will usually provide you with sufficient needs to start your installation. Any other outstanding items can be taken off the drawing and called for when you require them. Please remember to check your list against your job specification as this will guard against accidentally missing out specific equipment or out of the ordinary accessories which arise from time to time. Finally, remember all your fixing requirements.

Block diagrams

This is a very useful way to forward an idea without the need for too much technical detail. A block diagram will provide you with a means of focusing on a possible problem area and creates a gateway for more formal planning to be carried out at a later stage. Material requirements are easily removed for list forming at this stage as everything is so very basic.

Figure 7.10 illustrates such a diagram.

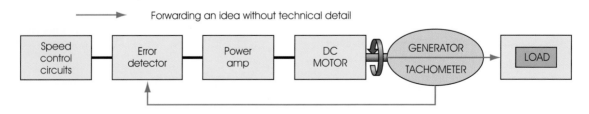

Figure 7.10 – *A block diagram is an ideal way to forward an idea without technical detail*

Location diagrams (also referred to as equipment location diagrams)

This type of diagram targets and identifies in great detail the location of the equipment to be installed within your installation. This information is often needed to allow for the structural design of the building to be developed, for example, where a large recessed mains distribution centre has to be positioned in a precise position or if a fire detection/alarm panel requires an exact location. The drawing comes as a large scaled engineering diagram with supplementary measurements and other details included. Usually one drawing occupies a single sheet of paper.

Circuit diagrams

Circuit diagrams are drawn in graphical detail again using BSEN electrical location symbols representing all components and their interconnections. The finished diagram will show how your circuit will operate without necessarily providing a physical layout drawing. Figure 7.11 shows the wiring as vertical and horizontal lines. The location symbols are those recommended by British Standards. Custom-made symbols must never be used if confusion is to be avoided.

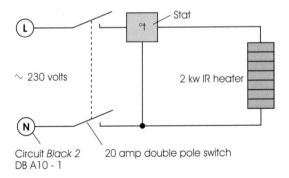

Figure 7.11 – *Circuit diagrams – point-to-point wiring is shown as vertical and horizontal lines*

Producing a formal contract

A contract between your company and a potential client is an important part of business life – it is a way to produce a formal, legal agreement in order to get work done. The common factor associated with all contracts is that there is an exchange of what someone wants. This, in legal terms is called a *benefit*. The payment for this benefit is known as the *consideration*.

Once signed it entrusts your company to carry out the agreed work as laid down and described within the pages of the signed contract.

A 'template' contract is usually drawn up by a company's solicitor and is often bursting with official terminology, often known as *jargon*. This is difficult to understand at the best of times. If you do not understand any of the official language, then you must ask your solicitor to explain its meaning using everyday language.

A word of warning! Never sign a contract unless you have both read and understood its contents – you could end up losing a great deal of money.

Contents in concise format

The basic contract features you can expect

Contracts do differ in involvement from a simple verbal or written site instruction to a far more comprehensive document offered to you for pricing.

The common characteristics associated with all contracts from start to finish are:

- Every contract contains rules of understanding which are known as *terms*.
- Both parties are bound by the terms of the contract.
- There is a 'benefit' and 'consideration' to each contract. In most cases the consideration takes the form of monetary payment.
- A contract starts off by an *offer* being made; in our language – a tender.
- Your tender has to obtain *acceptance* (approval) by your client.
- Once you are this far, it is just a short step to actually doing the job.

What would I expect to find within a standard contract?

Standard contractual documents are all rather similar and are drafted to include the following terms of reference:

- A description of the work you are invited to carry out.

- Naming the person who has the legal accountability for overall control of the contract.
- Terms of payment (once every month, three months, etc. or based on the amount of productivity).
- Ways in which productivity can be measured.
- The length of the contract in weeks or days.
- How much your company has to pay if the contracted work is not finished on time? This is known as *liquidated damages.*
- The amount of money held back and only given to your company on the acceptable completion of all contractual works. This is referred to as *retention.*
- The rules and regulations which have to be followed by both parties in case of dispute. This is legally known as the *claims procedure.*

A contract, by its very nature, can generate much worry, grief and loss of sleep if it is not tackled carefully. Read every word, every line and understand its contents like the back of your hand. You should be able to avoid problem areas by exercising cautious thought and, before signing, come to a sensible agreement with your client concerning its contents. Good luck!

Normally a tender will be accompanied by 'proposed installation drawings' together with an up to date specification and equipment schedule.

Ways we monitor the progress of the contract

Bar charts

A bar chart is one of the most common ways to show the intended progress of a project waiting to be carried out. A well designed bar chart must be worth a thousand words and can be presented in a horizontal or vertical format. A bar chart is like a picture-table and is widely used in the electrical installation engineering industry to present information at a glance. Figure 7.12 illustrates a simple bar chart, mapping progress from the start of first-fix lighting arrangements to completion.

Critical path networks (CPN)

Critical path networks are a useful visual office tool for keeping tasks, which are critical for the development and completion of works, on schedule. It is an excellent visual aid for planning a complete project is its embryonic stages – easy to understand and information can be retrieved at a glance.

First you must locate your critical works schedule and then fine tune it by going through your tasks with a fine-tooth comb to decide which ones can slide a little without necessarily affecting your completion date. This is a basic means of control which works and can prove extremely useful.

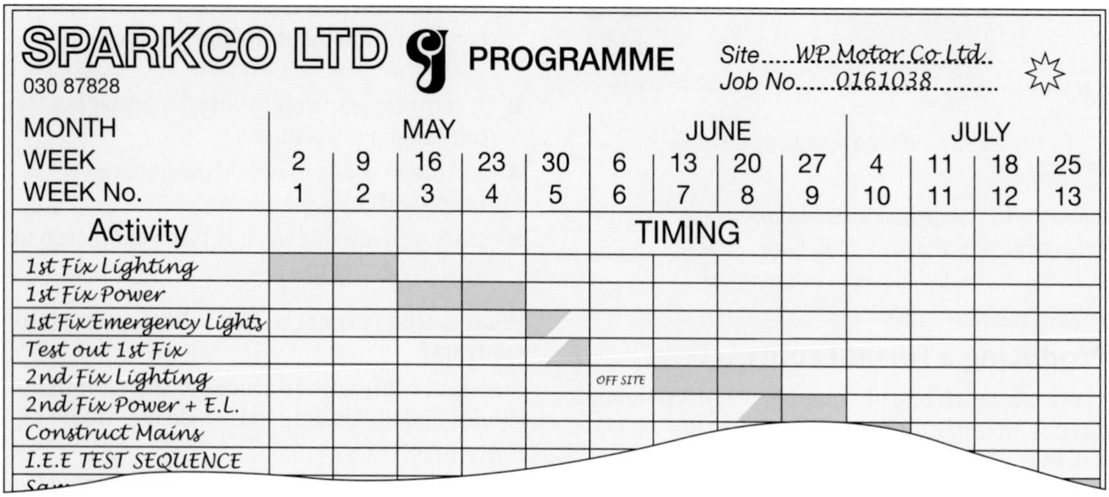

MONTH			MAY				JUNE				JULY		
WEEK	2	9	16	23	30	6	13	20	27	4	11	18	25
WEEK No.	1	2	3	4	5	6	7	8	9	10	11	12	13
Activity					TIMING								
1st Fix Lighting													
1st Fix Power													
1st Fix Emergency Lights													
Test out 1st Fix													
2nd Fix Lighting						OFF SITE							
2nd Fix Power + E.L.													
Construct Mains													
I.E.E TEST SEQUENCE													

SPARKCO LTD PROGRAMME   Site......W.P. Motor Co. Ltd.   Job No......0161038
030 87828

Figure 7.12 – Bar charts will provide guidance as to the sequence of work to be carried out

A small builder who has the task of erecting a single dwelling where all decorations, etc., have been arranged to be carried out by his client may bring together a critical path network similar to the one illustrated in Table 7.3

CPN serving an electrical installation

Table 7.4 provides a snap-shot view of how a critical path factor chart might appear if a large department store were to change hands and be redeveloped into a supermarket. Five days per week working arrangements apply.

Table 7.3 – *Critical path networks (CPN) seen through the eyes of a builder*

Tara Building Company – Critical path network for ZDR Project		
Ref. point	Assignment	Proposed activity dates
1	Clear the land and level	August 1st – 4th
2	Ground work, foundations and construction floor	August 5th – 19th
3	All walls and internal stud-work [SLIDE IF NECESSARY]	August 20th – September 20th
4	Roof timbers and felt	September 21st – 14th
5	Electrical – boxing out and first-fix wiring and other services	October 15th – 28th
6	Plastering throughout	October 29th – November 10th
7	Second-fix electrical and plumbing. Commission and test	November 11th – 16th
8	Land scaping [including garden lighting arrangements] [SLIDE IF NECESSARY]	November 17th – 29th
9	All internal works to be completed [allow for technical snags]	November 30th – December 4th
10	Roofing etc. to finish off	December 5th – 20th

Table 7.4 – *Critical path factor chart serving an electrical installation*

Tara Electrical Company Telephone 01794–000876 Critical Path Networks for Project #0008765 Site . . .Whiteparish A.G.												
Site Activity	Start Date	Finish Date	Extent of Work	November 1 Wk 1	8 2	15 3	22 4	29 5	December 6 Wk6	13 7	20 8	27 7	> > Next month
Site safety check	01/11	01/11	1 Day										
Set up site	02/11	05/11	4 Days	****								Xmas	
Temporary light/power	08/11	12/11	5 Days		*****							Break	
Carsacing cable cariers	16/11	10/12	20 Days			*****	*****	*****	*****			24/12 To	
Wire for sub-mains	13/12	9/12	7 Days							*****	**	2/1	
Construct main distribution	16/12	24/12	7 Days								** *****		
1st Fix lighting#	2/1	30/1	30 Days										***** > >

. . . and so on (# Can 'slide' if hard-pressed for completion)

Site records

This is very much an in-house task which we all do at some level throughout the duration of work we are asked to do.

Records and data can be gathered in many different ways and stored in plenty of diverse places. Please glance through the following list.

- *Your site day diary* – it is here you will record progress, the number of electricians you have available to you, staff who have called in sick, site meetings, delays caused through material items not being delivered, etc.
- *Site delivery note file* – this is where all your delivery notes are, which accompany the material items either you or your company have ordered from your local supplier. Once signed, add the delivery date and time for your own records and those of others.
- *List or photocopies of site variation orders or instructions* – this can make a difference to the duration of your job. If you have many, it could add many days, if not weeks, to the agreed completion date. Keep them safely in a ring binder file to make it easy to refer to when required
- *Site drawings (master and working location plans)* – for your personal records it is a good idea to pencil in all the 'as-fitted' circuits you have been responsible for, together with their circuit number. This is illustrated in Figure 7.13. Take time to record on your drawings the precise location of any additional work asked for by others so 'as-fitted' drawings can be prepared at a later stage.
- *Departures from BS 7671* – you need to verify in writing any departures from our electrical regulations. This is important if you are to cover yourself.
- *Bar charts and CPNs to monitor and keep check on progress and productivity* – these 'see-at-a-glance' graphic representations have been discussed in previous paragraphs.

Site diaries

Use your site diary to record things which are important to you at the site level. Some examples are:

- productivity, delays and occurrences,
- the number of staff you have on site,

- absenteeism,
- who is doing what,
- site visits by management or other company representatives,
- general site meetings you have to attend,
- future appointments or activities with respect to the job,
- whether your job is on target, in front of your expectations or lagging behind. If it is behind, then it is best to explain the reason – for example, it could be due to additional work you have had to do in the form of site instructions, variation orders or maybe delays caused by the non-delivery of material items you have ordered,
- the weather should be recorded within your site diary if it affects your productivity when work has to be carried outside,
- record variation orders given by word of mouth,
- any time off by operatives (dentist, doctor, professional appointment, etc.).

All these entries are important as it helps to provide an overview of the ups and downs we all have to bear under site conditions. At the end of the day, if you are the job holder you alone are held responsible as to whether your job has made a profit or not. If the profit margin is disappointing, then you will have at your fingertips all that is required in the form of written evidence from your site diary and other site records.

Site variation orders

Variation orders are written orders by others requesting work to be carried out. It may be an alteration or an addition to your original specification but whatever it is, the golden rule is to obtain the order in writing properly signed by a responsible person. Failing to do this could result in non-payment to your company.

Variation orders can originate from many a source; here are just a few:

- architect,
- your customer (this is possible, but rare, as usually fresh instructions are discussed with the architect),
- the electrical design engineer,
- the principal building contractor,

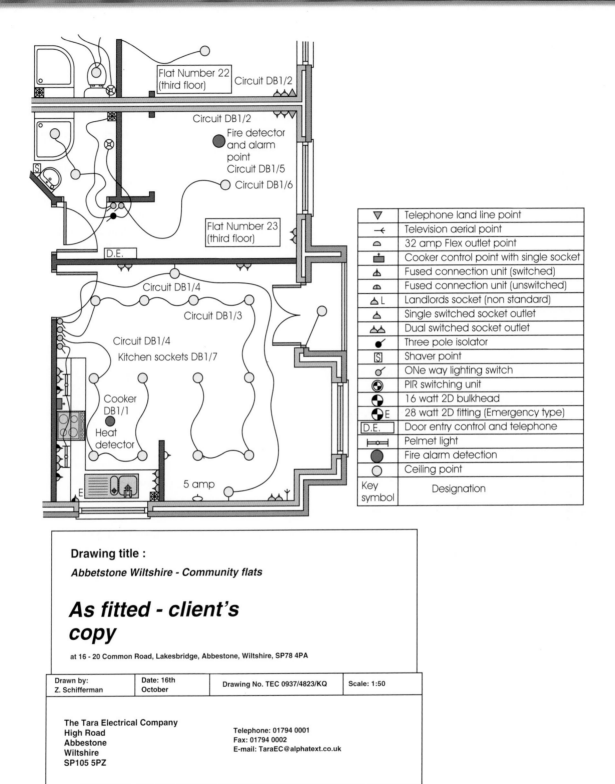

Key symbol	Designation
▽	Telephone land line point
⟶	Television aerial point
⌂	32 amp Flex outlet point
⊡	Cooker control point with single socket
⚥	Fused connection unit (switched)
⚥	Fused connection unit (unswitched)
⚥ L	Landlords socket (non standard)
⚥	Single switched socket outlet
⚥⚥	Dual switched socket outlet
●	Three pole isolator
Ⓢ	Shaver point
⟋	ONe way lighting switch
⊕	PIR switching unit
⊕	16 watt 2D bulkhead
⊕E	28 watt 2D fitting (Emergency type)
D.E.	Door entry control and telephone
⊢⊡⊣	Pelmet light
●	Fire alarm detection
○	Ceiling point

Drawing title :

Abbetstone Wiltshire - Community flats

As fitted - client's copy

at 16 - 20 Common Road, Lakesbridge, Abbestone, Wiltshire, SP78 4PA

Drawn by: Z. Schifferman	Date: 16th October	Drawing No. TEC 0937/4823/KQ	Scale: 1:50

The Tara Electrical Company
High Road
Abbestone
Wiltshire
SP105 5PZ

Telephone: 01794 0001
Fax: 01794 0002
E-mail: TaraEC@alphatext.co.uk

Figure 7.13 – *A small 'as-fitted' drawing showing the circuit wired together with the circuit number and over-current protection used, etc.*

- a fellow sub-contracting company,
- construction site manager or his agent.

These requests, known as variation orders, mean that someone wants you to carry out additional work, over and above what you are already doing. An example is shown as Figure 7.14.

If a written variation order is not possible, please log the instructions given within your site day diary but do not carry out any additional work until you obtain a signed, written notification. If you don't have the paperwork to back up what you have done, your company will find it difficult to receive payment.

An idea is to photocopy the variation order to include within your site records before you hand the original document to your office for processing. This way you can keep on top of everything.

Harmonizing your installation with other trades

It is important that only one trade is allowed to carry out work in one particular area. This rule can be ignored if the area to be worked in is large and spacious but not so if, for example, it is a small or confined space such as a bathroom or boxroom within a semi-detached house or bungalow. Imagine the chaos and disorder which would follow if you, the painter and the plumber were to be in the same bathroom at identical times. The site manager is responsible for ensuring that this does not occur.

PITBOND ENGINEERING COMPANY LIMITED   ℭℬ

Site Variation order   *No. E 0968*

Site address ....
To ....
Date .... Ref: ....

Pleae carryout the following works. A risk assessment and a permit to work will be required

1. *Alter your 17.5 metre run of steel conduit serving the basement final ring circuit. Repostion 300 mm from the construction ceiling level.*

2. *Change the existing one way basement lighting system into a two way switching arrangement. Site the switch 450 mm from the basement exit door.*

Signed... [Site manager]   Date...................................

Figure 7.14 – A site variation order will usually ask for alterations to be made to work which has already be done

The reality of variations

Most of us are in no way great supporters of variations to an original drawing and specification. We feel that in one way or another we are taking two steps forward and three back when carrying out our work or that the electrical design engineer should have got it right in the first place. Having said that variations (or the lack of them) can have the ability to either make or break a job. It is very important to get variations right first time – start initially with the paperwork.

Site variations are normally priced as 'time and materials' but when this does not apply you will be asked to price for the additional work. This may not seem at all logical as in theory, if your charge is too high, another electrical contractor could do the job and work alongside you.

Time and materials explained

After completing your variations and providing your company with all relevant paperwork your contracts manager will gather together all the wholesale costs of material items supplied and add a small percentage retail profit.

Companies vary as to how much they charge their clients for day work. This is known as the *day work rate* and usually works by targeting your hourly wage plus one and a half times this figure. So if you are paid, say £14 an hour, then your company will charge the client £29 an hour. Both figures (time plus materials) are added together – the total, together with VAT, is the cost to their client.

Alternatively a day work rate is formed when the contract is priced for at an earlier stage.

The importance of maintaining good relationships and successful communication

Customer and others

It is very important to maintain good working relationships with your own workmates and those of other trades. A happy, hassle-free site is often a smooth running and safe site. This must be our goal. Do remember, cooperation is better than conflict!

The main requirements for securing a good-natured and friendly atmosphere with your client, architect, surveyor, main contractor, other trades people and local authority representatives can be summed up by browsing through the following list:

- Patience and open-mindedness towards others.
- Don't fall asleep during a site meeting.
- Treat your visitors (local authority, etc.) with respect.
- Punctual completion of entrusted or important work asked for by the architect or main contractor.
- Carry out your duties in a professional manner.
- Keep to the agreed programme of work.
- Regular productivity to please the architect and surveyor.
- Avoidance of long periods of absence from site (this may not be your mistake but the fault of your company).
- Do not over-stay your break periods.
- Get to know your customer – remember names; this is important.
- Advise of disruption to your customer's activity.
- A serious breakdown in discipline must be reported to your management.
- Always respond to requests for information from senior visitors and other trades people.
- Keep jargon (the language of our trade) out of conversations with others who are not electrically minded.

Try to adopt a personal code of moral values when dealing with customers, potential clients or anyone associated with your project. A good image coupled with a high standard of professionalism can go a long way. It is no longer sufficient just to provide the required expertise and good workmanship asked for by your customer. You will not be judged by this alone. A mixture of old-fashioned manners with a few social graces will often be enough to generate the confidence to win a person over and form a good working relationship.

Listed in random order are thoughts to consider or discuss when dealing with people who are important to you:

- *Personal appearance* – don't be shabby.
- *Courteousness* – practice good manners.
- *Personal hygiene* – keep body odours away.
- *Attitude* – be business-like.
- *Image* – keep up a good standard.
- *Tact* – avoid gossip.
- *Respect for property* – care and consideration at all times.
- *Positive relationships* – treat both customers and co-workers in a manner which will result in greater understanding and encourage a positive working relationship.
- *A strained relationship* – choose an independent mediator to sort problems out.
- *When information is unavailable* – do your best; prompt action is required.
- *Confidential information* – keep it to yourself.
- *Identify your client's needs accurately* – provide appropriate information.
- *Check out your customer's technical awareness* – address them accordingly.

Voltages for supply and distribution

Supplying our power

Most of our electricity is now generated between 11 000 and 33 000 volts AC by machines known as alternators driven at high speed by steam turbines. This generated electricity is transformed in value to 132 000, 275 000 or 400,000 volts and fed into the National Grid, as illustrated in Figure 7.15. The National Grid was first considered in the 1920s. Since then it has been altered and modified into an interconnecting network of power lines, Figure 7.16, and now provides us all with an efficient fail-safe means for high voltage power transmission throughout the country.

Distribution to the customer

Substations, one of which is illustrated in Figure 7.17, built in principal locations, provide the means by which power can be drawn from the National Grid. The high voltage taken is reduced in value to 11 000 volts. In rural areas we deliver

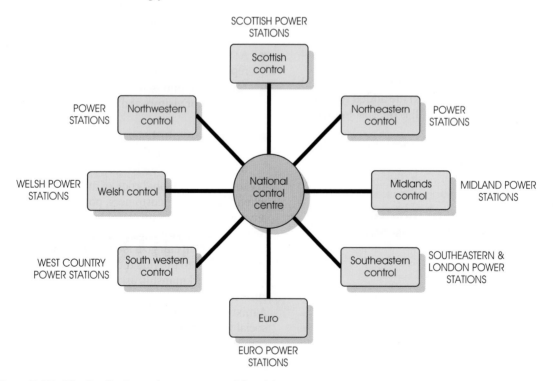

Figure 7.15 – *The distribution and management of electricity*

Figure 7.16 – *An electricity pylon serving the National Grid*

Figure 7.17 – *A National Grid substation in rural Hampshire, England*

our electricity to where it is needed by means of insulated conductors carried on high quality ceramic insulators fitted to tall wooden poles. The timbers are impregnated with an anti-rot preservative. These are routed from the local substation to supply energy to field sited community step-down transformers. The transformer is often cradled safely out of reach between two sizable wood poles as shown in Figure 7.18; smaller ones are housed on little platforms and supported by just one pole. At this point the electricity is reduced again to a conventional three phases and neutral, 400 volt system. Figure 7.19 shows how the supply is routed on single wooden poles evenly spaced through a small village in rural Wiltshire, England. This provides the means of dispensing both triple and 230 volt single phase supplies to the customer. In towns and cities power is distributed by means of a network of underground cables. Homes are supplied by alternative phases. For example, House 1 is supplied with power from the brown phase, House 2 from the black phase whilst House 3's power is via the grey phase and so on. This way the supply to the estate is more or less electrically balanced.

The three phase four wire system of distribution

The winding arrangement most often used for the distribution of power from a neighbourhood transformer to the community is known as the delta–star connection. The high voltage (11 kV) is placed across the delta windings as shown in Figure 7.20. The secondary windings are connected in star formation and it is these which provide us with the transformed energy we require in the form of a three phase and neutral supply.

Three phase and neutral (four wire) supplies are used for commercial and industrial purposes and are also found in large stately homes: Broadlands, for example, in rural Hampshire,

Figure 7.18 – *Community transformer serving rural requirements*

Figure 7.19 – *These are wooden power poles which supply electricity to village communities*

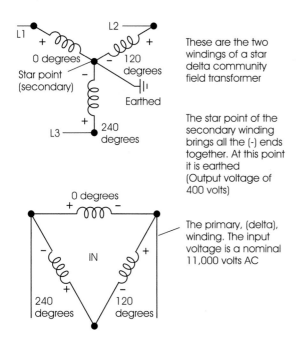

These are the two windings of a star delta community field transformer

The star point of the secondary winding brings all the (-) ends together. At this point it is earthed (Output voltage of 400 volts)

The primary, (delta), winding. The input voltage is a nominal 11,000 volts AC

Figure 7.20 – *The star–delta arrangement within a community transformer. The 11 kV primary supply is connected to the delta windings as shown*

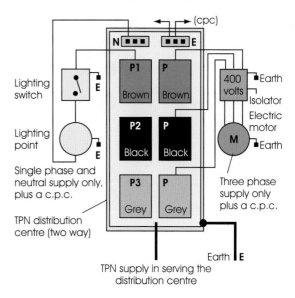

Figure 7.21 – *Triple pole and neutral (400 volt AC) distribution within an installation*

England. The supply comprises three phase conductors which are electrically separated from each other by 120°. These are known as L1, L2 and L3 – loosely described as the brown, black and grey phases. They are attended by one neutral conductor, colour coded blue. Figure 7.21 illustrates this.

Measuring between any of the three phases and neutral or earth will give you a nominal voltage of 230 volts, AC. A nominal value of 400 volts will be obtained when you measure directly between any combinations of phases. For example L1 and L2 will produce a nominal 400 volts as would L1 and L3 or L3 and L2. The word 'nominal' really means 'so-called' or 'supposed', so please never expect the recorded voltage to be exactly right as you will no doubt obtain variations either side of this figure.

Studying Figure 7.22 you will find the star point of the secondary side of the supply transformer – this is the point where all thee windings meet which is also the starting point of the supply neutral. This common connection is grounded by way of an arrangement of electrodes

to be found by the side of the supply transformer. This common connection also acts as an earth conductor to the system.

Individual three phase requirements to supply heavy electric motors and industrial machinery are wired from the load side of a power rated protective device from a suitably rated distribution centre. This could take the form of a 'B' or 'C' type miniature circuit breaker or a motor rated fuse. It is usual to use three different coloured conductors (brown, black and grey) to serve a three phase load. If this is not possible, then your conductors will be brown but in doing this please tag each one as L1, L2 and L3, respectively.

Some three phase equipment will require a neutral conductor such as a construction site crane. This neutral will be for any control circuitry and general lighting requirements within the crane.

The importance of a balanced load

In a balanced three phase system the current flowing in the neutral conductor is zero. However, when unevenly loaded it causes altered phase-neutral voltages and heavy currents will form within the neutral conductor. This must be

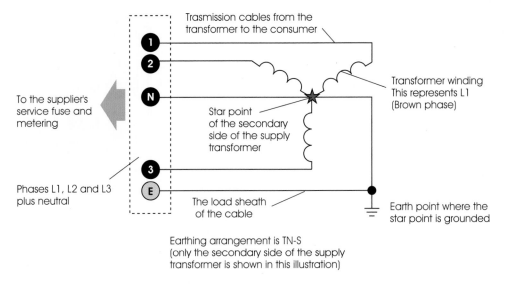

Figure 7.22 – *The star point of a community transformer marks the beginning of the neutral conductor*

corrected if other problems are to be avoided. An unstable three phase installation will be responsible for the following difficulties:

- an increase in neutral current within the neutral conductor,
- higher electricity bills,
- general energy loss,
- voltage dips when loaded suddenly,
- extra fuel is required to supply the generator,
- the 'phase angles' are not equal – they should be 120° from each other,
- your equipment will not be working as efficiently as intended,
- fuses and other over-current devices will not be as efficient as they should be,
- you will experience a certain amount of phase distortion throughout the system,
- losses will appear in distribution cables,
- an unfortunate neutral break or bad connection will result in high voltages to appear in single phase circuits – this could damage costly equipment,
- an unbalanced three phase system will cause uneven phase voltages.

The way ahead – getting it right

An unbalanced three phase and neutral system can be made technically better by the following methods:

- *By manually balancing the system* – this involves swapping over single phase circuits from one phase to another. It can be hard work and often difficult to get it right.
- *By use of a Phase and Neutral Balance System (PNBS)* – this equalizes the voltages in all three phases thus reducing the neutral current and doing away with wasted energy.

Single phase supplies

This type of supply is tailored for domestic and light-commercial use such as small shops and garages. When measuring the voltage between the phase and neutral or earth conductor (but not within a TT earthing arrangement) a nominal 230 volts will be recorded as described in Figure 7.23.

When in towns and cities, the main earth wire serving your distribution centre will originate from the protective lead sheath of the incoming supplier's cable. At the very end of this cable, which often has to travel many kilometres underground to a local community transformer, the metal sheath connects to the star point of the transformer as illustrated in Figure 7.24 – this connection is securely grounded by means of an arrangement of earth electrodes. This is known is a TN-S earthing arrangement.

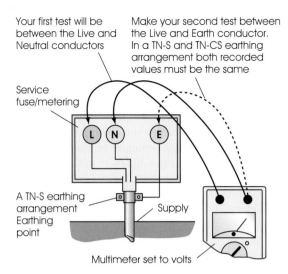

Your first test will be between the Live and Neutral conductors

Make your second test between the Live and Earth conductor. In a TN-S and TN-CS earthing arrangement both recorded values must be the same

Service fuse/metering

L   N   E

A TN-S earthing arrangement Earthing point

Supply

Multimeter set to volts

Figure 7.23 – Measuring your voltage across the phase and neutral conductors

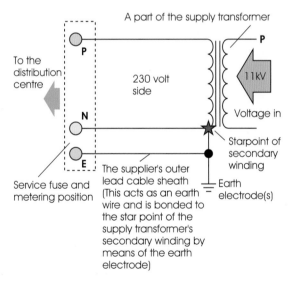

A part of the supply transformer

To the distribution centre

P

230 volt side

P

11kV

Voltage in

N

Starpoint of secondary winding

E

The supplier's outer lead cable sheath (This acts as an earth wire and is bonded to the star point of the supply transformer's secondary winding by means of the earth electrode)

Earth electrode(s)

Service fuse and metering position

Figure 7.24 – A TN-S earthing arrangement from the secondary star winding of a community transformer. A, the consumer's supply intake point; B, the supplier's cable sheath (earthed); C, one phase of the secondary winding of the community transformer; D, one phase of the delta winding served with 11 000 volts; L, live (phase conductor); N, neutral conductor; E, earth (the cable sheath is connected to the star point of the secondary side of the transformer and also grounded)

Most urban single phase distribution systems now have TN-C-S earthing arrangements. This is where an artificial earth is formed by connecting the main earth conductor serving an installation to the supplier's neutral conductor at the point of entry into the premise. A fault current occurring between the phase and earth conductor will have the same effect as though the phase conductor had touched the neutral. In either system the local circuit protective device would be triggered. Your single phase circuit to a connected load will always begin from the topside of your local fuse or miniature circuit breaker and will be accompanied by a single blue neutral conductor of the same cross sectional area.

Your protection against indirect contact

Whether a three or single phase installation, it is very important EEBADS is used for protection against indirect contact. These initials stand for Earth Equipotential Bonding – Automatic Disconnection of the Supply. The simple rule of thumb is that if you can touch anything connected to an EEBADS system or circuit protective conductor – join them together to maintain a common polarity of zero volts.

Isolation and switching

Isolation

Isolation and switching might appear to be the same but electrically they are very different. BS 7671 Regulation 460-01-01 asks for a means of isolation to be provided so that electrical work can be safely carried out on or near to live parts and equipment which would normally be energised.

The position of the switching contacts must be clearly visible or reliably indicated so you can physically tell whether it is 'on' or 'off'. If you are designing an isolation circuit never use a semiconductor device as a means of isolation – BS 7671 will not permit it.

Sometimes it might be physically difficult to install an isolator near to the equipment it serves. If you have to install it remotely then your isolator must be capable of being secured in the 'OFF' position. This will usually mean the use of a lockable isolator or an isolator which can be made safe with a padlock.

You must provide means of isolation for the following:

- *An electric motor circuit and any associated equipment* – at the origin of the installation (mcb) and adjacent to the electric motor (isolator).
- *Single phase domestic circuits* – at the origin of the installation, a main switch or circuit breaker and means to isolate the phase conductor serving each circuit (for example, a circuit breaker).
- *There must be a provision for disconnecting the neutral conductor on single phase circuits* – this can take the form of a neutral link housed within the distribution centre.

Switching

The term '*switching*' means you can safely interrupt the electrical supply whilst your circuit is still on load – Regulation 476-01-02 confirms this. BS 7671 will also allow you to use just one switch to switch off a group of circuits – an obvious example would be the mains switch serving a circuit distribution centre but a separate switch must be installed to control each individual circuit.

A means of switching or switching off for lamp replacement and electrical maintenance

You must have means to locally switch off your circuit for mechanical maintenance (for example, changing a spent lamp, replacing a worn out lamp holder or cracked ceiling rose, etc.). Our regulations demand that each switching device must conform to the following:

- Be so designed as to prevent accidental reclosure caused by mechanical vibration or shock.
- Your switch must be able to isolate the full load current of the circuit. Regulation 537-03-04 will confirm this.
- The switch you use must clearly show both the 'ON' and 'OFF' positions or have an externally visible switch contact gap so you can be clear that the device is either on or off. The use of a neon switch indicator is not encouraged by BS 7671 – they could fail.
- Your switch must be located close by its load and easy to get to – not, for example, two metres from the finished floor level in a

lockable cupboard! Regulation 462-01-02 confirms this.

- It must be indicated with a permanent label unless its function is clear – for example, a lounge lighting switch.

A brief look at emergency switching

Wire an emergency switch into your circuit where a possibility exists that someone might have to disconnect the power in a hurry as in the case of electrical mains equipment plugged into a final ring circuit serving a college workshop or science laboratory.

In practical terms red emergency stop switch buttons, all of which are wired in series formation, are scattered around the work room. Each button is so wired that when it is pushed it will break the circuit supplying electricity to the socket outlets. This circuit is usually a contactor coil, the contactor being the electromechanical switch which controls the final ring circuit.

Points to remember are as follows:

1 Never use a plug and socket outlet as a mechanism for emergency switching.
2 When designed to isolate by means of a contactor, the contactor must have fail-safe features. This means if it should fail, the circuit it controls should always switch to 'OFF'.
3 It must be within easy reach of the danger zone it monitors.
4 The device will switch both the phase and the neutral conductors when used on a single phase installation.
5 Your emergency switching device should have either a red handle or push button.
6 The current rating of your switching device must match or be rated greater than the full load current it switches.
7 Remember to install a fireman's switch for petrol pump or high voltage lighting installation. Please refer to Regulation 476-03-05 to 07 and also 537-04-06 of BS 7671.

Over-current protection

Over-current protection will protect your circuit against the following conditions:

- *Circuit overload conditions* – where a dangerously high current is flowing within your circuit.
- *Short circuits* – where both phase and neutral conductors are in direct contact with each other and a path of low impedance exists between them.
- *Earth fault situations* – where a path of very low impedance exists between earth and the phase conductor.
- *Protection against electric shock* – but only when your over-current protection takes the form of a RCBO to BS EN 61009-1

(the initials stand for Residual Current Breaker with Overload protection). It is a device which is part circuit breaker and part residual current breaker and can be obtained in various residual current ratings – see later in this section.

Table 7.5 reviews six of the commoner types of over-current protection devices together with how they are applied to an installation and relative comments.

There are advantages and disadvantages of using certain types of over-current protection. Table 7.6 reviews six of them.

Table 7.5 – *Common types of over-current protection devices*

Over-current device	Application and comment
Fuse (semi-enclosed) to BS 3036	Often referred to as rewireable fuse. Used in low cost electrical installations. Easy to see when the fuse element has ruptured. There are three types which correspond to three different prospective fault-current values. Not to be used in temperatures exceeding 35°C. It is easy to fit the wrong size fuse element
Small cartridge fuse to BS 1362	Fitted to UK 13 or 10 amp plugs to protect the flexible cable against over-current conditions. The larger version of this fuse (BS 1361) is used as a power supplier's fuse, and is usually rated at 100 amps in domestic installations
High breaking capacity fuse (HBC) to BS 88	This fuse is a large cartridge fuse with fixing lugs attached to either end-cap. The body of the fuse is made from a high grade ceramic material in order to withstand heat and the interruption of heavy circuit current when the element ruptures
Miniature circuit breaker (MCB) to BS 3871 and 60898	Used in most installations these days. Looking like a switch, it is very easy to see whether an over-current condition has occurred as the device will automatically switch to its 'OFF' position (switch pointing down). The supply to your circuit can be quickly restored. They are tamper proof and many different types and current ratings are available. Use type 'B' for domestic installations, type 'C' for industrial circuits and type 'D' for motor and large site transformer installations. Single or three phase models are obtainable
Moulded case circuit breaker (MCCB)	Often used to protect supply cables serving distribution centres. Similar but bulkier than a MCB and obtainable in three or single phase format. The MCCB can be selected to withstand a far greater circuit current and has a superior short circuit capacity. It is designed to uphold high levels of current flow indefinitely
Residual current breaker with overload protection (RCBO)	This device is part miniature circuit breaker and part residual current device. Obtainable in various current and residual tripping ratings the RCBO has two functions: (1) to protect against earth fault currents and (2) to guard against overload currents. Because short circuit sensing is only on the phase side of the device, you must make sure that both the phase and neutral conductors are properly connected

Table 7.6 – *The advantages and disadvantages when using over-current devices.*

Over-current device	Advantages gained	Disadvantages found
Semi-enclosed fuse	Cheap to buy. No moving parts. Easy to see when the element has ruptured. Inexpensive when replacing the element.	There is no quick way to replace the element. The fuse could blow in your hand if there is a fault within the circuit when replacing. Any size fuse can be installed. The fuse element will deteriorate with age. The fuse can cause damage if the short circuit is harsh
Cartridge fuse	Precise current rating. The fuse will not weaken with age. Usually small and easy to handle. No moving parts.	Not recommended for high current circuits. It can be replaced with an incorrect cartridge fuse. Far more costs are involved when replacement is needed. It is easy to short-out using tin foil or silver paper.
High breaking capacity fuse	It is straightforward to see when the element has ruptured. Very reliable. Distinguishes between heavy currents of short duration (for example: a 10 kVA site transformer being switched 'ON') and a genuine fault current.	These fuses are expensive to buy
Miniature circuit breaker	You can see at a glance which circuit is faulty. Values cannot be altered. Your circuit can be quickly re-established after it has auto-switched to 'OFF'. The device is tamper proof. Types 'C' and 'D' will only auto-trip on a continuous overcurrent condition. Lots of different current rating ratings are obtainable.	They have moving parts which can wear. Temperature variations can affect their electrical characteristics. Expensive to buy. They need to be regularly tested to ensure they are operating as intended
Moulded case circuit breaker	The same as MCB protection	The same as MCB protection
RCBO protection	The same as MBC protection. The device will protect against earth fault currents and sustained over-current problems. Can be obtained in a variety of different residual current values and switching current ratings.	The same as MCB protection

Earth fault protection devices

There are four principal types:

- residual current device (RCD),
- residual current circuit breaker (RCCB),
- residual current breaker – with over-current (RCBO),
- earth leakage circuit breaker (ELCB) – of which there are two types:
  - voltage operated (v),
  - differential current operated (c).

The residual current device

Residual current devices are divided into two categories:

- *Electromechanical* - this type uses the output energy generated from a supersensitive current transformer (sometimes called a sensing coil) to energise a highly responsive relay as illustrated in Figure 7.28.

Under a 'no-fault condition' the load current flowing both in and out of your RCD is totally balanced. Load current enters both phase and neutral coils wound around a laminated ring-like transformer known as a toroidal transformer. The two coils are wound in opposite ways which in a healthy, no fault circuit, would produce equal and opposite magnetic flux – no current would be induced into the current transformer. When an earth occurs within the installation – and this can be

Figure 7.28 – *An electromechanical residual current device (RCD)*

either a phase or neutral conductor leaking current to earth – your RCD is electrically thrown out of balance. This allows a tiny current to be induced in the fault sensing coil and it is this small output which provides the signal to the tripping relay which operates by simply cancelling the permanent magnetic flux generated in the windings of the sensing relay. Once energised, a mechanical device is triggered and your RCD is switched 'off', making the installation completely safe.

The sensitivity of the device is directed by the type of toroidal transformer and sensing relay that is fitted at the design stage.

RCD's are made in a wide range of sensitivities from 2.5 to 1000 milliamp. Variable load current ratings are also obtainable.

■ *Electronic* – this type of device puts to use its built-in electronic circuitry, powered from the mains supply, to detect earth fault current which, in turn, stimulates the mechanical tripping mechanism of the RCD. This works by electronically amplifying the output of the fault-sensing coil to operate the shunt trip. This effectively switches 'off' the current and makes your circuit safe.

In general, an electronic electrically amplified RCD is far slower than its electromechanical

equivalent and has the added disadvantage that it is voltage-dependent. An electronic RCD is prone to nuisance tripping because not only will it detect earth faults but will also monitor problems with the incoming supply.

I am afraid there are three other sub-categories for you to remember!

– type 'AC' for use for protection against alternating circuits,

– type 'A' for protection against AC and pulsating direct current circuits,

– type 'B' which will detect and protect against pure direct current earth fault circuits.

Residual current circuit breakers

This form of protection is a mixture of a mechanical switch and added residual current device to provide protection against hazardous earth fault currents. Sometimes found on distribution centres; Figure 7.29 illustrates this.

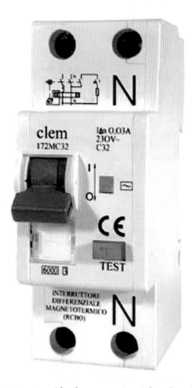

Figure 7.29 – *A residual current circuit breaker (RCCB)*

How it works

The device operates by comparing the current entering the RCCB to the current leaving it. When a circuit is working as intended the current supplied through your phase conductor is the same in value as the current returned to the system by way of the supply neutral. If a fault condition is identified, some of the fault current will escape by means of the earth wire and the amount routed through the neutral conductor will be reduced. The difference between these two values is known as the *residual current*. The sensitive winding, wired in series with the neutral conductor, becomes far weaker, magnetically, than its phase counterpart. When an iron rocker mechanism is triggered by the weakening neutral magnetic field it opens the RCCB's load switching contact which immediately isolates the supply from the load.

Line current ratings are 25, 40 and 63 amps. Residual current tripping values are 30,100, 300 and 500 milliamps.

The residual current breaker with overload current protection

An RCBO is an ideal device to protect against dangerous earth leakage current, short circuit or overload conditions – Figure 7.30 illustrates this. As it can be used to provide both forms of protection, it is ideally positioned to carry out tasks where additional equipment would

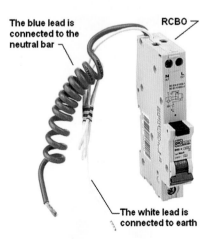

The blue lead is connected to the neutral bar

RCBO

The white lead is connected to earth

Figure 7.30 – *An RCBO will not take up too much room within your distribution centre. It will protect against residual currents to earth and over-current*

normally be required – for example, if all circuits connected to a TN-S distribution centre served by a TN-S earthing arrangement are lighting but one is required to serve an immersion heater or radial socket outlet. Additional protection can then be given without the expense of installing a suitable residual current device to monitor the complete distribution system.

Short circuit sensing is only applied to the phase conductor so it is very important to wire correctly as shown on the face of the device. Some are manufactured pre-wired, which is very handy.

Nuisance tripping

Nuisance tripping is not when a residual current device switches itself to the 'off' position when doing the job it was designed to carry out – it is irrational tripping for no apparent reason. There are many reasons why this should happen. Please find time to browse through a few of the more common reasons:

- switching 'on' or 'off' rows of fluorescent lighting,
- long runs of mineral insulated cable (MI) can act as a capacitor and activate the device,
- spikes or abnormalities in the supply voltage will set off the trip,
- isolating large electric motors can, sometimes, cause the device to act as though a fault condition to earth had occurred,
- using the wrong type of radio-frequency interference filter (RFI) will cause your trip to go out,
- lightning strikes targeted at your supplier's service cables will set off your RCD – unless it is fitted with a built-in time delay,
- vibration often causes RCD's to mechanically switch 'off',
- large current surges lasting only milliseconds can cause disruption.

Other problem areas can be caused by a deteriorating cooker plate, immersion or night storage elements after a critical temperature has been reached.

Lastly, a noisy residual current device often indicates a forthcoming fault or nuisance condition to happen.

Earth leakage circuit breaker (voltage operated)

How it works

Any fault current to earth is routed from the appliance through the circuit protective conductor to the common earth bar within your distribution centre. Figure 7.31 illustrates this. From this position it follows a route to the voltage sensing coil within the ELCB. Once it travels through the coil it discharges itself to the general mass of earth and on doing so both energises and activates the relay mechanism. This action effectively disconnects the incoming supply from the load.

This type of device has now been phased out but can still be seen in rural areas where a small amount of electrical installations have never been brought up to date.

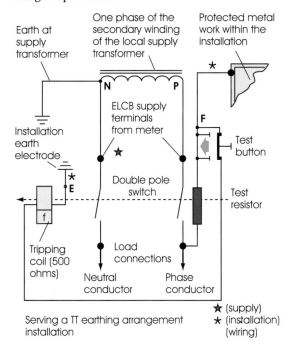

Serving a TT earthing arrangement installation

★ (supply)
★ (installation)
(wiring)

Figure 7.31 – *A voltage operated earth leakage circuit breaker*

Earth leakage circuit breakers – differential current type

How it works

The principle of the earth leakage circuit, current operated, circuit breaker is very similar to the residual current device.

Under normal working conditions the switching contacts are closed to allow for the passage of current to your load. Inside the device the main load bearing conductors pass through a round, doughnut shaped current transformer. These are known as CT's. As with the RCD, the current operated ELCB is fitted with a secondary winding. This is called the *sensory winding*. Under a 'no fault condition' situation current entering the device will be the same in value as the current leaving to return to the supply.

As the current generated is equal but opposite in both directions, a sensing voltage will not be induced in the secondary winding. However, under a fault condition, phase or neutral current leaking to earth will cause a measurable imbalance within the current transformer and this is sufficient to induce a small current into the secondary winding. If the current is above a predetermined leakage level it will cause the device to trip. This effectively isolates the supply from the load.

The dangers associated with the use of electricity

A few common sense thoughts

There are many dangers associated with the use of electricity. Ways of controlling these risks follow in random order:

- If possible, avoid working alone, especially when remote from others. If a problem arises there will always be someone there to help.
- Make sure the equipment you are using is in good working order. Get appliances tested regularly – it could save your life!
- Working when you are tired can be the cause of many a problem.
- Keep to 110 volts sources of power, even when working in an occupied premise.
- Keep physically away from metal pipes and ducting which are usually bonded to earth when working in confined spaces and lofts.
- Take great care when working with rechargeable batteries (secondary cells). The voltage will be low but the amperage can often be very high – more than enough to burn you or start a fire. Never smoke whilst working with secondary cells for fear of an

explosion, caused by free hydrogen evaporating from the plates and igniting.

■ Use a plug-in type residual current device to serve your 110 volt output transformer when working in someone's house.

Controlling risk with isolators and switching

You can reduce accidents involving electric shock by making a routine check of your circuit before making your final connection. Another way is to simply 'lock-off' the switch gear which provides electricity to your installation. If a locking facility is unavailable then either remove the switching bar from your isolator or take out the fuses within the local switch fuse. Whatever you take away, always keep in a safe place for when they are needed again.

Means of isolation must be provided within the following parts of your installation and these can be made good use of to help reduce the risk of getting an electric shock:

■ At the start of your installation, remembering that you must install a double pole isolator if the earthing arrangements are either 'TT' or 'TN'. When dealing with a three phase system your isolator must separate all three phases and the neutral conductor from when the earthing arrangement is 'TT' but within a 'TN-S' or 'TN-C' system, only the phase conductors need to be interrupted from the supply. Provision must be made for the neutral conductor to be mechanically disconnected. This is usually carried out by means of a removable copper link already built into the isolator at the manufacturing stage. Regulation 460-01-06 will confirm this.

■ Each of your circuits must have a means to isolate it from the electric supply. Examples: lighting arrangements within a room or a local accessible isolator serving an electric motor and any associated equipment. Please refer to Regulation 131-14-02.

Controlling the risk of direct and indirect contact with electricity

Direct contact

The term *direct contact* really means what it says – touching live conductive parts and this is something we all need to avoid. There are many things which we provide during the course of our installation work to maintain this safeguard. Please spare a few seconds to browse through the following list:

■ enclosures (insulated and steel),

■ barriers built during the course of the installation,

■ SELV and PELV systems.

Protection against direct contact within SELF and PELV system needn't be made available if the voltages you use do not go beyond the following laid out in Table 7.7.

Please note that non-sheathed single insulated conductors must be protected by cable trunking or conduit as demanded by Regulation 521-07-03. Alternatively, they may be placed in a suitable enclosure.

Indirect contact

We obtain our protection from indirect contact with electricity by limiting damaging earth fault voltages to a safer value and by making them far shorter in duration – after all, a 230 volt shock

Table 7.7 – *Protection against direct contact for SELV and PELV systems*

Locality of your circuit	SELV systems	PELV systems
Areas which are dry. Regulation 411-02-09 explains further.	25 volts AC/60 volts DC	25 volts AC/60 volts DC
Swimming pools, saunas and bathrooms. See also Regulations 601-03-2, 602-03-01 and 603-03-01.	You will need protection at all voltages	You will need protection at all voltages
All other areas. Regulations 471-01-01 and 471-14-02 explains further.	12 volts AC/30 volts DC	6 volts AC/15 volts DC

will do you more harm over a five second period than 40 milliseconds.

Other protections which add to our safety include the following:

■ The use of Class 2 equipment (where exposed metal work is not bonded to earth and supplementary insulation is used as protection).

■ Installation of residual current devices to allow for a far shorter disconnection time. BS 7671 recommends 0.04 seconds or less.

■ A complete harmonization of all protective devices and individual circuit impedances together with the bonding of all exposed and extraneous conductive parts. (As a memory jogger: *exposed* refers to the cladding you would find on an appliance such as a fridge or washing machine whereas *extraneous* is relevant to metal work not associated with your electrical installation, for example, water pipes and air ducting systems, etc.).

■ Separated extra-low voltage systems (SELV) – supplied from a safety isolating transformer which has no parts connected to earth or any other protective conductor and are placed within an insulated pocket added to their usual insulation. Regulation 411-02-07 confirms this.

■ Protected extra-low voltage systems (PELV) – this system is similar to SELV apart from circuits are not electrically separated from earth.

■ EEBADS systems within your installation.

■ Protected by supplementary equipotential bonding.

Getting your installation terminals right first time

Points of termination must be both electrically and mechanically sound when conductors are joined together. A loose-fitting connection will always give you problems when current is drawn from the circuit. Arcing will occur and once started it is only a matter of time before the accessory or joint is burnt or destroyed.

A few practical points to observe:

1 Over-tightening can cause the threads on the terminal screw to shear, thus making it impossible to tighten the terminal as intended.

2 Never leave too much copper exposed at the point of termination. It will cause problems when other conductors are added. Figure 7.32 shows the best way to terminate your conductors.

Aluminium conductors serving domestic installations

There is not much call these days for domestic, aluminium based cabling – but it is still available by order and can be seen from time to time in older installations. Aluminium is much softer than copper and will break easily if mishandled. Take care if you have to re-terminate older circuits as the terminal screw serving an accessory could snap the wire if too much pressure is placed upon it. Your error may not be noticeable until you switch back 'on' to re-establish your circuit.

Terminating flexible cables

Woven braided flexible cables often used for fancy decorative lighting arrangements regularly become frayed at the point of termination. This unsightly problem can be solved by one of the following methods:

■ by placing expandable rubber sleeving near to the end of each conductor, making sure that the frayed bit is completely hidden,

■ by the use of heat-shrink sleeving placed near to the end of each conductor. Shrink your sleeving snugly into place with an industrial hot air blower – never be tempted to use a flame,

■ bit mastic/wax can be smoothed over the frayed braid – a bit messy, but it works!

The method you choose will of course depend on the status and the amount of time and money allowed for your job.

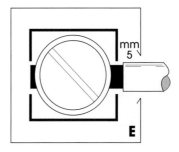

Figure 7.32 – *Expose just sufficient copper to terminate your conductor*

Simple joint box terminations

It might seem easy but it is possible to make mistakes so please browse through the following practical suggestions:

1 Tailor-make the length of your conductors to fit neatly within your joint box. Avoid masses of unwanted surplus cable within and around the point of termination.

2 Under no circumstances try to use conductors which are too large for the terminal you are working with. A 30/32 amp joint box (JB) will have room for 1.0, 1.5. 2.5, 4.0 and 6.0 mm^2 copper conductors.

3 All your terminals must be totally accessible and sound and be able to withstand the design current (the total load of your circuit). Twenty amp joint boxes are used for lighting circuits, 30 amps for power – not the other way around!

4 It is good practice to introduce about 4 or 5 mm of outer sheath into your joint box. The insulation serving the conductors must never be seen outside of the joint box as Figure 7.33 demonstrates. This rule also applies to socket outlets, ceiling roses, switches and fused connection units, just to mention a few.

5 When preparing your conductor, just remove enough insulation so that the bare copper will fit safely and neatly inside the brass terminal housing. If too much inner insulation is removed it could cause problems in the future. Don't forget to over-sleeve your protective conductor.

6 All your prepared terminals must be both electrically and mechanically sound. Loose terminals will lead to arching whilst on load.

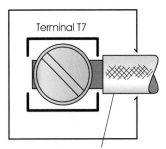

Terminal T7

Place about 4 to 5mm of sheath with your joint box

Figure 7.33 – *Place between four to five millimetres of cable sheath within your joint box*

Arching produces heat which will lead to a steady breakdown of the cable insulation, soften the copper conductor and destroy the termination. It is best to check the tightness of the joint before screwing your lid onto the base section of the joint box.

7 Make sure you have fitted the lid correctly and there are no unwanted cable ports visible.

8 It is good practice to fit a circuit destination label on the lid of your JB – this will help others who follow.

A guide to cable terminations

■ The size of a cable terminal reflects its current carrying capacity. This means a large terminal has the ability to carry a large amount of current and a small terminal, a small amount of current.

■ It is wise to know the rating in amps of the terminal so that any potential problem can be avoided.

■ Terminals rated too low for the current they carry will over-heat, become damaged and eventually burn out. More often than not the heat generated will destroy the insulation around the copper conductor,

Making a professional cable termination

Careless workmanship really can affect the quality of your termination. Here are a few points to consider:

■ Consider it bad practice to expose too much bare copper beyond the point of termination.

■ Under tightening a cable terminal will cause arching to occur on load.

■ Cable terminals can be damaged by using the wrong size screwdriver.

■ Blunt or damaged tools will not tighten terminals efficiently.

■ Heat softened terminations will lose their mechanical strength.

Terminations you will come across as an electrician

As you can imagine there are lots of different cable terminations used for plenty of different jobs within electrical installation engineering. To keep to the C & G scheme you will find them in future paragraphs within this chapter.

Dangers associated with the location of equipment and components

There are many things to consider when deciding where to position equipment and components and where to route your circuits within your installation. Here are a few to consider:

- *Basic requirements* – BS 7671 requires all the equipment you install to be correctly assembled and be capable of being inspected, tested and regularly maintained as far as it is practical.

- *The accessibility of your equipment* – remember to give proper attention to all your joints within any enclosures and the final terminations serving your equipment. Our regulations demand that they are accurately connected and are judged to be suitable in every way regarding your installation, even if these connections are within site mounted enclosures or compartments. This rule will not apply to an epoxy-resin cable joint where it would be impossible to open it up for routine inspection.

- *Positioning equipment used for outdoor use* – much of this is common sense; for example, it would be foolish to position a passive infra-red detector near to or above a gas central heating vent. The hot ventilated fumes would be constantly switching the device to 'ON.' Keep outside sockets away from potential stormwater or water taps. Make sure that all IP 56 rated equipment you install is located sensibly and is free from possible damage by impact. Make sure that all the components you use for your installation are fit for the job intended – for example, never fit domestic joint boxes within an outside or semi-outside installation. It is wise to protect your circuit with a 30 milliamp rated residual current device and keep to the published limiting value of earth fault loop impedance.

- *Protection of equipment against fire and/or explosion* – fit flame proof fittings within a potentially explosive environment and if you have to situate equipment within a location containing flammable liquid in excess of 25 litres please install within a fire resistant cabinet or chamber and ventilate this chamber to the atmosphere outside of the storage facility. Your builder must provide for adequate drainage to accommodate for any spilt flammable liquid. Fit extractor fans to battery charging rooms to prevent a potential build-up of explosive hydrogen gases.

- *Equipment within agricultural installations* – it is very important to keep equipment and components well out of reach of livestock. This rule is most important within a milking parlour where cattle can nibble at electrical cables and other electrical equipment. Keep powerful halogen lighting directed away from flammable stock such as hay and straw. Use only IP 56 rated equipment out of doors. Select and locate your luminaries sensibly in areas that are regularly hosed down.

- *The positioning of hot luminaries* – extra-low voltage lighting fittings must be positioned wisely. They produce a lot of unwanted heat and must not be directed at any potential fire risk. Place ceramic or special fabric covers over the luminaries when fitting to the ceiling – Figure 7.34 illustrates the fabric type. Please ensure that your fitting is not covered with thermal insulation as this will compound problems even further. Finally, check the tightness of the screws which serve the small connector block on the low voltage fitting and that your joint box is arranged to your satisfaction.

- *The need to position in a well ventilated location* – this is essential, especially when installing electric motors or designing an area to house battery charging equipment. It is

Figure 7.34 – *Where wooden flooring exists, fit thermal protection covers over extra low voltage and mains operated decorative spot lights; they are very hot!*

equally necessary to supply ventilation when the equipment you install has a normal working temperature exceeding 90°C.

■ *Accessibility of equipment and components* – this really means your equipment must be nice and easy to get at. Avoid installing consumer units within a tiny cupboard under the stairs. It is far better to mount within a full sized cupboard to be found in the hallway. Always provide a small trap to enable joint box connections to be checked or tested – never leave under a tiled floor.

■ *The environmental positioning of equipment out of doors* – make sure that it is IP 56 rated and free from any potential danger of damage through impact, collision or interference by animals. The use of a residual current device is preferred for additional safety. Make sure your equipment is well protected if it hasn't been designed expressly for outside use.

■ *Situating equipment with a view to regular maintenance* – when positioning your equipment remember that maintenance will be required from time to time so make sure it

is readily accessible and can be maintained in safety. For example, it might be considered unwise to install luminaries on a high ceiling serving a stairwell.

■ *Location of equipment within a bathroom or kitchen* – your bonding clamps must be readily accessible to enable maintenance and testing to be carried out. Position any unguarded lighting point well away from the zone nearest to the bath (zone 0) and position wisely any extractor fan or fixed heater you install to serve the bathroom. Keep socket outlets away from kitchen sink units. Extra-low voltage lighting units should be carefully located so heat damage can be totally avoided. Do not incorporate a lighting transformer within a flame proof protection cover or position very close to the lamp.

Wiring systems and enclosures for installations

Table 7.8 identifies appropriate wiring systems and enclosures for chosen installations.

Table 7.8 – *Wiring systems and cable carriers*

With reference to . . .	Suggested wiring system	Proposed enclosures	Comments
Using the structure of the building	1. Cable duct for power distribution cables 2. Steel infrastructure for installing steel wire armoured cables 3. Service risers for running power distribution cables – remember the fire barriers! (Regulations 527-02-01)	Steel enclosures are ideal for this type of installation – trunking and conduit. Reference is made to items 1, 2 and 3	1. Your cables will be drawn into the duct which forms part of the infrastructure of the building – but not fixed. Make sure that water cannot get in at the entry point of the cable duct. 2. Use 'Hit-Clips'® as a way of fixing to the steel work. 3. Use cable tray or steel trunking as a suitable cable carrier.
Environmental considerations 1. Ambient temperature 2. Where water and dampness is present 3. Pollution and corrosion problems 4. Mechanical stress 5. Solar radiation	1. Select your cable to serve the highest operating temperature likely to be present. 2. If your installation is within a wet environment then it must be designed to withstand such conditions. If damp, then ensure that all your termination points are damp proof. Best to use PVC-u trunking and conduit systems.	1. Select enclosures which are suitable to withstand the highest working temperature – steel conduit and cable trunking for example. If your installation is in a heavily dusty environment then your enclosures must offer a degree of protection to IP 5X 2. IP 56 enclosures but avoid aluminium conductors being placed in contact with brass or copper terminals unless	1. Mineral insulated (MI) cables for use in very hot environments – Artic grade cables for extremely cold conditions. See also Section 522-01 of BS 7671. 2. Avoid electrolytic contamination occurring. Never allow dissimilar metals to come into contact with one another. Provide drain-off points within a conduit or trunking installation

Table 7.8 – *continued*

With reference to . . .	Suggested wiring system	Proposed enclosures	Comments
	3. If at all possible, do not expose your installation to corrosive substances but if not practical select the type of system that would withstand a corrosive atmosphere. 4. All systems 5. Mineral insulated cable systems, steel cable-carrier and glass reinforced polyester systems are examples.	the aluminium is suitably plated. BS 7671 522-05 confirms this. Use galvanised or PVC-u cable carrying systems 3. PVC-u enclosures and cable carriers or glass reinforced polyester will be ideal. 4. Compatible with your wiring system. Epoxy-resin joints for SWA cable installations. Steel enclosures and cable carrying systems for industrial installations, etc. 5. Compatible with your chosen installation. If you are to use a plastic based cable carrier, please check that UV light will not cause problems in the future.	3. Materials which will attack include: magnesium chloride (found in floors and dadoes), unpainted walls which contain lime plaster and cement, corrosives salts, acidic wood such as oak and dissimilar metals such as copper and aluminium. 4. Support is required for your cables and carriers within vertical cable ducting. Do not exceed more than 5 metres without intermediate support. Please referrer to Appendix 4 of BS 7671. A 0.5 mm^2 flexible cable will support a maximum of 2 kg; 0.75 mm^2 a mass of 3 kg and 1.0 mm^2 a total mass of 5 kg 5. Ultra-violet light (UV) will discolour the sheath of PVC insulated cables and eventually crack and destroy the insulation. Any other PVC products used within the electrical industry will suffer the same outcome.
Current demand	All systems	Compatible with your chosen installation	BS 7671 demands that all equipment installed must be suitable for the maximum power requirements by the current using equipment. This will mean paying particular attention to the cross sectional area of your conductors. When calculating the size you will be required, please take into account the four correction factors (ambient temperature, thermal insulation, cable grouping and the factor we use for a semi-enclosed fuse)
Over-current protection Please refer back to Table 7.5 where you will find details concerning the various types of over-current protection reviewed within the last section of this table	All systems – general overview	Compatible to your chosen installation	1. Semi-enclosed fuse (BS 3036) 2. Cartridge fuse (BS 1361 and 1362) and equivalent 3. High breaking capacity fuse (HBC) (BS 88), see Figure 7.35 4. Miniature circuit breaker (MCB) (BS EN 60898) 5. Residual current breaker with over-load protection (RCBO), see Figure 7.36 6. Moulded case circuit breakers (MCCB)

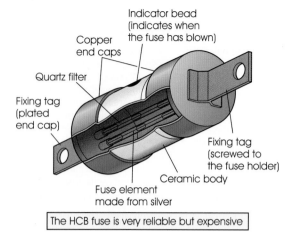

Indicator bead
(indicates when
the fuse has blown)

Copper
end caps

Quartz filter

Fixing tag
(plated
end cap)

Fixing tag
(screwed to
the fuse holder)

Ceramic body

Fuse element
made from silver

The HCB fuse is very reliable but expensive

Figure 7.35 – *An HBC fuse*

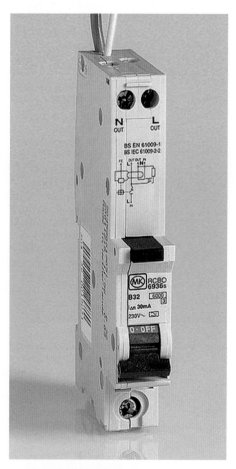

Figure 7.36 – *An RCBO device*

The advantages and limitations of common wiring systems and cable carriers

There are many wiring systems to choose from but it must be right for the job. For example, it would not be wise to use PVC sheathed cable clipped directly to a surface to serve an industrial installation nor would it be sensible to bury this type of cable underground as a way of providing temporary power to a construction site. If you are a position to have to an option of choice then think carefully – it could save you a lot of money in the long run.

Table 7.9 provides you with an overview of the advantages and limitations of using a chosen wiring system.

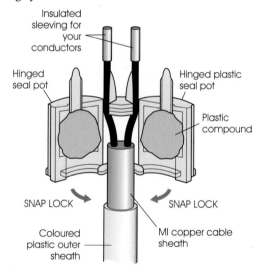

Insulated
sleeving for
your
conductors

Hinged
seal pot

Hinged plastic
seal pot

Plastic
compound

SNAP LOCK

SNAP LOCK

Coloured
plastic outer
sheath

MI copper cable
sheath

The cable is prepared as normally. The plastic pot is
fitted around the prepared end as shown. When locked,
a clear fitting sleeve is placed over the pot

Figure 7.37 – *A mineral insulated cable termination from British Insulated Calendar Cables (BICC)®*

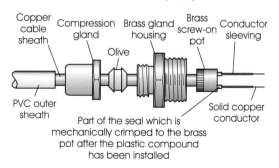

Copper
cable
sheath

Compression
gland

Brass gland
housing

Brass
screw-on
pot

Conductor
sleeving

Olive

PVC outer
sheath

Part of the seal which is
mechanically crimped to the brass
pot after the plastic compound
has been installed

Solid copper
conductor

Figure 7.38 – *An MI cable termination*

Table 7.9 – *The advantages and limitations of wiring systems*

Wiring system	Main advantages	Disadvantages	Comment
PVC sheathed cable installation	Inexpensive compared to other systems. Quick to install. Basic skill requirements needed	Difficult to add to when wiring modern flat and houses. Open to accidental damage by other trades. Must have all points of connection accessible when the installation is complete	It is important that you plan your circuits before you start your first fix. Remember it is wise to have all the kitchen appliances wired to their own final ring circuit – but not supplementary protected by a RCD but ideally a Type 'B' 32 amp MCB
Mineral Insulated cable installation	A good all round cable which will resist impact damage and can be installed in very hot environments (a temperature up to 343 K). Many sizes and core numbers available. A simpler cable termination is available from BICC Pryontex® – Figure 7.37 illustrates this	Expensive and a high degree of skill required to carry out the installation. Much planning is needed when wiring. Time consuming. Many accessory boxes are necessary. Special tools are needed to carry out your installation	A good knowledge of MI wiring systems is essential. Figure 7.38 Illustrates a MI cable termination and Figure 7.39 how the copper sheath is removed in preparation for the cable termination. Figure 7.40 illustrates how the copper can be neatly removed when using basic side cutters
PVC/SWA cable installations	Robust and versatile. Ideal for power requirements in industrial and commercial work and temporary installations serving construction sites. Easy to terminate smaller sizes. No special tools are required	Can be heavy to install, sometimes requiring two operatives or more. The steel wire armouring is not strong enough to withstand accidents involving heavy construction plant. When buried in the ground it requires the additional protection of sand together with a yellow warning tape (Hep-tape), see Figure 7.42, placed some 140 mm above the cable or alternatively, cable tiles may be placed above the installed cable as shown in Figure 7.43 When wiring for temporary installations it is wise not to rely on the steel wire armouring for your principal earth conductor – use a spare core instead	Always dispense cable from the top of the wooden drum. Cable terminations are carried out as illustrated in Figure 7.41. Use a hot air blower to soften the outer PVC sheath in very cold weather. This will allow for safer cutting. Remember to sandwich a brass earthing tag between the armoured cable gland and the enclosure. This will allow for a supplementary circuit protective conductor
PVC insulated single core cable installations	Ideal for cable trunking, conduit and panel wiring installations. Easy to install. Colours and sizes available to suit the needs of your installation. Your installation can be added to, rewired and mixed sizes of cable can be used	Errors can occur with cable identification. Conductors can become trapped or friction-burnt when installing. Too many conductors installed within a cable carrier will sometimes cause warming to occur when the circuits are on load. You must think well ahead if wiring errors are to be avoided. A high skill factor is required	This is an ideal wiring system but you will need to build on your experience before carrying out such an installation

Table 7.9 – *continued*

Wiring system	Main advantages	Disadvantages	Comment
Fire retardant cable installations	An ideal cable to use for fire alarm and central battery emergency lighting circuits. It can also be used for other sensitive circuits which require protection from potential fire. It is light and easy to install and its red sheath colour identifies it from other cables. The insulation serving the cable cores is made from silicone rubber which, under conditions of extreme heat, chemically converts into silicon dioxide. This then becomes a first rate insulator. Lower smoke, zero halogen cable can also be used.	The cable can be expensive. You must be very careful when removing the outer sheath not to snag the insulation around the conductor – it is easy to tear. It can have a mind of its own if not unwound from the drum correctly. If your emergency lighting cable is clipped to the surface you must separate it from other cables by a minimum distance of 300 mm or you can use MI cables for your installation. Fire retardant fire alarm cables can be installed within a cable carrier especially reserved for fire alarm circuits but if this is impractical they must be separated from other circuits by a minimum of 300 mm	For additional information regarding the segregation of fire alarm and emergency lighting (central battery system) cables to cables serving other purposes, please refer to Table 7.4 in the On Site Guide. Some brands of fire retardant cable use a small ferrule at the end of the cable. This is to protect the insulation from potential damage caused by the metallic sheath – Figure 7.44 shows this. Self-contained emergency lighting units are not considered to be a part of an emergency lighting installation and can be wired as you would a lighting point

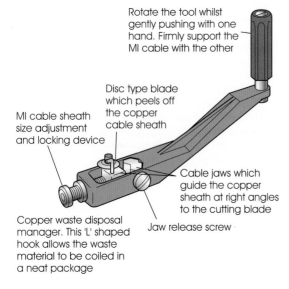

Figure 7.39 – *Removing the copper sheath from MI cable*

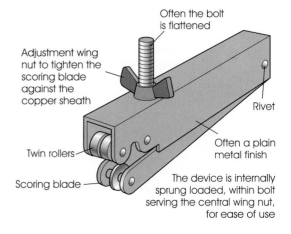

Figure 7.40 – *Ring around the MI cable sheath when using a pair of side cutters or an old fashioned 'stripping rod' to remove the copper sheath. Scoring a light groove around the copper cable sheath with this device will allow you to neatly cease your stripping task once arriving at the furrow*

The pros and cons of cable carriers and enclosures

Table 7.10 provides a brief overview of the advantages and disadvantages of various cable carrying systems.

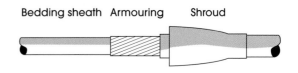

Figure 7.41 – *Installing an SWA cable gland*

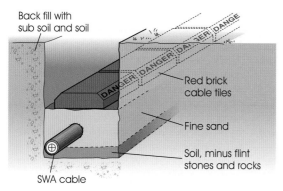

Back fill with
sub soil and soil

DANGER DANGER DANGER DANGER

Red brick
cable tiles

Fine sand

Soil, minus flint
stones and rocks

SWA cable

Check out the BS 7671 for the depth of the cable
(The depth will depend on the voltage carried)

Figure 7.42 – *Plastic hep-tape can be placed above the buried cable instead of tiles*

Figure 7.43 – *Red brick cable tiles*

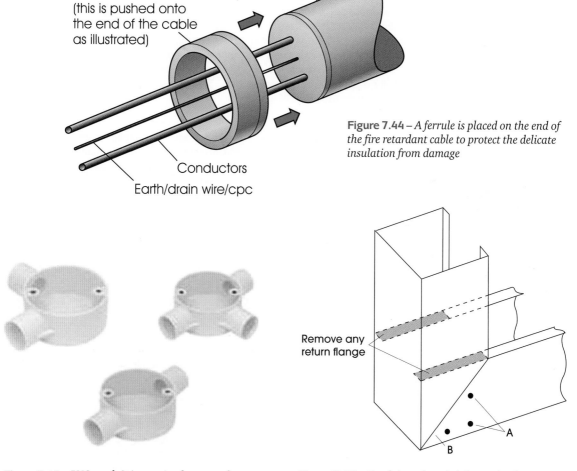

Cable

Ferrule
(this is pushed onto
the end of the cable
as illustrated)

Conductors

Earth/drain wire/cpc

Figure 7.44 – *A ferrule is placed on the end of the fire retardant cable to protect the delicate insulation from damage*

Remove any
return flange

A

B

Figure 7.45 – *PVC conduit inspection boxes – a few examples*

Figure 7.46 – *Conduit and conduit inspection boxes cast in concrete*

Table 7.10

Cable carrier/enclosure	Advantages	Disadvantages
1. PVC-u conduit systems	1. It is light and easy to use. The skill fact or for installation is considerably lower. PVC-u conduit will not rust. It is far quicker to install than steel conduit. Thread making tools (dies) are not required. All joints are glued into position. Bends and sets can be made quickly. Not so many tools are required to do the job. Can be easily altered if required once installed to first-fix status. Plenty of inspection boxes are available as shown in Figure 7.45	1. PVC-u conduit requires a circuit protective conductor. It will warp and bend with heat if safeguards are not carried out. This system can be over-accommodated with cables. Plastic conduit is not flame proof. PVC can be affected by UV light over a long period of time. No good to use in hot environments. Often requires additional support. PVC-u conduit is not as strong as steel conduit systems
2. Steel conduit systems	2. Can be rewired or added to. An excellent protective conductor. An option of four finishes – black stove enamelled conduit, galvanised, aluminium and stainless steel. Galvanised is ideally suited for external use. Your system can incorporate various sizes of conduit. Can be wired using many different sizes of cable. Can be cast in concrete as shown in Figure 7.46. Can be used in hazardous situations. Steel conduit is strong. An excellent choice of screw on inspection boxes, similar to those used for PVC-u condui	2. High material and labour costs. Additional man hours are required to carry out the task. Careless or defective joints can cause large current drains should a fault condition to earth occur. Bushes and couplers can work loose during the lifetime of the installation – maintenance is required. If broken or damaged, steel conduit can be a less effective circuit protective conductor. Stove enamelled conduit will corrode gradually if not maintained. A high skill factor is required. Can be a dirty, messy job
1. PVC-u cable trunking	1. PVC high impact trunking is light and easy to handle. Quick to install. The risk of personal injury is somewhat reduced. Plenty of sizes available – many with optional dividing fillets. Many trunking accessories as shown in Figure 7.47. Lids designed either to be screwed or snapped on. Aluminium screening can be fitted if required to protect sensitive circuits. Accessories are designed to be fitted by means of plastic rivets or permanently stuck. PVC-u trunking is made in many profiles – square, delta shaped, rectangular, etc	1. The disadvantages follow a similar vain as PVC-u conduit. If over-accommodated with cables (Figure 7.48 illustrates the space factor which is demanded), the trunking will warm up when the circuits within are on load. Plastic trunking will sag and distort in warm environments if not secured properly. CPCs will need to be included as plastic is not a conductor of electricity
2. Steel trunking systems	2. Strong – ideal for commercial and industrial use. Many manufactured accessories available but it is possible to form your own. It is made in various finishes – painted, stainless steel and galvanised are examples. Ideal for large installations. Provides for a good circuit protective conductor when copper earth links are fitted throughout the system. Can be worked accurately.	2. Accessories and trunking must match. Cable straps must be used for vertical trunking runs. You will require additional tools. Cables can become trapped between the lid and the body of the trunking causing problems. It is possible to over-accommodate your trunking with cables. You need to ensure that it is well and truly bonded throughout its entire length. Takes longer to install compared to plastic or non metallic trunking

Table 7.10 – *continued*

Cable carrier/enclosure	Advantages	Disadvantages
Cable tray and ladder systems	Cable tray is ideal for small to medium SWA and mineral insulated cables. Very large cables can be routed on ladder systems of which two types are available – heavy and medium duty. The non-metallic variety is lighter and quicker to install. There are many accessories to choose from. Figure 7.49 illustrates an external cable tray bend. Continuity bonding is not required. Polyester cable tray is ideal suited for use in high temperatures. It is also good for damp situations or sensitive circuits. There are plenty of accessories to choose from in a cable ladder system – Figure 7.50 illustrates these	Some non-metallic cable trays have a poor chemical resistance to ammonia and nitric acid – check with your wholesaler before obtaining if in doubt. It will buckle at very high temperatures if expansion has not been allowed for. Non-metallic cable will not tolerate site made accessories. It is hard to cut cable ladder with a standard hacksaw – best use power assisted means. You must leave a gap of about 2 mm between each length of cable ladder to allow for the effects of expansion; this also applies when fitting glass reinforced polyester cable ladder
1. Cable duct systems	Steel wire armoured cables laid in open (ground level) ducting systems are easy to install. Providing there is plenty of room, SWA cables can be laid within an underground ducting system reasonably easily although two or more operatives might be required to carry out the task. MI cable and steel conduit installations can be installed by fixing on to the ducting walls, preferably at a higher level	Can be over-accommodated with cable making it difficult to add more. Some poorly maintained ducting systems are open to flooding and vermin and other disgusting pests. Ducting systems are seldom illuminated nor have power point facilities within their length. They also require to be ventilated
2. Cable ducting systems	This is a system where medium to large armoured cables are laid within a manufactured enclosure made from metal, wood or an insulating material which is not cable trunking or conduit. Cable ducting is not really intended to be part of the structure of the building – this is the fundamental difference between the two ducting systems. The advantage is it can be site custom-built to your own specifications to accommodate all your power requirements	2. Things you must consider: Fire barriers need to be fitted if your ducting passes through a fire resistant structure. All joints serving your ducting system must be mechanically sound – metallic ducting can be used as a circuit protective conductor. Make sure your ducting is well supported and that any route changes will accommodate the radius of the largest size cable. For safety, site constructed ducting must be free from sharp edges
Cable basket systems	It makes installing heavy duty cables and lighter control cables a fairly easy experience. Widths are from 60 to 600 mm, so it can be adapted for a range of many applications. The square mesh wire sides are made in three heights –– 35, 60 and 100 mm. The 3 m long wire mesh lengths of basket are obtainable in an industrial paint finish, hot galvanised, pvc coated or a stainless steel finish depending on what the system is to be used for. Various fittings are available. Straightforward to install with experience.	Can be expensive. Experience is required when you are installing this type of cable carrier. You need to know what accessories are required for your installation – how many and the names of these accessories. It is difficult to install with just one operative – it is much easier with two people. Planning your route and listing the types of accessories you want is very important. Do not guess – take time to research first and you will find your job will turn out better for it!

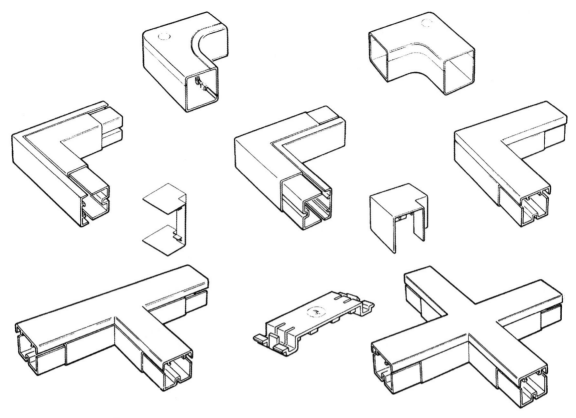

Figure 7.47 – *PVC trunking accessories*

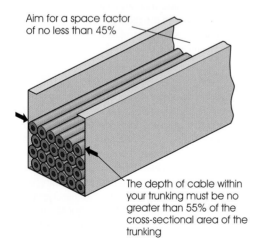

Aim for a space factor of no less than 45%

The depth of cable within your trunking must be no greater than 55% of the cross-sectional area of the trunking

Figure 7.48 – *A space factor is essential for trunking installations*

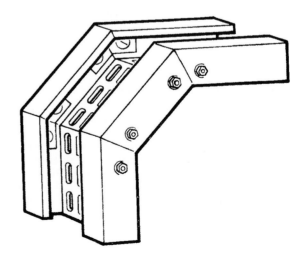

Figure 7.49 – *External cable tray bend*

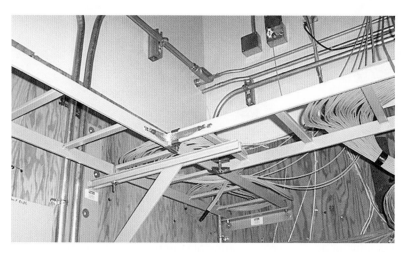

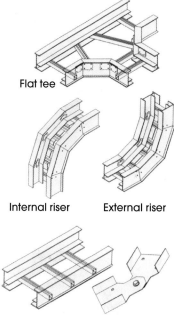

Flat tee

Internal riser   External riser

Ladder   Ladder hinge

Figure 7.50 – *Accessories for cable ladder*

Reliable cable terminations

An electrical overview

Your terminations and joints must be suitable for the type, size and insulated to the working voltage you are to use – horses for courses! Remember also to take into consideration the maximum current which will flow within your circuit and that all cable cores must be protected within a suitable enclosure. Your termination must be able to withstand the design current of the circuit. Make your compression joints with a reliable tool, or one which has been the subject of a test certificate as prescribed in BS 4579.

Other things to consider are as follows:

- You must always sleeve mineral insulated cable cores with a temperature similar to that of the seal.
- A cable gland serving steel wire armoured cable must be secure without damaging any part of the cable.
- Cable couplers (stuffing/compression glands) must be used for flexible cable or cord.

- Avoid using conductors which are too large for your termination point.
- Figure 7.51 shows how best to terminate sheathed cable into a joint box whereas Figure 7.52 illustrates how not to.
- Make sure your terminals are secured – loose terminals will cause arching.
- Label your finished joint box.
- Do not cram unnecessary conductors within your joint box – it will hinder.

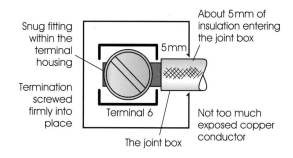

Figure 7.51 – *Pin and bullet (crimp-type) cable terminations*

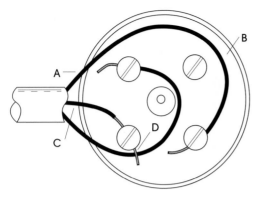

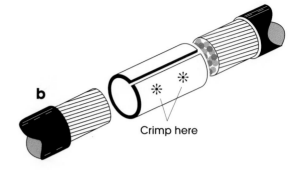

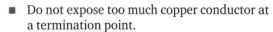

Figure 7.52 – *How you shouldn't terminate PCV sheathed cable in a joint box*

Figure 7.53 – *An uninsulated tunnel (crimped) cable termination*

- Do not expose too much copper conductor at a termination point.
- Damaged tools will lead to damaged terminal points.
- Always secure your termination enclosure to a wooden joist or masonry block.
- Protect against mechanical damage and vibration – be wise where you site your connection.
- Avoid placing strain on your jointed conductors.
- Choose a suitable size adaptable box for your joints and connections to be housed.

A simple way to check whether your joint is electrically sound is to use a reliable milliohm meter:

1 Touch the two test leads together and 'auto-zero' the display.
2 Target your probes leads firmly on either side of your joint.
3 Pressing the test button should present a value of just above zero but no greater than 0.05 ohms.
4 If higher you must remake your joint and retest.
5 As a final check, touch your test leads together to make sure that your instrument is working as intended.

Types of termination which we use

There are many different types of cable terminations which are used in our industry but to avoid over-current problems you must be aware of the current limitations and maximum voltage rating of your terminal. Here are a few examples:

- *Crimped tunnel terminations* – some are insulated, others are not, as Figure 7.53 illustrates. The insulated variety are colour coded in three separate colours: red up to 1.5 mm^2 conductors, blue will acccept 2.5 mm^2 and the yellow type 6 mm^2 conductors.
- *Bus bar terminations* – these are uninsulated and are fitted to the live copper bus bars to enable medium to heavy conductors to supply your installation needs – Figure 7.54 illustrates this.
- *Crimped cable lug terminations* – these are crimped to the ends of conductors to serve

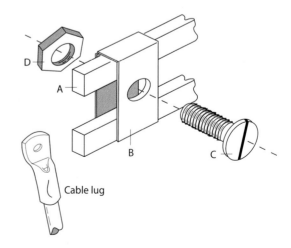

Figure 7.54 – *Typical bus bar terminations*

Figure 7.55 – *Crimped cable terminations*

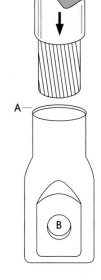

Figure 7.56 – *Soldered cable-lug termination*

large switch fuses and isolators and are made from soft tin copper as shown in Figure 7.55.

- *Soldered cable-lug terminations* – Figure 7.56 provides an illustration of a typical solder-type cable lug. These are not used so much these days but they are often seen in older installations. Made from tinned copper, solder is melted into the top of the lug. The conductor is then tinned and pushed into the cable lug whilst very hot and topped up with solder. The completed lug is wiped clean of solder and allowed to set. This system is ideal for terminating conductors serving bus-bars and larger switch gear.

- *Tag terminations (crimped type)* – insulated with hard PVC and colour coded into three groups:
  - red from 0.75 to 1.5 mm^2 conductors,
  - blue from 1.5 to 2.5 mm^2,
  - yellow, the largest of the three, from 4.0 to 6.0 mm^2 conductors.

- Obtain the best possible joint with a hand-held crimping machine; Figure 7.57 illustrates a larger type which is used for heavier cables.

- *Block terminations* – these mechanically screwed terminations are very common within the electrical industry. They are served with insulation made from polyethylene, nylon or rigid Bakelite® and are for general use.

- *Pin and bullet crimped terminations* – colour coded in three colours: red 0.75 to 1.5 mm^2, blue up to 2.5 mm^2 conductors and yellow will acccept conductors totalling a cross sectional area of 6mm^2. The male crimped termination (the pin) is pressed firmly into the female bullet termination. These are illustrated as Figure 7.58 with other examples of crimped termination hardware.

- *Miniature circuit breaker termination points* – there are two: one at the bottom of the device which serves the common distribution bus bar and the other is used as a take-off point for your outgoing circuit. Although it is *necessary* to ensure tightness, please steer clear of using a battery operated drill as a means of securing the fixing screws. When testing is carried out on your final ring circuits it might be a problem releasing the conductors using an average cross-slotted screwdriver.

Figure 7.57 – *Hand held crimping tool*

The size of your termination point

It is generally accepted that the physical size of your cable terminal reflects on its current carrying ability. In short, the larger it is the more current it is capable of bearing. One of the requirements of our regulations (BS 7671) is that the joints between flexible and rigid conductors must be easily reachable for inspection and the resistance across the joint must not exceed 0.05 ohms. But, with this in mind, inspection access is not relevant to the following joints:

1   Compression-type joints enclosed within a box made from fire-resistant material.

2   A steel wire armoured joint buried in the ground.

3   Any joint which is enclosed within fire proof building fabric but not mechanically clamped.

4   A joint which is formed by soldering, brazing or welding.

Preventing corrosion

Corrosion and pollution will cause problems to electrical installations and the longer it is left the worst it will get. Here are a few practical tips on how to avoid or control pollution – some you will have experienced yourself.

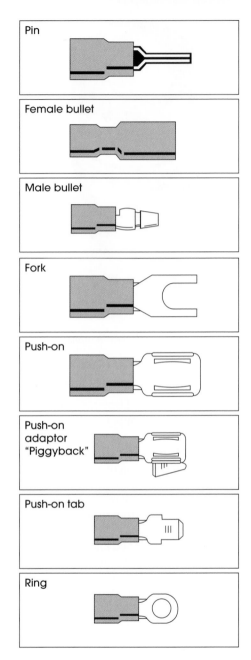

Figure 7.58 – *Pin and bullet terminations are used for extra low voltage applications*

- Never secure an aluminium conductor onto a steel, copper coated earth electrode as sooner or later corrosion will occur.

- Paint plain copper earth links when fitting to steel/galvanised trunking systems. Use 'cold galvanising' paint for this task.

- Aim to use mineral insulated cable with an overall PVC covering and plastic shrouds when planning an outdoor installation, underground or installing within concert ducts. This will greatly reduce the risk of corrosion occurring – especially when IP 56 non-metallic accessories are used in conjunction with your installation.

- Some soldering fluxes are naturally acidic or corrosive. Only use these if you have ways to reduce this effect.

- Only use steel wire armoured cables with a plastic covering when planning an outdoor or underground installation.

- Rusting will only occur in the presence of water and oxygen and will speed up from the result of contaminations within the iron and the presence of acid rain. Prepare untreated ferric metal (that is, metal containing iron) with a suitable finish if rusting is to be avoided within your installation.

Electrolytic attack and corrosion are found within damp situations where two dissimilar materials are touching each other. There are many materials which will cause this – here are some of them:

1 corrosive salts (sometimes found in plaster undercoats),

2 acidic timbers – the best known is oak,

3 dissimilar metals touching each other causing an electrochemical reaction – for example copper and zinc galvanising,

4 freshly plastered walls and solid brass accessories,

5 floors and dadoes which contain magnesium chloride (a white salt-like compound).

Choosing the size of your conductor

The design current of your circuit

Gone are the days when you could say that a 1.0 mm^2 conductor would support 'so many amps'. There are many considerations you must think about before making your final choice. In summary format, these are:

- The *design current* of your circuit – the number of amps which will be flowing.

- The *ambient temperature* – hot, cold or average room temperature.

- The *installation method* you will be using – for example, laid within trunking, drawn through conduit or clipped to the surface of a wall, etc.

- Will your installation be *grouped together*? This is known as bunching.

- Will there be a large voltage drop once you have switched 'ON' – is the size of your cable suitable for its proposed length?

- What type of over-current device do you want to use? – there are many types to choose from.

- Will correction factors need to be applied? Using correction factors correctly will help determine the size conductor you will use. There are four groups for you to consider:

  - Cg: This is the symbol for cable grouping, where cables are grouped or bunched together,

  - Ca: use this for the ambient (surrounding) temperature,

  - Ci: this is used for thermal insulation (for example, roof space lagging). If just one side of your cable is in contact with the insulation, the factor is 0.75 but when totally surrounded the factor is decreased to 0.5.

  - Cr: this term is used for the type of over-current protection you wish to use. Choosing a semi-enclosed fuse will provide a correction factor of 0.725 but not if your circuit has been wired using mineral insulated cable.

- *It* is the tabulated current carrying capacity of the selected cable.
- *In* is the current rating of the over-current device.

How do we apply correction factors?

As an example, please consider the following:

A 230 volt, 3000 watt industrial machine is planned to be installed using PVC insulated copper single conductors drawn into steel conduit and protected by a 15 amp semi-enclosed fuse rated at 15 amps. The installation is planned to be installed in an area where the maximum temperature could reach 35°C. The circuit will be drawn in alongside of two other single phase circuits (see Figure 7.59) and all three circuits will support about the same load. There is no thermal insulation to consider.

We must now find the minimum size conductor that can be used to act in accordance with BS 7671.

- First you must calculate the current which flows in your circuit – the design current: Ib = 3000 watts/230 volt = 13 amps.
- Next you must find the correction factor (Ca) for a maximum ambient temperature of 35°C. BS 7671 supply a factor of 0.94.

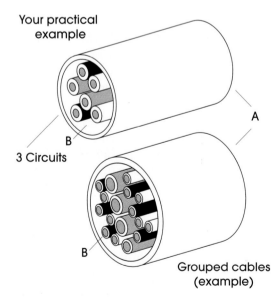

Your practical example

A

B

3 Circuits

B

Grouped cables (example)

Figure 7.59 – A correction factor example – six conductors drawn into a conduit

- The grouping factor (Cg) is the next to find for cables enclosed in conduit. The factor provided by the Regulations is 0.7.
- The last correction factor you will have to find is for a semi-enclosed fuse (Cf) which is 0.725.
- In this example there is no thermal insulation to consider, so Ci will not apply to our calculation.

You must now calculate the minimum current carrying capacity of your proposed cable – this is called *It*:

$$It = \frac{\text{rating of the protective device in amps}}{\text{temperature} \times \text{grouping} \times \text{fuse factor}} = \frac{\text{rating of the protective device in amps}}{Ca \times Cg \times Cf}$$

(7.1)

Thus:

$$It = \frac{15}{0.94 \times 0.7 \times 0.725} = \frac{15}{0.477} = 31.4 \text{ amps.}$$

Turn now to BS 7671 and under *Installation Method 3* select the minimum size cable required for your circuit. This will clearly confirm that a 4 mm² size conductor having a current rating of 32 amps would be the most cost effective size to use. This is ideal for a reasonably short cable length where volt drop need not be considered.

Thinking about volt drop

A large volt drop within a system will have a direct effect on other current-using equipment. This is very clear when your home is the very end house supplied with electricity from the local village community transformer during a harsh winter! Fluorescent lighting will start to flicker and go out, your television picture will shrink in size and incandescent lamps will fade.

BS 7671 is satisfied when a voltage drop occurring between your supplier's terminal cut-out fuses and fixed current using equipment does *not* exceed 4% of the nominal voltage ($U0$) of the supply. Volt drop must always be taken into consideration when selecting the size of cable required for your circuit. As a practical example, please think about the following. Correction factors can be ignored for this illustration:

A 20 metre, 230 volt, single phase circuit is planned to provide power to an industrial machine. It is to be wired in single PVC insulated cables within steel conduit. Your job is to calculate the minimum size cable which you can use to comply with BS 7671.

Given that:

Permitted volt drop = $0.04 \times$ voltage (7.2)

substituting for figures gives:

Permitted volt drop = $0.04 \times 230 = 9.2$ volts.

The way to calculate the actual volt drop within your circuit is by use of the following expression:

Actual volt drop = $\dfrac{mV/A/m \times I \times L}{1000}$ (7.3)

where milliV/A/m is the millivolt drop per metre, I is the current flowing within your circuit and L is the length of the conductor in metres.

Now substitute for all known values:

$9.2 = \dfrac{mV/A/m \times 27.75 \times 20}{1000}$.

Now cross-multiplying the above expression:

$9.2 \times 1000 = mV/A/m \times 555$.

Then dividing each side of the equation by 555:

$mV/A/m = \dfrac{9200}{555} = 16.57$ mV/A/m.

Choosing the size of your conductor

By checking out BS 7671, Table 4D1B, the correct size of conductor can be chosen for your circuit which has a volt drop of less than 16.57 millivolts per amp per metre. By casting your eye down the column of figures you will find that a 6.0 mm^2 conductor will meet your needs, having a volt drop of 7.3 mV per amp per metre.

This is an ideal means to find the cable size you require for your circuit –but please remember *correction factors*: they are equally important!

Shock protection – direct contact

To protect ourselves and others against direct contact with live parts the following basic defensive measures have been put in place for our safety:

- *Installation of barriers and enclosures* – but you must provide proper safety measures to prevent others (people, livestock, etc.) from accidentally touching the live conductive parts. Enclosures must be either lockable or access gained by the use of a tool
- *Protection by insulation of live parts* – the insulation must be appropriate for applied voltage.
- *By placing out of reach* – for example, bare overhead lines for power distribution between buildings. These are always situated at high level but bare conductive parts which are *not* part of an overhead power distribution system must not be within arm's reach and this is without any help from a ladder or tool. If they can be accessed by any other means they must not be within 2.5 metre of any exposed conductive part (the steel body of an appliance, etc.), any extraneous conductive part (steel central heating pipes, etc.) and the bare conductive parts of other circuits.
- *Obstacles (usually at high level)* – this is to protect operatives from live traction rails found within heavy industry or from the unintentional contact with live parts when operating machinery and other equipment.

Shock protection – indirect contact

This item has been regularly reviewed within the last few paragraphs. To summarise briefly:

- *Electrical separation* – this will take the form of an isolating transformer in which the secondary winding is not grounded to earth. Additional details from BS 3535 413-06-02.
- *Earth equipotential bonding with the automatic disconnection of the supply (EEBADS)* – use RCBO's, residual current devices, etc., with a tripping rating of 30 milliamps.
- *A non-conducting location* – this is where all exposed conductive parts are installed in such a way that it will be impossible for a person to come into simultaneous contact with two exposed conductive parts or with an exposed conductive part and an extraneous conductive part. In this type of installation there are no protective conductors whatsoever. For

additional information please refer to
BS 7671 413-05.

■ *By using Class 2 equipment or similar
insulation* – that is when the insulation is
either double or totally reinforced or if low
voltage switch gear and control gear is used
with total insulation. Please refer to
Regulation 413-03-01 for additional data.

■ *Earth free local equipotential bonding
installations* – this is a very specialised subject
which is not fully covered in this syllabus.
Briefly, all equipotential bonding conductors
must connect all simultaneously accessible
exposed conductive parts and all extraneous
conductive parts but the protective conductor
is not grounded to earth. For more details
please check out BS 7671, Regulation 413-32
to 34.

Choosing your disconnection times

For everyday run-of-the mill circuits the correct
disconnection time is obtained by using the
following:

■ a well chosen protective device,

■ keeping within the maximum permitted
impedance ($R_1 + R_2$) of your circuit.

Many of us will fit a 'B' type 32 amp miniature
circuit breaker to serve a cooker circuit within
a TN-S earthing arrangement. This is fine
providing the total length of your circuit together
with your earth loop impedance value is no
greater than 1.2 ohms – Figure 7.60 illustrates
this. If the maximum impedance is much higher
than can be tolerated the quickest way forward
is to remove the Type 'B' breaker and fit an RCBO
of the same value in its place. The maximum
impedance value allowed will then be 1666
ohms.

A disconnection time of 0.04 seconds is needed
for circuits which serve the following:

■ socket outlets which are used for any sort of
portable equipment,

■ fixed equipment placed outside the
equipotential zone where you are able to
touch or handle exposed conductive parts –
for example, a garden, aluminium clad, socket
outlet,

■ any hand held appliance which has a metallic
case and requires an earth conductor,
powered from a socket outlet,

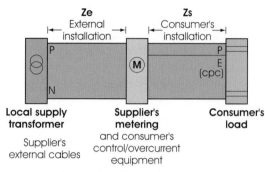

Remember, **Ze + Zs** to be equal
or be less than 1.2 ohms

Figure 7.60 – *When a type 'B', 32 amp miniature circuit
breaker is used to serve a circuit, the total earth loop
impedance of the circuit must not be more than 1.2 ohms*

■ a radial socket outlet designed to supply
power to portable equipment outdoors by
means of a flexible lead,

■ socket outlets installed within a room
containing a shower cubicle.

After completing your installation, check out the
disconnection time in seconds with a suitable
meter. You will find that values will vary
depending on your maximum recorded
impedance.

There are time/current characteristics within
BS 7671 which are very useful to establish the
disconnection times you will require for your
circuit.

What size conduit will I need for my installation?

Finding the size of conduit you will require is
reasonably straightforward but you will have to
consult a couple of tables. The figures you extract
from the tables are referred to as factors and this
is the way to proceed:

■ First obtain a cable factor from Table 7.11 for
each cable which is to be drawn into your
conduit.

■ Once you have collected together all your
cable factors, add them together and compare

Table 7.11 – *Cable factors – choosing the correct size of conduit*

Type of cable	Cross-sectional area of the conductor in mm²	Cable factor
Stranded or solid copper, PVC insulated	0	16
	1.5	22
	2.5	30
	4.0	43
	6.0	58
	10.0	105
	16.0	145
	25.0	217

the following single core PVC insulated copper cables – Figure 7.61 illustrates this:

- three 2.5 mm² cables,
- four 1.5 mm² cables,
- ten 1.0 mm² cables.

To find the size of conduit required first take a look at Table 7.11.

this figure with the cable factors given in Table 7.12.

- You must also take into account the length of your conduit installation and the number of bends that will be incorporated. For this you will have to plan ahead.
- The minimum size of conduit you will require is the size which has a cable factor either equal to or greater than the total sum of your cable factors.

An example

Imagine a small installation comprising one 9 metre run of steel conduit served with a single right angled bend. The conduit has to cater for

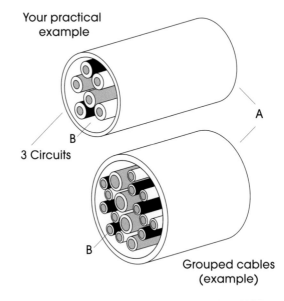

Figure 7.61 – *The size of conduit needed for an installation – an example*

Table 7.12 – *Selecting the size of conduit by means of cable factors*

	A straight run of conduit				Conduit with one bend			
Diameter (mm) of conduit	16	20	25	32	16	20	25	32
Length of conduit (m)								
3.5	179	290	521	911	162	263	475	837
4.0	177	286	514	900	158	156	463	818
4.5	174	282	507	899	154	250	452	800
5.0	171	278	500	878	150	244	442	783
6.0	167	270	487	857	143	233	422	692
7.0	162	263	475	837	136	222	404 .	720
8.0	158	256	463	818	130	213	388	692
9.0	154	250	452	800	125	204	373	997
10.0	150	244	442	783	120	196	358	643

- The cable factor for 2.5 is 30 and there are just three cables = 90.
- The cable factor for 1.5 is 22 and there are four of them = 88.
- The last factor to consider is for 1.0 which is 16, of which there are ten cables = 160.
- Next add up all the cable factors: 90 + 88 + 160 = 338.
- Now refer to Table 7.12 and look along the horizontal row of factors serving the 9 metre run incorporating one bend. The factor nearest to your calculated figure of 338 is 373.
- Look at the topmost number in this column and you will find that a 25 mm diameter conduit is recommended for your installation.

Life for the average electrician is not quite as easy as having a 9 metre run of conduit incorporating just a single right angled bend. The vast majority of schemes are designed to include many sets and right angled bends throughout the installation. If too many bends and sets are incorporated within your circuit, overlooking the use of inspection boxes, wiring would be impracticable.

As a rule of thumb the following points should be noted:

1 The sum of your total conduit angles should not be greater than 180° before an inspection box is installed – Figure 7.62 illustrates this.
2 Add an inspection box in your installation after the following:
   - two return sets,
   - a mixture of one right angle and two 45° sets,
   - every second right angle,
   - after every tenth metre of straight conduit.

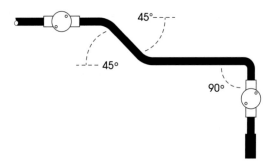

Figure 7.62 – *The sum of your total conduit angles must not exceed 180° before installing an inspection box*

Never undervalue the importance of inspection boxes especially if you have incorporated tight knuckle bends or factory made bends within your installation.

Earthing and bonding explained

Earthing

The functional earth conductor, colour coded cream, forms part of the electrical distribution system inasmuch that it leads the route back to the local community transformer's star point via the sheath of the suppliers underground cable. Alternatively, in a 'TT' earthing arrangement this conductor can be connected to a locally staked earth electrode where a fault current will find its way back to the source of energy, the transformer, by way of the general mass of earth – Figure 7.63 illustrates this.

In a 'TN' earthing arrangement system, where the size of the neutral conductor is less than 35 mm^2, the cross sectional area of the main earthing conductor has to be 16 mm^2. Conventionally, the electrical potential of earth is always taken as zero.

At the transformer the secondary winding is grounded at the star point – it is connected to the general mass of earth – see Figure 7.64. Any fault condition to earth would, in effect, be a fault between the phase and neutral conductor. The breaker would come into play.

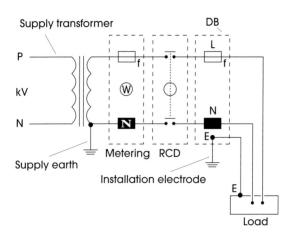

Figure 7.63 – *In a 'TT' earthing arrangement, stake your earth electrode as near to the source of the installation as possible*

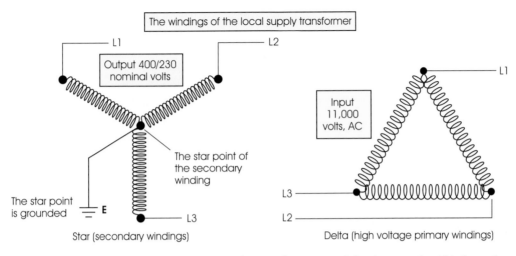

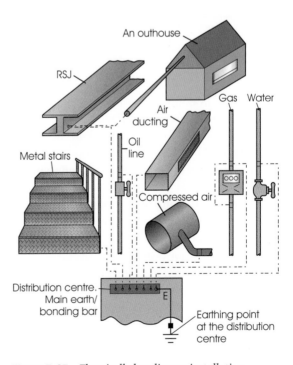

Figure 7.64 – *At the community transformer the secondary winding is grounded at its star point. This forms the system earth*

Bonding

Protective bonding conductors are colour coded green–yellow and are connected to all extraneous conductive parts, every exposed conductive part and to the main earthing terminal. Further supplementary equipotential conductors can be used but in general they are only required in special risk locations such as swimming pools, bathrooms or in very damp areas as contact with water will lower your body's electrical resistance to produce lethal currents.

Bonding will reduce the potential difference which under a fault to earth condition could appear between two conductive parts – for example, a length of copper water pipe and the metal structure of an appliance.

Main equipotential bonding

This is dealt with by Regulations 413-02 and 471-08. This requires all metal work not forming a part of your installation (extraneous metal work) and all exposed conductive parts, for example, the steel housing of a washing machine, to be bonded together and then to an electrical earth as illustrated in Figure 7.65. In practice these are all made common at the main earth/bonding connection point. The illustration shows a series of radial bonding conductors serving an assortment of extraneous parts. If you wish to design your installation looping one bonding conductor from part to part, you can, providing the bond is not cut as shown in Figure 7.66.

There are many extraneous conductive parts in an average installation. Here are just some of them:

- The main copper/steel water and gas supply – bond within 600 mm of the water or gas meter on the consumer's side of the installation or where it enters the building.

Figure 7.65 – *Electrically bonding an installation*

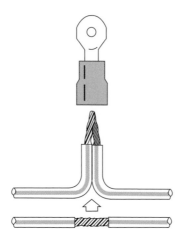

Figure 7.66 – *The bonding conductor must not be cut – wire as illustrated*

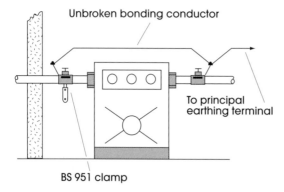

Figure 7.67 – *Bonding the gas meter*

It is wise to shunt the gas meter as illustrated in Figure 7.67 to prevent fault currents passing through the mechanical mechanism.

- Metal cladding forming both doors and walls.
- The infrastructure and any guard rails within a dairy.
- The metal studding serving partitioning.
- Central heating pipe work and air conditioning ducting.
- Exposed steel joists and metal stairs.
- Lightning protection arrangements.

You must serve all extraneous conductive parts with a bonding conductor which has a cross sectional area of no less than 50% of your main bonding conductor (in an average dwelling your main bonding conductor will be 10 mm² when the supply cable to your distribution centre is

25 mm²). Alternatively, a cable of no less than 6 mm in cross sectional area can be used. All bonding termination points must be accessible for testing. This is a requirement of BS 7671 526-04-01.

Supplementary bonding

You need only supplementary bond where there is an increased risk of shock in areas containing a fixed bath or shower, a swimming pool or anywhere it could become damp or wet.

In a bathroom you must make common all protective cable termination points serving each circuit delivering power to Class I and II equipment in zones 1, 2 or 3. Please connect everything which is made from metal together. These will include the following:

- metal gas and water pipes,
- central heating pipes and radiators,
- metal stairs, fixed heaters and luminaries as shown in Figure 7.68,
- air conditioning ducting and grills,
- metal baths, steel shower basins and metallic waste pipes. Please note that metal waste pipes in contact with earth should be bonded and the protective cable taken to the principal earthing point at your electrical mains.

There is no need to connect supplementary bonding to bath or basin tapes supplied with plastic water pipes nor is it necessary to supplementary bond a steel bath which is not connected to any extraneous metal work. Other areas where supplementary bonding is not required are:

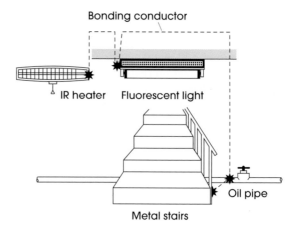

Figure 7.68 – *Supplementary bonding*

- metal furniture placed in kitchens or bathrooms,
- kitchen sinks, draining boards and pipe work,
- wash hand basins which are not part of a bathroom suite, together with the associated metal pipe work.

What size of cable may I use for supplementary bonding?

All green/yellow colour coded supplementary bonding conductors require to have a minimum cross sectional area of 2.5 mm when they are mechanically protected from harm – for example, channel strip placed under the plaster line, mini/micro trunking, PVC-u conduit, etc. Alternatively a single 2.5 mm^2 protective cable may be used if placed in an inaccessible position. Use 4.0 mm^2 supplementary protection if the installation is to be unprotected – BS 7671 547-03 confirms this.

Earthing arrangements

We have several different types of earthing arrangements to serve our supply systems in the UK. BS 7671 Regulation 312-03-01 provides details for the following groups:

- TT,
- TN-S,
- TN-C-S,
- TN-C.

TT systems

If you live in rural Britain you will be very familiar with this type of earthing arrangement. A 'TT' earthing arrangement requires the consumer to supply an effective means of earth leakage protection. This protection is either achieved by the installation of a series of copper earth plates or by installing a suitably rated residual current device (RCD) accompanied by a steel/copper plated electrode staked firmly into the ground.

The phase and neutral supply is provided by two single overhead service cables which originate from the local community transformer. The star point of the transformer's secondary winding is grounded to earth and is completely independent from the consumer's earthing arrangements.

Wiring throughout the installation is carried out in the traditional way and all circuit protective conductors are terminated within the main earth/ bonding terminal bar serving the consumer/distribution unit. A fault current to earth will unbalance the RCD and isolate the circuit/installation from the supply.

Figure 7.69 illustrates the schematic design associated with the type of earthing arrangement whereas Figure 7.70 shows the system in illustrative form.

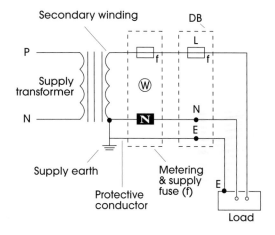

Figure 7.69 – A 'TT' earthing arrangement – schematic layout

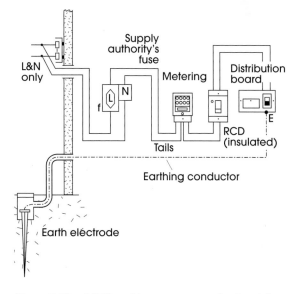

Figure 7.70 – A 'TT' earthing arrangement in pictorial form

TN-S systems

This system is found in towns and cities, especially in older properties, where the lead sheath of the underground service cable acts as a protective conductor. The service cable originates from the local community transformer where the transformer's secondary winding is grounded to earth at the star point along with the lead sheath of the supply cable. An installation is protected by connecting a suitably sized conductor from the sheath of the service cable to the earth terminal point serving the distribution centre. As with the 'TT' earthing arrangement, any voltage leakage to earth will be the same as a

direct short circuit between phase and neutral. Figure 7.71 explains the basic wiring schedule for this type of earthing arrangement whilst Figure 7.72 provides you with the same arrangement in pictorial form.

TN-C-S systems

A TN-C-S earthing arrangement can be found in both rural and urban settings. The vast majority of new installations are provided with this type of virtual earthing.

The service cable to your installation is called a PEN cable – a cable in which the earth and neutral (E and N) are combined and are formed from copper stranded armouring wrapped around an insulated phase conductor (P). The cable is completed with a layer of insulation around the stranded copper.

After the supplier's cable has been safely terminated at the service fuses, a suitably sized earth conductor (colour coded cream) is connected from the supplier's neutral terminal block to the principal earthing terminal serving the distribution centre. In an average house, say a three bedroom semi, where the double-insulated supply conductors are 25 mm^2 and are served with an over-current device rated at 100 amps, the principal earthing cable will be 16 mm^2. A fault condition to earth within a TN-C-S system would be the same as a direct short circuit from phase to neutral.

Figure 7.73 illustrates this system schematically, Figure 7.74 pictorially.

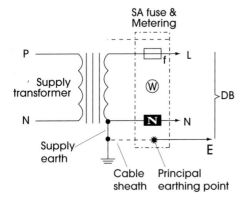

Figure 7.71 – *A 'TN-S' earthing arrangement – schematic*

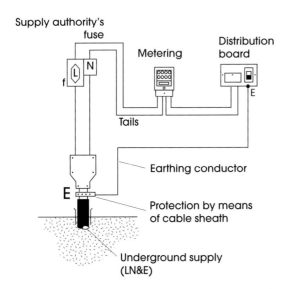

Figure 7.72 – *A 'TN-S' earthing arrangement – pictorially shown*

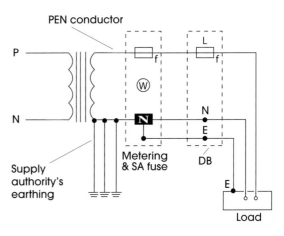

Figure 7.73 – *A 'TN-C-S' earthing arrangement – schematic*

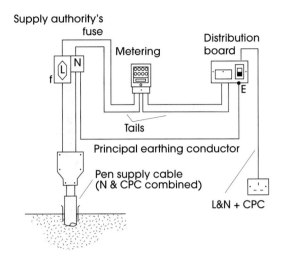

Supply authority's fuse

Metering

Distribution board

Tails

Principal earthing conductor

Pen supply cable (N & CPC combined)

L&N + CPC

Figure 7.74 – A 'TN-C-S' earthing arrangement – pictorially shown

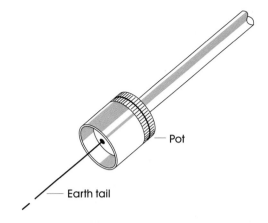

Pot

Earth tail

Figure 7.75 – A special MI gland pot is used for an installation served by a 'TN-C' earthing arrangement. This illustration shows the positioning of the special flexible copper earth/neutral fly-lead

TN-C systems

When your installation draws its power from a privately owned isolating transformer (where the input voltage equals that of the output voltage) supplied by a PEN cable and the installation is wired throughout in copper sheathed cable, the system is known as a 'TN-C' earthing arrangement.

Your installation will be wired all the way through using mineral insulated cables (MI) and fitted with a special gland-pot. A simple lighting circuit would be wired in single core MI cable. As the cables are so thin, it can be quite difficult to gland-off properly first time – it takes practice! The special gland-pot has a flexible copper lead which has been embedded into the brass pot by the manufacturers – Figure 7.75 illustrates this. This acts as a combined neutral and protective conductor. Figure 7.76 provides an additional view of the special gland-pot in cross section.

This type of earthing arrangement is not in general use but you may come across it from time to time. As with most other power arrangements, the secondary side of the privately own transformer is grounded to earth. Figure 7.77 illustrates this schematically whilst Figure 7.78 shows it in pictorial form.

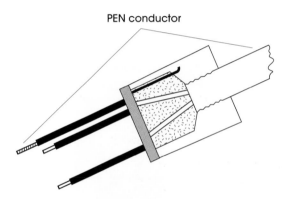

PEN conductor

Figure 7.76 – The special MI gland pot used within 'TN-C' installations shown in cross section

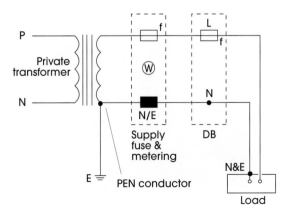

P

Private transformer

N

W

f

L

f

N

N/E

Supply fuse & metering

DB

E

PEN conductor

N&E

Load

Figure 7.77 – A 'TN-C' earthing arrangement shown schematically

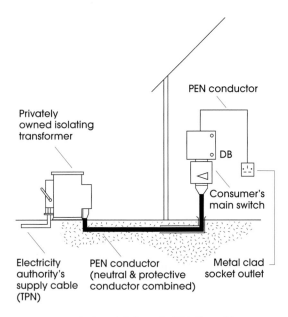

Figure 7.78 – *A pictorial view of a 'TN-C' earthing arrangement*

Figure 7.79 – *It is possible for an unskilled person to replace the fuse element with any value of fuse wire or whatever comes to hand!*

Circuit protective devices

Over-current protection is available in many shapes and sizes and is designed to defend our circuits automatically. Table 7.13 reviews nine of the common most over-current or thermal protection devices we have available to us.

Figure 7.80 – *A high breakiing capacity fuse (HBC)*

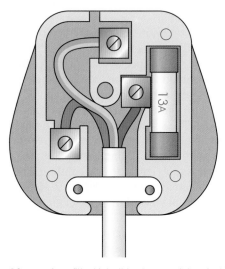

The 13 amp fuse fitted into this plug was intended to serve a small 100 mm extract fan. A 3 amp fuse would be far more suitable

Figure 7.81 – *A standard plug fuse to BS 1361*

Table 7.13 – *The application and limitations of protective devices. (Please note: a miniature circuit breaker can maintain an over-current condition of about 25% before tripping. Between 60 and 70% is needed to rupture the element of a HBC fuse and nearly 100%, or over, to melt the element of a rewireable fuse.)*

Overcurrent device	Application	Limitations
1. Semi-enclosed fuses – BS 3036	1. Low cost overcurrent protection. OK for TN-S earthing arrangement installations	1. The fuse element cannot be replaced quickly and can be repaired with the wrong size fuse. The element will blacken and deteriorate with age. It can be re-inserted in the circuit whilst the fault is still there. There is a lack of discrimination when selecting this type of over-current protection – Figure 7.79 illustrates. Gives poor protection under a small overload conditions and ruptures far too quickly under much heavier overload conditions.
2. Cartridge fuses – BS 1361 and BS 1362.	2. BS 1361 – as service fuses used by the supplier; the miniature version is still used in older type consumer units. BS 1362 – widely used as over-current protection within UK 13 amp plugs. The body of the fuse is made either of glass or low grade ceramic material which can be filled with fine sand. There are metal end caps, used as contact points	2. More expensive to replace than a rewireable fuse and can be shorted out using tinfoil paper. They are not appropriate for high current circuits. Can be replaced with the wrong size cartridge – Figure 7.81 illustrates.
3. High breaking capacity fuses (HBC) – BS 88	3. Ideal for motor circuits and circuits which draw high currents. The fuse comprises a body made from high ceramic material. The end caps are tagged as illustrated in Figure 7.80.	3. Expensive but very reliable and dependable in operation
4.Miniature circuit breakers [MCB] – BS EN 60898	4. For tamper proof over-current protection for use in domestic, agricultural, commercial and industrial installations – Figure 7.82 illustrates. A fault is easily recognised and the supply quickly restored	4. They are expensive. They have mechanical moving parts which require testing every now and again. Their tripping characteristics are affected by the ambient temperature – if it is very hot they will trip before the rated current has been reached. Use type 'D' for site transformers and motor circuits. A type 'B' MCB will often trip when an incandescent lamp comes to the end of its working life
5. Residual current breaker with overload protection (RCBO) – BSEN 61009–1	5. Used where a residual current device would take up too much space. Ideal to serve a special outdoor radial circuit in order to protect against earth/current leakage and over-current problems or where a particular circuit requires monitoring – Figure 7.83 illustrates.	5. They are expensive. Because short circuit sensing is only on the phase side of the device, mistakes can be made when the RCBO is connected to the distribution centre. Mistakes can also be made when selecting the current and residual current rating of the device. The screwed connection points need to be firmly tightened
6. Moulded case circuit breaker (MCCB). Figure 7.84 illustrates	6. Generally to protect supply cables or as a means of isolation for part of an extended installation	6. They are expensive. A choice of two types: • magnetic • thermal magnetic. Prior knowledge of the characteristics and category of duty of the device are required before a choice is made

Table 7.13 – *continued*

Overcurrent device	Application	Limitations
7. Miniature bimetallic devices	7. For thermal and over-current protection. Applied to electric motors rated under 0.37 kW. Two varieties are available: • with an integral heating element – Figure 7.85. • without an integral heating element as shown in Figure 7.86	7. Some models will automatically snap back into their working position, once cooled; others need to be manually re-set after they have switched themselves 'OFF'. They can become polluted with contamination and lodge themselves in the open position
8. Motor rated overloads	8. Fitted to electromechanical starters to protect electric motors against overload problems. The device is screwed into the starter unit where the outgoing load is connected from – Figure 7.87 illustrates.	8. It is possible to select the wrong type of motor overload unit – the tripping current could be ranged either too high or too low. The overload relay must be compatible with the starter assembly it serves – different models seldom fit. They can develop internal open circuit problems and are subjected to ambient temperature fluctuations. This will provide false tripping problems during periods of very hot weather. The air ventilation slots need to be free of dust and general contamination
9. Thermal links for heating appliances	9. Are found wired within older night storage heating appliances. If, for any reason the appliance overheats, a soft metal thermal link will melt. This acts as a switch and isolates the current from the element – Figure 7.88 illustrates	9. They are very sensitive –sometimes even melting when clothing is accidentally left on top of the night-store heater.

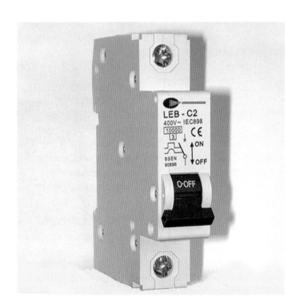

Figure 7.82 – *A typical miniature circuit breaker*

Figure 7.83 – *Most RCOBs are the same dimensions as a miniature circuit breaker*

Figure 7.84 – A typical MCCB

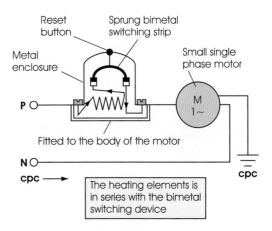

Reset button

Sprung bimetal switching strip

Metal enclosure

Small single phase motor

P

M 1~

Fitted to the body of the motor

N

cpc →

cpc

The heating elements is in series with the bimetal switching device

Figure 7.85– *Miniature bimetal overcurrent devices are fitted to electric motors which are rate under 0.37 kW (370 watts). This illustration shows a device with a built-in heater element*

Installing your customer's switch gear

The arrangement of your customer's switch gear will depend on the type of earthing arrangements you have.

Let us suppose that your installation is to be supplied from a TN-S system where the supplier's cable, comprising a singe phase and neutral lead sheathed/armoured cable, travels underground from the local community transformer to your client's home.

To recap on earthing matters: your main earthing conductor will be clamped firmly to the lead sheath of the supplier's cable and then routed directly to the earth connection point serving your distribution centre. This will effectively provide a continuous path back to the local supply transformer's secondary winding at the star point. At this point it is grounded to earth.

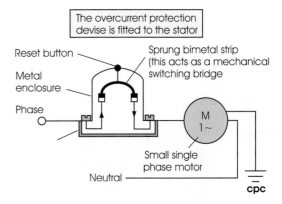

The overcurrent protection devise is fitted to the stator

Reset button

Sprung bimetal strip (this acts as a mechanical switching bridge

Metal enclosure

Phase

M 1~

Small single phase motor

Neutral

cpc

Figure 7.86 – *A miniature bimetal overcurrent device without a heating element*

Doing it by the book

You should check out the following features of the electrical supply provided. This is to avoid dangers and inconveniences as a result of a fault condition occurring:

- *The type and current rating of the supplier's fuse* – usually between 80 and 100 amps and supplied by a BS 1361 Type 2b fuse.
- *The frequency of the supply* – the nominal frequency in the UK is 50 hertz.

Figure 7.87 – *An overload set has to be mechanically fitted onto a starter to provide protection for a motor circuit*

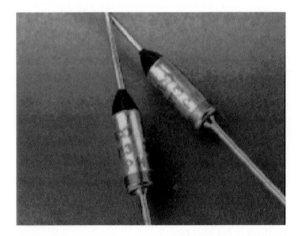

Figure 7.88 – *Thermal links are fitted in some older night storage heaters*

- *The prospective short circuit current at the origin of the installation* – it must be able to rupture your highest fuse rating.

- *The voltage of the supply* – it would be a poor supply if your installation was the last to be served from the community transformer providing just 200 or so volts during the winter months!

- *The maximum demand, in amps, of your proposed installation* – it must be less than the rating of the supplier's fuse.

- *Will the installation requirements be suitable with what has been surveyed?* – will you have complete access to all equipment? Placing everything within a cramped cupboard under the stairs is not the answer!

- *A check must be made that you have a suitable external earth loop impedance value (Ze)* – anything between 0.1 to 0.8 ohms within a TN-S system will be fine.

Fitting your equipment and switch gear

You will require:

- *The supplier's service fuse and neutral block* – they should be there and ready for you.

- *Metering arrangements (credit or prepaid)* – supplied by the metering company including the double insulated 'tails' from the service cutout to the meter.

- *Your distribution centre (your best option for an installation from a TN-S system would be to incorporate both MCB and RCBO protection)* – supplied by you.

- *Double insulated tails from the supplier's meter to your distribution centre, by way of an optional 100 amp main switch (or two switches if you are installing off peak heating circuits) together with cable ties* – supplied by you.

- *You are allowed a maximum of 3 metres of double insulated tails but if you require longer, a double pole switch fuse rated between 80 and 100 amps must be incorporated in-line and near to the supplier's metering equipment* – equipment supplied by you.

- An earth clamp together with a measured length of earth wire connected to the lead sheath of the supplier's cable and attached within the distribution centre. For a TN-S system when both phase and neutral conductor sizes are either 16, 25 or 35 mm^2 you must use a 16 mm^2 copper earth wire. Your main copper equipotential bonding conductor will be 10 mm^2 for all three sizes of cable – material supplied by you.

- *Some electricity companies might require you to install a larger size earth conductor* – if in doubt, check with your supplier.

- If you are asked to bury your main earthing conductor and do not have plans to protect it from mechanical damage or potential corrosion, it must be at least 25 mm in cross sectional area.

Figure 7.89 illustrates a typical electrical mains layout in pictorial form. The two 100 amp miniature isolators are optional but should be used if your distribution centre has been designed without a main switch or if your service tails are over 3 metres in length.

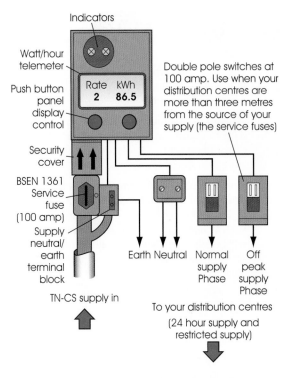

Figure 7.89 – *This is a typical electrical mains layout in pictorial form. Designs and layouts will vary throughout the UK*

Figure 7.90 – *A typical site induction sticker*

Handy hints

- Pen-dispensed white correction fluid is ideal for identifying temporary site power cables at the point of distribution.

- When tracing a fault make a habit of removing all the neutral conductors from your distribution centre. Replace in the order they were removed.

- Site induction programmes are vitally important – never avoid them. Always tell your site manager if you have any medical problems such as diabetes or if you are an epileptic. You will be asked to sign a completed form and often provided with a sticker – Figure 7.90 illustrates.

- Steel and concrete can mask mobile telephone signals. Clear yourself from the building when making a call and advise your caller that you are working in a poor signal area.

- It is best not to leave tools on the topmost platform of a pair of steps – sooner or later they will come crashing down.

- It is wise not to shorten circuit-detail on destination charts within distribution boards. You and others who work with you will know what 'Imm' and 'OSL' means (immersion heater and outside lights) but will the people who have to live there know?

- It can be far more straightforward to attach a bonding conductor to an earth clamp during the first-fix period of your installation but recheck the connection before the second phase is completed.

- Corrosion will set in if you use an aluminium bonding conductor in a damp situation when connection is made to a dissimilar metal.

- Remember – a socket outlet must be at least 3 metres from a shower cubicle.

- The service side of a water main can never be relied upon to supply a good electrical earth. Many old water mains which were once steel have been changed to high performance thermoplastic pipe work.

- Smear a little petroleum jelly on the machine screws serving outside lighting fittings – it will make it easy when lamps have to be changed.

- It is wise to keep a first aid box on your company vehicle in case an emergency arises.

- Control the build-up of condensation by drilling a couple of 6 mm holes in the bottom of an inspection box serving an outside installation. Carry this out at the lowest point.

- A warm to touch switch serving an outside installation can indicate the build up of condensation within the accessory box. Once checked out, drill a couple of drainage holes at the base of the switch to avoid future problems. Reference is made to Regulation 522-03-02.

- Identify your tools with a couple of laps of coloured PVC tape. It will not prevent someone pinching them but it will provide you with a means of instant recognition if they are left unattended or end up in someone else's tool box!

Summary so far . . .

- Ensure your work area is safe before starting work.

- Install your temporary site accommodation with care.

- Get to know the role others play within your working environment.

- A contract must never be signed unless you are fully aware of its contents.

- A bar chart is a simple way to show your intended progress.

- It is very important at site level to maintain a good relationship with your workmates and those of other trades.

- Electricity is generated at 11 000 or 33 000 volts AC. Local distribution from the sub-station across the countryside is usually 11 000 volts. Community transformers reduce this again to a nominal voltage of 400/230 for delivery to homes and industry.

- The term 'switching' means you can safely disconnect the supply whilst your equipment is still on load. 'Isolation' is a term we use for when maintenance work is required and we need equipment to be electrically removed from the installation.

- Terminals must be tight to avoid arcing. Arcing produces heat and heat destroys!

- Cable duct is ideal for power distribution when large cables are used.

- The physical size of a cable termination point generally reflects its current carrying capacity.

- Over-accommodated cable trunking will warm up when on load. This could cause a breakdown within the insulation, leading to problems.

- Corrosion and pollution will cause problems within an electrical installation – the longer it's left the worse it will get. Act accordingly.

- Remember to apply correction factors when choosing the size of your cable. Potential volt drop problems should also be considered.

- Remember not to exceed the maximum circuit impedance allowed by BS 7671 if disconnection times are to be met.

- Use Tables 7.11 and 7.12 to determine the size of conduit you will require for your installation. Remember: an adequate number of inspection boxes are essential if you are to wire your installation without problems.

- The term 'earth' refers to the return path back to your community transformer's secondary winding at the star point connection, where it is grounded to the general mass of earth.

- There are four types of earthing arrangement systems in the UK – TT, used mainly in the countryside, TN-S, widely employed in towns and cities, TN-C-S found both in urban and rural settings and TN-C where privately owned generators or transformers are used.

- Semi-enclosed fuses provide poor protection under small over-current conditions and rupture far too quickly under much heavier overload conditions.

- The arrangement of your customer's switch gear will depend on the type of earthing system available to serve your proposed installation. In a TN-S system the lead sheath of the supplier's cable will act as the principal earth. In a TT system, you will have to provide your own earthing arrangements.

Review questions

1 Name three basic temporary accommodation requirements when site work commences.

2 Briefly state the role of your supervisor when site visits are carried out.

3 Describe the function of a critical path network.

4 Emergency switching buttons are wired as follows:
   (a) in parallel formation,
   (b) in series formation,
   (c) in series/parallel formation,
   (d) in parallel formation but only to the first appliance.

5 State three earth leakage protection devices.

6 Explain the difference between switching and isolating a circuit.

7 Describe the type of installation needed for a corrosive atmosphere.

8 Suggest three different uses for fire retardant cable.

9 Discuss how best to ensure a joint is electrically sound.

10 List four ways of preventing corrosion from occurring.

11 Express, using your own words, why cable factors are important in determining the size of conduit needed for a particular circuit or installation.

12 List three different methods of preventing direct electric shock.

13 Suggest a couple of ways in which 'D' type miniature circuit breakers may be used.

14 Calculate mathematically, the size of steel conduit required for the following small installation:
   – a 10 metre run of steel conduit,
   – one right angle bend will be included,
   – the PVC insulated single core cable to be drawn through the conduit amounts to:
     • four 2.5 mm^2 cables,
     • two 1.5 mm^2 cables,
     • five 1.0 mm^2 cables.

Work out using the tables provided within this chapter.

15 Describe the difference between earthing and bonding.

Testing and commissioning electrical installations

Introduction

Derived from Outcome 2 of Unit 2 (installation (buildings and structures) inspection, testing and commissioning), this chapter has been written to explain the principles and techniques of testing and commissioning your installation after completion and why you need to do this. For further details, please refer to the *CG 2330 Scheme Hand Book*.

Your practical activities at college

These will include the following:

- deciding the correct measures to take for inspecting and testing your installation, components and equipment,
- demonstrating safe and efficient methods of carrying out electrical testing,
- showing your assessor that you are able to select and use the instruments to test and commission your installation.

Why do we bother to test a new installation?

A precise and detailed examination of your new installation is both necessary and essential. Always provide official recognition, in the form of documentation, to show your customer that no hazardous or dangerous conditions exist, which could result in death or injury to an unsuspecting individual.

BS 7671 (Regulation 713-02 to13) has laid down guidelines for the *visual* and *instrumental* inspections of a new installation and these you will find within the pages of this chapter. Try to be serious and demanding when testing your work – a quick check with an installation tester followed by a swift 'Mickey Mouse' certificate is just not good enough. The figures alone will give you away. Completely avoid any potential damage to both people and property by keeping to the rules and regulations – this is most important and is good common sense.

BS 7671, Regulation 713-02 to 12 has provided all the testing advice together with the practical suggestions you will find within this chapter.

What type of fault would I expect?

Listed at random are a dozen practical examples that you can relate to during your period of testing:

1 When *scuffed cable insulation* exposes bare copper to a metallic cable carrier–steel conduit/trunking and steel accessory boxes.

2 *Trapped or pinched conductors* within steel cable carriers serving an industrial installation.

3 A conductor pierced by an accessories fixing screw.

4 *Snapped conductors* within an accessory caused by using side cutters to prepare cable ends. Indents created on the copper wire, originating from the pressure of the stripping tool, produce a weakness within the conductor. Aligning the prepared accessory into place can cause the weakened wires to snap, snagging their bare ends onto an earthed metal accessory box.

5 Wood screws driven through concealed cables will cause problems.

6 Cables *snagged on steel channelling* forming the infrastructure supporting a plaster board partitioning.

7 *Faulty manufactured cable*. Often the phase or neutral conductor melts into the circuit protective conductor (cpc) at the manufacturing stage.

8 Conductors which occupy the *wrong terminals* – when, for example, unfamiliar accessories are second fixed, less experienced people sometimes make wiring termination mistakes.

9 Confusion with wiring arrangements caused by the introduction of *new EU colour codes*.

10 *Cable burn* within steel conduit or inside timber access holes drilled in wooden joists.

11 Lost *final ring circuit continuity* caused by accidentally plastering over a first-fix socket by others. This can also cause bad insulation values at your distribution centre from the phase and neutral conductors to earth and between your two current carrying wires.

12 *Masonry nails* driven into hastily positioned cable runs protected by steel or PVC-u channelling, sometimes referred to as capping.

Initial confirmation procedure

Ask yourself the following questions:

- Does your installation conform to BS 7671, including the recent amendments? If not, list the departures from the regulations within the 'cell' provided on the electrical installation certificate.
- Has the work carried out met with the job specifications? You can find this information out by checking through the most recent job specification sheet.
- Is your installation safe to use? This you can provisionally tell by taking the insulation readings on all circuits leaving your distribution centre, together with the principal earth-loop impedance value.

Now think about the following issues:

- What safety precautions should you take concerning safe working practices? Will you have to use a high voltage insulation tester whilst working from a ladder, or test, for example, in very cramped conditions? Is it possible you will have to muddle through any thermal-couple effects? These are electric currents generated by temperature differences. These are controlled by reversing your meter's test probes once you have obtained your first reading, then averaging out the two readings into one.

- Remember to have your instruments calibrated! Their accuracy will depend on this factor. Keep in mind that the values you obtain will also be affected by the condition of the instrument's battery, the surrounding temperature, even the angle you hold your instrument or if your meter slips from your grasp and crashes to the floor!

Low resistance ohmmeters must be able to read values down to 0.01 ohm. It is handy if you can obtain an instrument which you are able to 'null' the leads to read 000 ohms before carrying out tests.

- Will you require a *work permit* to carry out your testing procedures? No, if the site you are testing is small but a permit will be required on larger sites.

- Are you aware of the purpose of your installation, and able to recognise all of your circuits? You will require this knowledge for recording data. Check your drawing and specification sheet.

- Are you familiar with the order you have to carry out your testing and the correct methods to apply? This, you will find on page 246 of this chapter. Is your test equipment in good working order, undamaged and within the last calibration date?

- Have you all the information you will need for your test? This will enable you to carry out labelling and data recording accurately. Check out your job file and works specification for additional detail.

- Are you in contact with your customer, architect, works department and contracts manager? If not, log in their mobile telephone numbers into your own cell phone. You may require additional information for documentation and testing purposes.

Periodic inspections – purpose and conditions

Periodic inspections are very similar to completion tests carried out at the end of your job. The difference is that there is provision for you to add personal observations, recommendations and actions concerning the electrical safety of the installation you are inspecting and you will find you have a lot more freedom. For example, the percentage of 'sampling' you wish to carry out is at your discretion – but a sample of less than 10% is not recommended. A practical rule of thumb would be between 20 and 30%.

Contrary to popular belief, your testing need *not* be carried out in the order listed in *Guidance Notes Number 3*. It is for you to decide what is both practical and manageable for your inspection and testing programme. An earth loop impedance meter may be used to confirm the continuity of the protective conductor at socket outlets, accessories and easy-to-get-to current consuming equipment but please remember that *some* earth loop impedance testers will trip a residual current device if wired within the circuit under test.

Never insulation-test electronic equipment, it could cause a major breakdown. Always disconnect this type of equipment from your installation before testing – play safe!

Whenever you find a significant difference between previous test results and the values you have recently recorded, you must investigate as to the cause of this inconsistency and amend your documents accordingly. The test sheets are formatted in a slightly different way to the standard electrical installation certificate – it is a little longer but virtually the same or very similar information is required for both sets of documents.

Your tests and inspection must be carried out in a manner as to reduce general disruption to the installation and disturbance to others who might be working there. Remember to seek out permission if you need to isolate any power circuits for testing purposes – there might be computers on-line.

Periodic testing you will be required to carry out where it is practicable within an existing installation

Remember – choose the order of testing which is suitable and practical for you:

- *The continuity of your protective conductors* – test between the main earth terminal bar within your distribution centre and the following exposed conductive parts: the earthing point serving each socket outlet and fused connection unit and exposed conductive parts attached to current consuming equipment and accessories.

- *The continuity of bonding conductors* – this applies to all bonding conductors and all necessary supplementary bonding conductors.

- *The continuity of final ring circuits* – this test may not be necessary if you have formal written documentation of previous tests. If there is physical evidence that the ring circuit has been altered at some stage, then a test must be carried out to confirm that the modified ring is continuous and as intended.

- *The insulation values of the conductors* – if tests are to be made; then between the live conductor (with both phase and neutral joined together) and earth. This must be repeated at all distribution centres; at all sub-main and main distribution centres whilst these are isolated from the electrical mains.

- *The polarity of your installation* – perform your tests at the following positions: the origin of the installation, at the distribution centre(s), all reachable socket outlets and at the business end of each of your radial circuits (cooker, immersion heater and lighting circuits, etc.).

- *Earth electrode resistance test* – test each earth rod/plate or group of rods individually with the test link removed at the distribution centre and your installation switched off from the source of the supply. Record the value obtained in ohms.

- *Earth fault loop impedance testing* – carry out your tests at the following locations within your installation: at the origin of the installation (where the supply authority's fuses are sited); at each distribution board; at accessible socket outlets; at the business ends of all radial circuits (cooker, panel heating point, extractor fan, etc.).

- *Functional (practical) testing* – test the timing of your residual current device in milliseconds with an RCD tester; then test the operation of the manual test button to check that the device will switch off the circuit as intended within the time specified. For additional information please refer to BS 7671, Regulation 713-13-01 or the current edition of the *IEE On Site Guide*.
- *Functional testing of circuit breakers, isolators and other switching devices* – physically switch the device both 'ON' and 'OFF' to ensure that it disconnects and reconnects the electrical supply as intended.

You must decide which of the tests listed above are suitable for your needs. This can be done by drawing from your own experience of the installation, your knowledge of inspection and testing and by checking out past records which have been handed down to you.

When to carry out a periodic inspection report

1 If the original electrical installation is damaged through fire or by mechanical means.
2 If a building has a change of use from, for example, an industrial premise to a commercial outlet.
3 If alterations and additional wiring is made to the original electrical installation.
4 During a change of ownership of a domestic dwelling – this is often requested by the participating estate agent.
5 After a qualifying period – ten years for a private house or flat, three months for construction of a site and one year for a cinema and theatre, etc.
6 Request from a company's insurance broker

Useful information to help you with inspecting and testing

- *BS 7671 – the Wiring Regulations* – obtainable, by order, from most High Street book sellers; mailed direct from the IEE or bought from your local wholesaler.
- Amendments to bring BS 7671 up to date – obtained directly from the IEE and sometimes downloaded from their web site free of charge.
- The latest edition of the *'On Site Guide'* – often, electrical suppliers have this publication in stock but, if not, it can be swiftly acquired by post directly from the IEE.
- Your *contract specifications* – these will be in your site office, kept with your general site manager or, alternatively, held by your contracts manager at your company office.
- Construction and wiring diagrams – there should be two site copies available for your use.
- Manufacturers' handouts and instruction leaflets – these should have been filed and left safely within your site office. Unfortunately, few ever see the inside of a file!
- *Related legal rules and regulations* – for example, it may now be required to install additional mains supplied smoke alarms within a certain type of dwelling or to use harmonised EC colour coding.

Testing on completion – your visual inspection checklist

The visual inspection (taken from BS 7671, Regulation 712-01-02) forms two separate parts:

1 *The British Standards Inspection*: check your installation to make sure all the electrical equipment you have installed complies with proper British Standards or is an EC equivalent to that standard, is not mechanical or electrically damaged, and that the complete system acts in accordance with the conditions set down within BS 7671. Please refer to Regulation 511-01-01 for further details.
2 *Principal visual inspection*: there are many, in fact nineteen of them, of which some will not be appropriate to your installation.

1 *Connections* – make sure all your connections are both electrically and mechanically sound. Ensure good conductance; correct and suitable insulation – Regulation 526-01-01.
2 *Identify your conductors* – carry this out by using coloured conductors, tags, discs, and

sleeves or by number/lettering. For conduit systems, apply a distinguishing orange band around the conduit complying to BS 1710: 1984 (1989). Please see Regulation 514-02-01 for additional guidance.

3 *Check the routing of your cable* – check the routing of your cables, keeping in mind where they go, and whether they are installed inside or out; Regulation 512-06 and 522 confirms this. This will include cable runs in the following areas:
    – airing cupboards,
    – bathrooms and kitchens,
    – within an agricultural location,
    – caravans,
    – boiler houses.
    Please also consider the effects of the following:
    – ionisation,
    – the ambient temperature,
    – the effect of wind and storm.

4 *Sizes of random conductors selected* – choose an assortment of conductors for both voltage drop and current carrying capacities. Check your cable sizes to make sure that the current carrying ability is not less than the design current of the circuit. For example, a 1 mm^2 conductor would not be at all suitable for a 3 kW immersion heater circuit – even though it will carry 16 amps when clipped direct. It just isn't good practice! Confirm that voltage drop will not exceed 4% of the nominal voltage. See Regulation 525-01-02.

5 *The polarities of all switch devices* – test all single pole switching devices to ensure that isolation occurs in the phase conductor only; Regulation 131-13-01 confirms this.

6 *The polarity of all socket outlets and lamp holders* – switches serving sockets must isolate the *phase conductor* only. Connect the switched phase conductor serving older Edison type lamp-holder to the middle terminal – see Regulation 553-03-04. Most modern Edison lamp holders do not require polarity testing. Regulation 713-09-01 refers to the older type.

7 *Protection against thermal effects and the presents of fire barriers* – this includes protection offered between floors and walls against the thermal effects from over-heating and arcing within trunking and tray-work; see Regulation 527-02. Consider the positioning of fixed luminaries when near to heating pipes/unit heaters and when distribution centres have been fitted without a back – please regard this as a potential fire risk.

8 *Protection against direct contact* – meet this requirement by use of the following methods:
    – by the insulation of live conductors,
    – using barriers or enclosures,
    – by the use of obstacles and barriers,
    – by placing out of reach, as with bare overhead conductors.
    Refer to Regulation 412-02 to 06.

9 *Protection against indirect contact* – meet this requirement by one or more of the following measures. Please refer to Regulation 413-01-01 for general detail:
    – automatic disconnection of the supply by the use of a RCD,
    – supplementary equipotential bonding,
    – class 2 equipment,
    – main equipotential bonding,
    – protective conductors,
    – earthing arrangements,
    – protection using a non-conducting location,
    – see Regulation 413-04,
    – main earthing conductor,
    – protection through electrical separation,
    – not to exceed 500 volts; see Regulation 413-06,
    – Earth-free local equipotential bonding.

10 *Prevention of common damaging influences* – select and erect the equipment you install in a way to avoid potentially damage or harm from non-electrical installations. A heater placed near to a plastic water pipe is an example. When this proves unworkable, both systems must be set apart from each other. See Regulation 515-01-01.

11 *Isolating and switching arrangements* – this includes breakers, fault current protective devices and emergency switching and monitoring equipment. Check that they are suitable and correctly rated for the circuits they control. Inspect that switches and other

switching devices are unable to re-close due to mechanical vibration or shock and are correctly positioned. Please refer to Regulation 537 for further details.

12 *The presence of under voltage protective devices* – meet this requirement by one or more of the following examples, which monitor equipment liable to damaged resulting from a drop in voltage:
   – incorporate a time delay mechanism in circuit,
   – if anticipated, then give ample warning to whoever. For further details, please refer to BS 7671 Chapter 45, Regulation 451-01-01 and 451-01-06.

13 *The choice and setting of protective monitoring devices* – a 30 milliamp residual tripping rate provides protection for circuits serving socket outlets, whereas over-current protection for this type of circuit is rated at 32 amp. In a TT, earthing/supply arrangement, look for a maximum disconnection time of 0.4 seconds when the nominal voltage is between 220 and 277 volts AC. Refer to Regulation 471-08 and 473-01.

Lighting arrangements can have over-current ratings of 6 or 10 amps, cookers 32, whilst immersion heaters have a 16 amp breaker.

14 *Labelling of circuits, switches, fuses and terminals* – carry out the following labelling as Regulation 514-13-01 requests:
   – earthing and bonding arrangements,
   – distribution centres,
   – double pole 20 amp switches and isolators,
   – central joint box terminal arrangements,
   – central heating control wiring arrangements,
   – buried cables ('Hep-tape'® underground; notices above),
   – position of caravan inlet points,
   – fuses and circuit breakers,
   – emergency and maintained key-switching devices.

15 *Equipment and protective measures to external influences* – the term, 'external influences' means one or more of the following:
   – the surrounding temperature (Regulation 522-01),

   – external heat sources (Regulation 522-02),
   – presence of water and high humidity (Regulation 522-03),
   – accompanying solid foreign bodies (Regulation 522-04),
   – corrosive or polluting material (Regulation 522-05),
   – mechanical shock or impact (Regulation 522-06),
   – vibration and shuddering (Regulation 522-07),
   – mechanical stress (Regulation 522-08),
   – presence of flora, i.e. plant life (Regulation 522-09),
   – presence of fauna, i.e. animals (Regulation 522-10),
   – solar radiation, i.e. radiation from the sun (Regulation 522-11),
   – building design (Regulation 522-12).

Consider both the type and location of your cable and accessories when carrying out your installation. For example, it would be hardly suitable to install domestic accessories out of doors or to use in an industrial location. It is the case of 'horses for courses' – nothing more, nothing less.

16 *Suitable access to switch gear and equipment* – there are many practical reasons why we must have complete access to our electrical switch gear. So often, the electrical mains serving a commercial building is within a small dedicated room, which, over the years, becomes a company junk room. It becomes a daunting task to get anywhere near the equipment you need to service! Other reasons, which spring to mind, follow:
   – the need to switch off in case of an emergency,
   – periodic inspection and testing activity,
   – an electrical breakdown in a section of the installation,
   – planned maintenance work,
   – additional work, adding to the original installation.

For additional information and data, please refer to Regulation 562-04-01. Other examples to consider follow:

- access to cable joints (Regulation 513-01-01),
- access to earthing and bonding connections (Regulation 543-03-03),
- access to emergency switching, for example, eed fire-fighters' switch or switch or emergency lighting test key switches.

17 *Danger notices* – visually, make sure there are warnings or danger notices, on or near to the electrical equipment under inspection; see Regulation 514-09-01. Normally this means you must label equipment where the supply voltages *exceeds* 250 volts or where the phase serving any accompanying control circuit differs from that of the phase serving the equipment.

18 *Diagrams and notices* – ideally, provide diagrams, charts, and tables mounted within large document frames and fit to the wall next to main switch and control gear. This will allow easy access for information gathering. This method of presentation keeps the drawings and documents in an ideal condition. If this method is not forthcoming, then check for the availability of documents relating to the installation. Formal detail is obtainable by reading through Regulation 514-09-01. Mains room documentation will include one or more of the following:

- voltage warning notices,
- details/destination of outgoing circuits, Figure 8.1,

- method of wiring, size and type of conductor, Figure 8.2,
- protective devices installed (RCD, MCB, MCCB, etc.),
- earthing arrangements (PME, i.e. TN-CS, TT or TN-S),
- notice recommending when to re-test the installation; Figure 8.3 will illustrate,
- data providing detail of the number of points installed,
- notices, advising not to switch off a circuit specified.

Examples: fire alarm, intruder alarm, mainframe computer and CCTV circuit,

- fault-operated protective device (RCD) warning notice; Figure 8.4 illustrates,
- caravan site notices.

19 *The erection/assembly of equipment* – always install equipment in accordance with appropriate requirements and British/EU Standards. If equipment other than EU is used, then you will have to check to prove that the equipment is suitable for one or more of the following examples:

- the nominal voltage (110, 230, 400 volts, etc.),
- the maximum design current (100 or 500 amps, etc.),
- the frequency of the supply – 50 hertz within the EU,

Tara Electrical Company PLC

Circuit destinations No. 195 Newton Close Elmore

Number	Destination of the circuit
1	Hob [kitchen]
2	Appliance final ring circuit [kitchen]
3	Ground floor final ring circuit – including the kitchen
4	First floor final ring circuit
...and so on	

Figure 8.1 – *A typical circuit destination chart serving a small consumer unit*

Electducation Ltd Installation address	Baalbck, Tenbury Wells DB1		
Circuit description	Cable size	Cable type	Protective device
Cooker	6 mm	pvc/pvc	32 A Type B MCB
Shower	10 mm	pvc/pvc	40 A Type B MCB
Garage	6 mm	pvc/swA	32 A Type RCBO
Upstairs lights	1.5 mm	pvc/pvc	6 A Type B MCB

Figure 8.2 – *This is a chart showing the method of wiring, the size of cable used and the over protection provided*

IMPORTANT

THIS INSTALLATION SHOULD BE PERIODICALLY INSPECTED AND TESTED AND A REPORT ON ITS CONDITION OBTAINED, AS PRESCRIBED IN BS7671 (FORMERLY THE IEE WIRING REGULATIONS FOR ELECTRICAL INSTALLATIONS) PUBLISHED BY THE INSTITUTE OF ELECTRICAL ENGINEERS.

DATE OF LAST INSPECTION.. *10th July 2008* ....

RECOMMENDED DATE OF
NEXT INSPECTION.. *July of 2018 [10 year domestic]*

Figure 8.3 – *The notice is attached on or near to your distribution centre. It advises the consumer as to when the installation needs to be retested*

Important notice

THIS INSTALLATION, OR PART OF IT, IS PROTECTED BY A DEVICE WHICH AUTOMATICALLY SWITCHES OFF THE SUPPLY IF AN EARTH FAULT DEVELOPS. TEST QUARTERLY BY PRESSING THE BUTTON MARKED "T" OR "TEST". THE DEVICE SHOULD SWITCH OFF THE SUPPLY AND SHOULD THEN BE SWITCHED IN TO RESTORE THE SUPPLY. IF THE DEVICE DOES NOT SWITCH OFF THE SUPPLY WHEN THE BUTTON IS PRESSED, SEEK EXPERT ADVICE.

Figure 8.4 – *Warning notice concerning the automatic disconnection of a circuit under a fault to earth condition*

- the power characteristics, sometimes difficult to find,
- how well matched will your non-EC equipment be?
- are there any conditions you are likely to come across?
- will there be ease of access for inspection and maintenance?

Please refer to Regulation 510-01-01 to 513-01-01 and 120-02 for finer detail regarding this visual inspection directive.

Choosing your test instruments

A review of the following test instruments will follow:

- the low-ohm continuity meter,
- the basic insulation tester with a megohm range at 500 to 1000 volts DC,
- the standard clamp meter, known also as tong-testers to measure current drawn from your circuit in amps or milliamps,
- the earth-loop impedance meter,
- the residual current device, timing meter (in milliseconds),
- the prospective short circuit current tester measuring your supply characteristics in kilovolts (kV),
- the standard kilowatt hour meter (kWhmeter),
- socket polarity tester.

1. The low-resistance continuity meter

You have the option of three types:

- a stand-alone low resistance ohmmeter – these are best to use for testing,
- a digital multi-meter with a low ohms scale. This sometimes is considered bad practice in some schools of thought,
- a combined insulation and continuity tester, Figure 8.5.

Whatever option you take your test current must not be less than 20 milliamps, supplied by a minimum of 3 volts, to a maximum of 24 volts,

AC or DC. It is sensible to choose an instrument that is able to deliver a minimum value of 0.01 ohm. This is most helpful, especially when you are faultfinding or need to record the exact value of a conductor for your records. Please avoid the temptation of using a bell and battery set to carry out continuity checks – yes, it works, but it is not professional.

2. The insulation tester

Take care when using this type of instrument. It will generate a vicious electric shock, which although not lethal, can cause heart fibrillations (this when your heart beats erratically). Take extra care when you are using your instrument standing on a stepladder as reaction to shock could cause your departure from the steps rather quickly!

The instrument you choose must be able to deliver the following test voltages:

- 500 volts DC for circuits rated up to 500 volts,
- 250 volts DC for extra-low voltage circuits,
- 1000 volts DC for circuits supplied by voltages in excess of 500 volts but no greater than 1000 volts.

Your insulation tester must be able to deliver an output current of one milliamp at the voltage

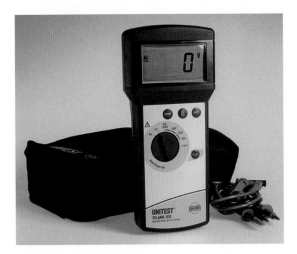

Figure 8.5 – *This is a combined continuity and high voltage insulation tester*

range of your choice and have an accuracy factor of about 2%. In practice, expect about 1.5%, which is fine. This will enable your meter to work as intended. Some instruments do not meet this standard and these will be the ones to avoid. If in doubt, check out the technical specifications on the side of the packaging – Figure 8.6 will illustrate a typical site insulation tester.

3. AC clamp meter

A clamp meter is an instrument for measuring current flow within a circuit. Working by capturing the electromagnetic field that surrounds a load-bearing conductor within its jaws then processing it into a readable value, this type of instrument is very straightforward to use.

The clamp meters you will come across in electrical installation engineering will be for use with alternating current (AC) circuits but direct current instruments are readily available. Some are a hybrid and can be used for both but if you have a choice, select a digital model. They are auto-ranging and multi-sensing, which means that a selector switch is not necessary. They are ideally suited for testing installations, stable loads, motor starting currents and other peak current and surges we experience in electrical work. The accuracy of this meter is typically ± 2% of the display value.

To test: just open the hooded insulated jaws by squeezing the mechanical catch shown on the top right hand side of Figure 8.7 then place around the conductor. Either test just one conductor, the phase or the neutral but never place both conductors within the jaws at the same time, otherwise it will not work as intended.

4. The earth-fault loop impedance meter

Figure 8.8 illustrates an earth-fault impedance tester. This type of meter operates an AC test current of between 20 and 25 amps for about 20

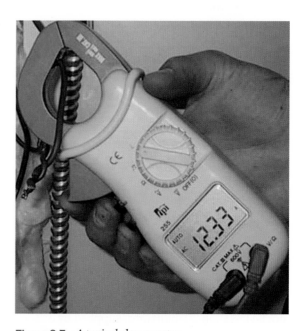

Figure 8.6 – *Some electricians prefer analogue insulation testers*

Figure 8.7 – *A typical clamp meter*

milliseconds with an accuracy of about ± 2%. The lowest resolution you can expect is about 0.01 ohm but this can vary from model to model. A typical earth-loop impedance value for an installation served by a TN-CS earthing arrangement would be between 0.05 and 0.3 ohm, depending how near you were to the supply transformer. For an installation, served by a TT arrangement, expect a much higher reading but this will depend on the electrical resistance of the soil. When very wet, values of between 2 and 5 ohms can be expected, but dry soil will produce far higher values – again this will depend on just how far away the point of measurement lies in relation to the supply transformer.

How to use your earth-fault impedance meter

The test instrument has three coloured test leads. The phase and neutral leads are in the form of insulated probes and are coloured brown and blue. Firmly place the probes on the phase and neutral connection points at the source of the supply in the correct order. This instrument is polarised; it is important to place the leads correctly. The green/yellow lead has an insulated crocodile clip attached to the end – attach this firmly to the earthing bar at the source of the supply. Now follow these simple instructions:

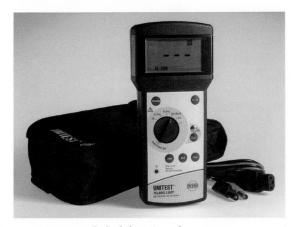

Figure 8.8 – *Earth-fault loop impedance meter*

- Check that you have correct polarity in relation to your coloured leads.
- Confirm that the indicator lights serving the instrument verify that it is OK to proceed with your test.
- Select the desired range (maximum of 20, 200 or 2000 ohms, depending on your supply systems earthing arrangement).
- Press the 'test button'. The value of the earth-fault loop impedance will promptly appear within the instrument's liquid crystal display (LCD) window in ohms.

You can use an earth-fault loop impedance tester to review values at any point throughout your installation and to measure phase to earth impedance values of exposed metalwork using an external earth probe. Please refer to BS 7671, Regulation 413-02-07 for additional information.

5. The standard multi-meter

Figure 8.9 shows a typical digital multi-meter – an ideal instrument for general site work. They are easy to use and are applied to trouble shooting problems within electrical installations during the commissioning period.

This type of meter has a large LCD panel in which individual values appear directly on screen. Designed for general-purpose use, test instruments such as these have many functions, such as:

- voltage measurements from millivolts to 400 volts, AC or DC,
- often frequency up to 2000 hertz (Hz),
- diode test facility,
- low resistance scale from 0.01 ohms,
- bar graph indicator within the display panel,
- capacitance measurement from typically 1 microfarad to 1000 microfarads,
- transistor test facility,
- a resistance scale from 40 ohms to 40 megohms,
- continuity measurement (audio or visual),
- current measurement from 400 microamps (a microamp is one thousandth of one milliamp) to 10 amps, AC or DC.

Figure 8.9 – *A site multi-meter will prove to be a very useful tool*

The low resistance scale is particularly useful when phase to earth fault finding is carried out as values as low as 0.01 ohm can be measured. One disadvantage of using a digital multi-meter is that it is possible to mistake the position of the decimal point and read the values incorrectly.

The accuracy of this type of meter is very good, having a value of between ±0.3 and 2% of the reading displayed.

6. The residual current device, timing meter

Designed to automatically trip out a residual current device during test conditions, this type of meter provides precise information as to the length of time this operation takes. An integral

selector switch provides a full range of test currents to suit the tripping value of the device under test including fast/normal-trip facilities and a phase angle selector switch for positive and negative half cycle testing. The in-service accuracy factor for this type of instrument is 10% and takes in the effect of local voltage differences that may occur. Figure 8.10 illustrates a RCD tester made by Robin Electronics.

7. The prospective short circuit current tester

This type of instrument, see Figure 8.11, will record the exact prospective (probable) short circuit value (PSC) at the source of the supply. If a direct short circuit were to occur at the source of the supply, the current generated would be the PSC value in amps. In practical terms, if the largest fuse serving your installation was 500 amp and the value of your prospective short circuit was only 0.4 kA (just 400 amp), your 500 amp fuse would never rupture in the event of a serious over-current problem. The PSC (or PFC as styled on BS 7671 forms) value must *always* be greater than the largest value of over-current protection and adequate for the prospective fault current at that particular point within the installation.

To test: the test instrument has three coloured test leads representing the phase, neutral and

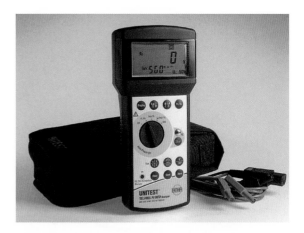

Figure 8.10 – *A timing meter for a residual current device – read in milli-seconds*

earth. Connect the *brown* lead to the phase of the circuit, and the *blue* and *green/yellow* leads to the neutral conductor. This will guarantee that the test will take place between the phase and neutral conductors. Press the 'test button' and your PCS value will appear within the instrument's LCD window in kiloamps (kA). Typical values follow:

- industrial, urban areas – 2 to 6.8 kA,
- rural residential areas with TN-CS earthing arrangements and served by overhead cabling – 1 to 2.25 kA,
- rural residential areas with TT earthing arrangements and served by overhead cabling – from 0.9 to 0.25 kA, depending on the distance to the supply transformer,
- industrial rural areas with TN-CS earthing arrangements served by a local sub-station – 2 to 6 kA,
- the fifth supply system is called an IT earthing arrangement and it is so called when the source of the supply has either no earth whatsoever or is earthed via high impedance. Variable values will apply depending on the arrangement.

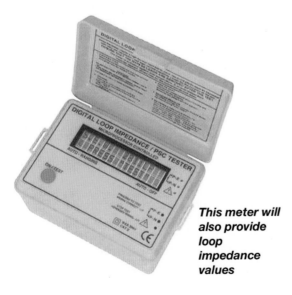

This meter will also provide loop impedance values

Figure 8.11 – *A prospective short circuit test meter – values in kA*

Please measure your PCS values at the source of your supply. Disregard the supplier's cutouts and neutral terminal block. There will be no access at this point, but gain admittance to your supply via your distribution centre. Avoid measuring the prospective short circuit from a socket or from an outlet, within a cooker point. This method provides false and unacceptably low values.

8. The kilowatt-hour meter

We are all familiar with the electric credit meter. There are two basic types, but many different designs and models. Today, digital meters have over taken the original dial variety and come in many different shapes, sizes and internal design technology.

Dial type

This type of credit meter is still readily obtainable from your electrical wholesaler. They are often fairly basic in construction and are used mainly for private metering or construction site management where the source of the temporary electrical supply is not from the public network.

How it works

The kilowatt-hour meter is, in reality, a skeletal electric motor in which the armature (the part that rotates) consists of a non-ferric (any metal which is not iron or steel) disc. Current consumed within your installation passes through the windings of an electromagnet, which is built-in to the meter. This provides power to rotate the disc. The disc, geared to an arrangement of interconnecting cogwheels, provides the means of registering the amount of electricity used. The speed of the disc is directly proportional to the current flow within the installation. A large permanent magnet encompasses a small section of the disc providing a magnetic break to enable fine control.

The way the dial works

One complete revolution of the first dial, divided up into units, will advance the second dial, registering 'tens', by just one division. When the second dial completes a full cycle it mechanically moves the third dial, which measures 'hundreds' by a single division and so on.

One complete revolution of the smallest dial, usually coloured red, corresponds to one unit of electricity. This is the test dial and its value is not normally included within a meter reading.

What is a unit of electricity?

The term *unit* has been fashioned for costing your electricity bill and is the product of time, in hours, and the power consumed in kilowatts. If, for example, you had a 2 kilowatt (2000 watts) electric fire burning for 10 hours, then the amount of units you would have to pay for would be 20. If the cost of each unit were to be 9 pence, then your bill for warmth generated would amount to £1.80.

When working in *Système International d'Unités (SI units)*, one unit represents 3.6 megajoule. A joule corresponds to 'the amount of work done per second by a current of one amps flowing through an impedance of 1 ohm'.

Reading a dial type kilowatt-hour meter

Figure 8.12 illustrates a typical dial face arrangement serving a single phase kilowatt-hour meter. The dials are divided into units of 10 000, 1000, 100, 10 and 1 to record the amount of electricity used. The small dial, often coloured red with white figures, is for testing purposes. Ignore when reading for costing.

First, check out the dial recording the highest value (in our case, it is the left hand dial in the illustration, calibrated in divisions of 10 000 units). Note the position of the dial indicator and select the *lower* of the two figures either side of the pointer. Our illustration shows that it has just past 1. Now go for the next highest value dial, to be found next to the left hand dial and take the smallest value of the two

figures ranged either side of the dial indicator. In this case, it is 3. Review the last three dials by repeating the procedure, making sure you record all values in a strict dial value sequence. The kilowatt-hour meter will then read 13 641 units.

Digital display meter

This type of meter is easy to read, as the amount of electricity used comprises just one single row of numbers read directly from screen for your records.

When a restricted (off peak) tariff is used within the same meter, two independent rows of figures are applied; one for the unrestricted day tariff units and the other serves all restricted circuits such as immersion heaters, night storage heaters and commercial ice banks supporting milking parlour bulk milk tanks. A touch of a button will change from one tariff to the other – known as radio telemeters.

Figure 8.13 illustrates the basic components serving a radio telemeter. The meter has a miniature FM radio receiver switch that responds to signals generated from radio transmitters at preset times of the day. The signal received switches 'ON' the restricted section of the telemeter to provide cheap electricity for the consumer. After a set period, the meter receives another signal; an order to switch 'OFF' the restricted supply. The twenty-four hour unrestricted tariff is constantly in circuit, recording whenever we use current from the supply. The liquid crystal display (LCD) panel, ranged in the centre of the meter, provides the following data at the touch of a button:

- tariff cost per unit,
- units recorded tariff by tariff,
- the time of day – usually this is in Greenwich Mean Time,
- the calendar date,
- status.

Clockwork/electric time switches provided in some older installations carry out the same function as the modern radio telemeter switch. They are still there because the supplier has not caught up with changing them yet!

10,000   1000   100   10   1

The arrows indicate the direction of the dial
Read off the number which has been past

Figure 8.12 – *The dial arrangement serving a kilowatt-hour meter*

Figure 8.13 – *The radio telemeter is designed to receive radio signals which will either switch 'ON' or 'OFF' the restricted supply or the customer's night storage equipment*

Figure 8.14 – *A compact voltage, continuity, resistance and current tester*

9. The socket polarity tester

This pocket meter is a very useful aid to check out the following but do not use this tester as a means to record your data:

- the correct polarity of a socket outlet,
- the status of the earth connection (whether sound or missing),
- the reversal of the phase and neutral conductors,
- some have a residual current device tripping facility up to 300 milliseconds,
- some are equipped with a small touch pad enabling the detection of raised earth voltages, equal to or above 50 volts,
- some, but not all models, have a built-in audio sounder notifying the wiring condition.

The device is easy to use. Just place it in the socket under test and it will automatically provide data for your records.

10. Other types of test instrument you might require

- An all-in-one compact voltage, continuity, resistance and current tester as shown in Figure 8.14. These are known as general purpose multi-meters and they are also able to check your diodes and transistors.
- Steel trunking and conduit continuity tester (value read in ohms).
- Applied high voltage tester.
- High voltage insulation tester – 1000 volts DC and up to 5000 volts AC.

Important issues

Confirming the correct test instruments

Before you start your testing procedures, verify you have the correct instruments to use. For example, it is not a good idea to use a high-scale multi-meter to measure the electrical insulation resistance of your installation nor, equally, is it wise to measure the external earth loop impedance (*Ze*) with a low scaled digital ohmmeter. Under these conditions, values obtained are totally unreliable – so keep to the rules!

Your minimum test instrument requirements will be as follows:

- continuity/insulation meter,
- socket polarity tester,
- earth fault loop impedance instrument,
- prospective short circuit tester,
- possibly an earth resistance tester,
- RCD auto-isolation timing meter graduated in milliseconds,
- a multi-meter.

Why we need to calibrate our instruments

Using your test equipment regularly will gradually weaken their reliability/accuracy factor that you have learnt to depend on. To keep your test instruments in tip-top condition make sure they are *calibrated* every so often to national standards. This will maintain their reliability within the specification of the instrument and keep you within the requirements of BS 7671.

A technically qualified engineer will standardise your instrument and restore to its former glory. This work is carried out in an environmentally controlled laboratory by experienced service engineers, equipped with the latest problem-solving tools and equipment.

Calibration service facilities are available from the following:

- large, nationwide electrical wholesalers,
- small, local, specialised businesses,
- most instrument manufacturers such as Robin® or Fluke® have an in-house calibration service.

Regular calibration is now a requirement of all approved bodies which recognise ISO 9000 standards – don't miss out!

The use of documentary evidence

Once you have completed your test certificate, the data you have collected may be used in many different ways. Many of the uses are summarised within the next paragraph; other uses can be as follows:

- The external earth-loop impedance can judged to be satisfactory, or otherwise, in relationship with the type of earthing arrangement provided. For an example, there would be a serious problem if the external earth-loop impedance you measured from a TN-S earthing arrangement were to be between 10 and 15 ohms.
- The date of the next inspection can be recorded. This is helpful, especially when the installation is a temporary arrangement which serves a construction site.
- The cross sectional area of the circuit conductor and rating of the protective device will advise your client whether he/she can upgrade the connected load. Example: from a 6 kW shower to a shower rated at 8.5 kW.
- The electrical installation certificate will clearly indicate whether the maximum permitted earth-loop impedance has been breached on any circuit installed. It is not a good to have a 32 amp circuit served by a 'B' type mcb with a earth-loop value of say, 6 or 7 ohms, when clearly it should be 1.2 ohms or under.
- Your customer, or their technical advisor, will be able to decide at a glance whether the installation is satisfactory or has borderline cases which might materialise into more serious problems. Example: insulation values of 0.5 megohms or slightly above and protected by a 30 milliamp residual current device, or whether, for example, too many lighting points have been added to a particular lighting circuit.
- Technically aware people will be able to judge whether any RCD in circuit trips within the time and requirements laid down in BS 7671.
- Customers who buy a property for renting will require this document.
- Your builder will also require the electrical installation certificates for his/her records.

Compiling your test certificate

After you have completed your *periodic report* or *test and inspection certificate* you will find it mainly comprises a mass of figures and snippets of text here and there. But what does it all mean? The following paragraphs will briefly explain and is based on the *NICEIC Domestic Electrical Installation Certificate* in accordance with BS

7671: 2001. BS 7671 also has forms; please ensure you know what form you are dealing with as it could cause confusion. Please also be careful to avoid confusion with the Building Regulations Document 'P' 2002 and 2004 Amended requirements.

Page 1

1   *Client/address* – the address of the person who is paying for the work done.

2   *Installation address* – where the work has been carried out.

3   *The extent of your testing and inspecting* – what you have tested. For example: all fixed wiring served by 230 volts, AC.

4   *Whether it is 'New', 'An addition' or just 'An alteration'* – tick the appropriate box.

5   *Particulars of the installation* – fill in each box as appropriate, according to the design of your installation.

6   *Design, construction, inspection and testing* – this is the section in which you sign, print your name and add the date when the installation was tested. There is a text box in which you can add details of any departures from our regulations. If all is correct, then just write the word, 'None'. A small window, to record written detail of the design, construction and the BS 7671 amendment worked to, will also be present. A typical entry would be: 'Amendment number 2: 2002'.

7   *Particulars of the approved contractor* – just fill in what is required.

8   *Next inspection* – for a private dwelling it will ten years, a church five years and a temporary site installation, just three months.

9   *System type, earthing and bonding arrangements* – tick the appropriate box: TT, TN-CS, or TN-S earthing system. Give details of the location of any earth electrode installed and the value of your measured external earth-loop impedance (Ze). There are also boxes available for the type and size of main protective conductors used and which extraneous conductive parts have been bonded. These must be filled in.

Page 2

1   *Schedule of items inspected* – there are about thirty-four small boxes in which to tick or write the abbreviation 'N/A' within. It is important that all boxes are written up; none must be over-looked.

Most of the information required is straightforward but some of the box instructions are slightly difficult to figure out. Here is a small example:

– *Erection methods* – how the installation has been built and formed.

– *Particular protective measures for special installations and locations* – this is to do with bathrooms, kitchens and other wet areas such as shower rooms and possibly milking parlours.

– *Choice and setting of protective and monitoring devices (for the protection against indirect contact and/or over-current)* – this refers to the type and current rating of the mcb/fuse protecting the circuit under review. For example, it would be very unwise to install a type 'B' 32 amp mcb to serve a 10 kVA site transformer circuit. What form and type of over-current protection would you suggest?

– *Functional testing of assemblies* – do all the appliances and current consuming assemblies operate as intended? These you will have to check.

– *Proximity of non-electrical services and other influences* – this means you must make sure your cabling and other electrical services such as television aerial systems and telephone wire are routed well away from hot surfaces or steam pipes, etc.

I am sure there must be other instructions you wish had been written in a more understandable way. Please ask your tutor to explain in plain words what you wish to know.

Circuit details

In this section you must provide the following details:

■   *The destination of your circuits* – cooker, ground floor lighting, etc.

■   *The type of wiring used* – this column is entered as a code, for example, pvc/pvc cable will be annotated as 'A' whereas pvc cables drawn into steel conduit will be entered as 'B', etc. Cable code details are printed on the side of the test certificate.

- *Reference method used* – this is written as a numerical code which you will find complete details of in Appendix 4 of BS 7671.
- *Number of points served* – just count the number of socket outlets that have been installed, remembering that a twin socket will count as two points. Other items you will have to count are the number of lighting points, associated extractor fans and shaver sockets which are served by the circuit under review; the number of fixed wall heaters and how many smoke detectors have been installed, etc.
- *The cross sectional area of the conductors* – provide details as installed. For example, 2.5 mm^2 for final ring circuits and immersion heaters, 1.0 or1.5 mm^2 serving lighting arrangements, etc.
- *The maximum permitted disconnection time in milliseconds allowed by BS 7671.* For circuits which have sockets incorporated, the time is listed as 0.4 second. For other circuits such as lighting arrangements and intruder alarms etc., the value prescribed is 5 seconds.
- *Details of the over-current device serving the circuit under review* – please provide the BSEN number of the device (BS 60898 for mcbs), the type that has been used (Type 1, 2 or 3, B, C or D) and the current rating of the device (6, 10, 16, 20 or 32 amp, etc.).
- *Details of any inline residual current device that has been used* – for example, 30 milliamps to protect socket outlets, 100 milliamps to serve lighting arrangements, immersion heaters and central heating control circuits.
- *The maximum Zs impedance permitted by BS 7671* – values for many frequently used over-current protection devices are given in Tables 41B1, 41B2 and 41D of BS 7671. The NICEIC or BS7671 electrical installation certificate provides us with limiting values of earth-loop impedance when measured at ambient temperatures of up to 20°C. These figures are about 80% of the values provided by our regulations but are rounded down. A selection of maximum *Zs* impedance values follow for various types of over-current circuit breakers:

Type 'B' (circuit breaker to BSEN 60898 or BS 3871 at 0.4 and 5 seconds disconnection time)

- 6 amp, maximum of 6.4 ohms
- 10 amp, maximum of 3.84 ohms
- 16 amp, maximum of 2.4 ohms
- 20 amp, maximum of 1.92 ohms
- 32 amp, maximum of 1.2 ohms

Type 'D' (circuit breaker to BSEN 60898 or BS 3871 at 0.4 and 5 second disconnection time)

- 6 amp, maximum of 1.60 ohms
- 10 amp, maximum of 0.96 ohms
- 16 amp, maximum of 0.6 ohms
- 20 amp, maximum of 0.48 ohms
- 32 amp, maximum of 0.3 ohms

Semi-enclosed fuse to BS 3036 at 0.4 seconds disconnection time

- 5 amp, maximum of 8.0 ohms
- 15 amps, maximum of 2.14 ohms
- 20 amps, maximum of 1.48 ohms
- 30 amp, maximum of 0.91 ohms.

Test results

The last section of the test certificate has a dozen columns in which installation data must be provided. Details follow:

Circuit impedances in ohms

- *The resistance of a final ring circuit when the 2.5 mm^2 phase, then neutral conductor is measured from end to end.* For an average size one bedroom flat this would be between 0.65 and 0.86 ohms for each conductor measured.
- *The resistance of the circuit protective conductor when measured from end to end.* When working alongside the sample figures above, expect values of between 1.08 and 1.43 ohms. A rule of thumb is to multiply the impedance of the phase conductor by a factor of 1.67.
- *$R_1 + R_2$ values (R_1 is the resistance of the phase conductor whereas R_2 represents the resistance value of the circuit protective conductor).* When applying our sample figures to this example, expect values between 0.43 and 0.57 ohms. To obtain this $R_1 + R_2$ value, apply the following expression:

$$\frac{R_1 + R_2}{4} \qquad (8.1)$$

- *The value of R_2, the circuit protective conductor, when measured from end to end.*

Providing you have completed the $R_1 + R_2$ column, there is no need to provide data for this column. R_1 and R_2 are not just related to final ring circuits – any circuit can have an $R_1 + R_2$ value.

Insulation resistance values throughout the installation

Insulation values

There are three columns to complete here. Values obtained will be in megohms. First obtain the insulation resistance between the phase and neutral conductors. Then target your test between the phase conductor and earth. Finally acquire the insulation value between the neutral conductor and earth. Aim for high values. Although 0.5 megohms is acceptable for a standalone circuit, it is not good practice to have too many circuits at this level. As all circuits are in parallel formation with each other from the distribution centre, it stands to reason that the total value of the installation will be considerably lower than this figure. If a value of less than 2 megohm is recorded, a problem may well exist and is worth looking in to. See *On Site Guide* for more detail.

Figure 8.15 – *A light emitting diode (LED) voltage tester*

- *Polarity of the accessories and appliances fitted.* Use a plug-in polarity tester to test your socket outlets and a simple LED probe tester to check your other accessories – Figure 8.15 illustrates.

- *Maximum measured earth-loop impedance.* Check out the maximum earth-loop impedance you obtain from every circuit listed on your test certificate. As a rule of thumb the total impedance for each individual circuit will be more than the $R_1 + R_2$ values obtained within the installation plus the value in ohms of the common external earth-loop impedance value.

- *RCD operating times.* Use a RCD disconnection/timing meter to obtain the disconnection value in milliseconds. First test at the normal setting of, say, 30 mA and then again at a faster setting using the five times IΔN switching mode. A typical value to expect would be 18.9 ms on a normal 30 mA test setting and 6.3 ms when using the heavier test

current. The maximum disconnection time permitted by BS 7671 is 40 ms (0.04 seconds) for circuits which have sockets incorporated and 5 seconds for other circuits such as immersion heaters and central heating controls, etc.

The last section of your electrical installation certificate is designed to record the serial numbers of your test equipment such as continuity meters, insulation testers, instruments which measure earth-loop impedance values and so on.

Other types of certification available

- Electrical Contractors Association (ECA).
- National Association of Professional Inspectors and Testers (NAPIT).
- ELECSA, for work which is covered by Part P of the Building Regulations 2002 and 2004 Amended. For this, you must be registered with them.

Methods of testing

Continuity

There are many ways to carry out this test; this is just one of them.

BS 7671 requires you to check that all circuit protective conductors are electrically sound and securely connected – whether they form part of a multi-core cable or routed independently.

The way it is done

To test, securely attach a long stranded or flexible insulated lead to the common service terminal of your continuity meter. The accompanying lead will be the instrument's own test lead and represents the business end of the procedure. You now have a long and a short lead attached to your continuity meter.

Connect the bared end of the longest lead firmly into the earthing terminal serving your distribution centre. This will let you wander freely to test all circuit protective conductors within the radius of the longest lead as shown in Figure 8.16. If you are unable to 'NULL' the resistance of your test leads then you must measure the total resistance offered by the leads and deduct this value from future test readings – Figure 8.17 illustrates.

A point to remember

The resistance of a conductor is directly proportional to its *length* but inversely proportional to its *cross sectional area*. For this reason, the further you travel from your distribution centre, the greater the resistance value of the protective conductor under test will be.

Testing

Start near to your distribution centre and test the protective conductors serving each electrical accessory within the installation. Remember to test all independently routed conductors which serve water pipes and extraneous metal work such as RSJs, air ducting or the steel channelling constructed for wall partitioning, etc. Disconnect independently installed protective conductors before you test to remove any parallel earth situation. This will then provide you with a better reading.

When you record higher than expected continuity values

Higher than expected continuity values must be treated with suspicion and the cause investigated at once. The problem could be caused by one or more of the following:

- test meter problems (fault, loose service leads),
- contamination,
- loose connection at the accessory,
- connection overlooked (not made),
- site damaged/broken connection,
- an accidentally severed conductor by others,
- conductor severed through over-tightening,

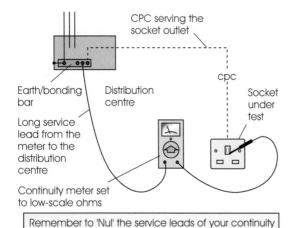

Figure 8.16 – *Continuity testing with a meter and long lead*

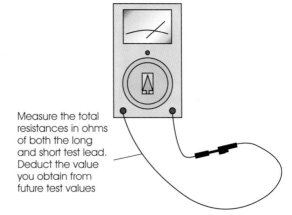

Figure 8.17 – *Finding the value of your test leads if you are unable to 'null' them to zero*

conductor snapped resulting from preparing the bare end with side cutters.

Please make sure all your protective conductors and supplementary bonding conductors, linked to other parts but not necessarily to do with the installation, are sound and electrically secure before you commission your installation.

When your continuity testing is complete, re-check your instrument by placing the leads together to obtain a *zero reading*. If you have been using an analogue meter, it might be necessary to re-adjust the instrument's pointer to zero via the adjustment screw below the display panel.

Insulation resistance

This test will make sure that the insulation serving the copper conductors and electrical accessories are free from any fault condition between your current carrying conductors and earth.

The way ahead

- First, remove or disconnect all lamps and current-using equipment. Isolate any electronic gear to avoid permanent damage to printed circuit boards.
- Leave all switches in the 'ON' position but switch off all the over-current devices. This will allow you to carry out tests on each circuit without fear that one circuit could influence another. It is best to disconnect all your circuit neutral conductors from the neutral bus bar for fear that one circuit could influence another.
- Testing is carried out in four easy stages. Let's imagine a 400 volt, three phase and neutral installation serving a machine is ready for examination.
  - *Stage 1:* place the test leads into your instrument, turn the selector switch to the 500 volt range and test your instrument by touching both leads together to ensure it works as intended. Measure the insulation resistance between all current carrying conductors. Start your test on the load side between L1 and neutral, making sure that the neutral conductor is disconnected from the supply and then follow the procedural advice given within the last illustration.

  - *Stage 2:* next, test the insulation resistance between each individual current-carrying conductor and the main protective conductor as shown in Figure 8.18. The value you obtain must not fall below the value prescribed by BS 7671.
  - *Stage 3:* finally, measure the collective resistance between all current-carrying conductors and earth. The conductors can be made common by linking them out as illustrated in Figure 8.19. The value obtained must not be less than the accepted value.
  - *Stage 4:* all tested connections must be identified to meet specified requirements. Remember to place a 400 volt warning label on the front cover of the switch gear before leaving.

Acceptable values to attain

The total resistance recorded must not fall below 0.5 megohms (500 000 ohms) when the nominal voltage is no greater than 500 volts. Above 500 volts and up to 1000 volts, the minimum

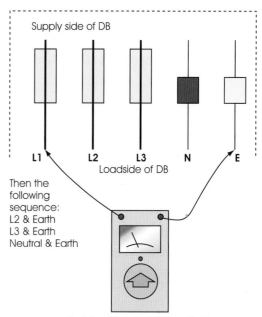

Switch off breakers before testing

Figure 8.18 – *Obtaining the insulation resistance between the conductors and the circuit protective conductor*

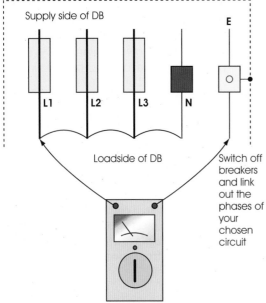

Test between the linked phases and earth

Figure 8.19 – *The insulation resistance between all phase conductors and earth*

insulation resistance must not fall below 1 megohm (1000 000 ohms). When SELV or PELV circuits are tested, the minimum insulation resistance is 0.25 megohms. See Regulation 713-04-04 for details.

When you obtain unacceptable values

If the values you obtain are below those which are prescribed by BS 7671, then a check must be made to find the reason for the fault condition. Never walk away from this problem; it must be remedied to bring the insulation resistance up to an acceptable level. Page 227, at the beginning of this chapter, lists several reasons why fault conditions occur within a completed installation.

Your instrument's output voltage

A typical digital insulation tester has a test voltage output of 500 and 1000 volts DC. Some models often have three voltage ranges: 250 volts to serve SELV and PELV circuits, 500 volts for low voltage circuits and 1000 volt DC to provide test facilities for high voltage circuits. The accuracy factor for such a meter is about ± 1 digit or ± 1.5% of the value obtained.

Polarity testing – with no electrical supply present

A simple way is by use of an extra low-voltage bell and battery set with one very long lead attached to the phase bus-bar of the distribution centre. The other lead will be short and so wired that, when both leads are placed together, the bell will ring. With all lighting switches placed in the 'ON' position and lamps removed, just unscrew the cover of the ceiling roses, or unscrew a switch if you prefer, and place the bare end of your shorter cable on the brown phase conductor serving the accessory. If the circuit under test is wired correctly, the bell will ring. If you prefer, use a purpose made polarity tester.

To polarity-test socket outlets with the electricity switched 'ON', just plug in a polarity testing instrument as shown in the previous paragraphs and test each socket outlet served by your chosen circuit.

Earth-fault loop impedance tester

There are three leads attached to this instrument. They are colour coded: *brown* for the phase conductor, *blue* to serve the neutral connection and the *green/yellow* lead is attached to the principal earthing bar. Attach the coloured leads from the instrument to the correct supply conductors feeding the main switch within your distribution centre. Clip the green/yellow lead to the principal earthing bar. As all the leads are polarised, it is important they are correctly connected. Once connected, press the 'TEST' button and read off the value in ohms. Additional information is obtainable from previous paragraphs within this chapter.

Earth electrode resistance meter

This type of meter, illustrated in Figure 8.20, calculates the electrical resistance of the installation's earth electrode. It is quite a cumbersome instrument to carry around as it comes with many attachments which includes the following:

- a long test lead set,
- two auxiliary earth-test spikes,
- a carrying case for the instrument and accessories,
- an instruction manual,
- the earth electrode resistance meter.

Most earth resistance testers offer three test ranges and are able to detect, and measure, the presence of residual earth voltages.

When your installation is served by a 'TT' earthing system or your installation is supplied from a private generator you are obliged to provide your own earthing arrangement. This will take the form of a steel, copper coated electrode or copper earthing tape. Once installed, the earth electrode resistance must be measured and the value logged within an appropriate document. To operate effectively the average value obtained should not exceed 100 ohms.

The *IEE Guidance Notes 3* suggests a couple of methods to test an earth electrode – both will be reviewed.

How it is done

Method one

- Drive the two auxiliary earth spikes, supplied as part of your kit, into the ground at distances recommended by the manufacturer.

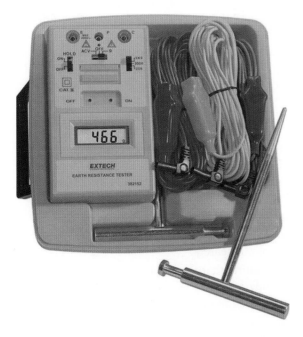

Figure 8.20 – *Earth electrode resistance meter. This test can also be carried out with an earth-loop impedance meter*

As a rule of thumb, this is usually 10 to 5 metres between each spike and the principal electrode under test – Figure 8.21 illustrates using a hand cranked instrument.

- Disconnect the earth electrode conductor at your distribution centre. This is very important.
- Connect the electrode under test, together with the supplementary and auxiliary test spikes, to the meter. Please refer to the maker's instruction manual for precise connection details. Usually, the test lead serving the auxiliary spike is connected to meter terminal 'C', or if you are using an older model, 'C2'. The supplementary test spike is attached to meter terminal 'P' (in older versions, 'P2') and finally the test lead serving your earth electrode is coupled to meter terminal 'P1'.
- Select a suitable resistance range (start at the highest rating if you are not sure) and press the meter's test button to reveal the resistance value of your earth electrode in ohms. During the winter months, when the soil has had its fair share of rain, expect lower values than you would, had you carried out the same test during summer time.
- To confirm your first reading the central supplementary test spike must be removed and placed, first further and then nearer to the electrode under test, keeping the distance travelled from the central point the same in both directions. Each time you reposition the supplementary test spike you must make a test.
- Once you have made three successive tests, calculate the average value and record appropriately.

Method two

- As an alternative, a phase earth-loop impedance tester may be used to obtain the value of your earth electrode.
- Isolate all outgoing circuits from your distribution centre and remove the outgoing green/yellow conductor serving the earth electrode.
- Your earth-loop impedance tester has a set of three, plug-in, flexible test leads. Connect the

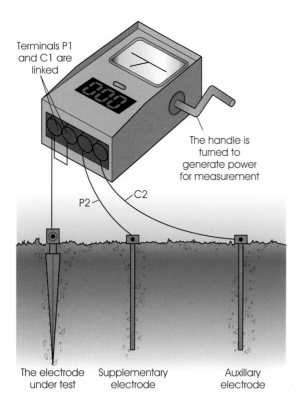

Terminals P1 and C1 are linked

The handle is turned to generate power for measurement

P2

C2

The electrode under test

Supplementary electrode

Auxillary electrode

Figure 8.21 – *Earth electrode testing using spikes driven into the ground*

Testing your residual current device

RCDs must be both electrically and mechanically tested and their auto-tripping values recorded in an appropriate document. To mechanically test, just press the small test button marked 'T' on the face of the device. Disconnection should take place more or less within 0.2 seconds but the test should not last longer than 1 second. Regulation 713-13-01 will confirm.

Test an inline RCD serving a final ring circuit by plugging your test meter into one of the mainline sockets; turn the selector switch to a suitable test current, then press the test button. A design feature incorporated within some models makes it essential to hold this button down in order to read the tripping value in milliseconds. Fast/normal-trip facilities and a phase angle selector switch for positive and negatve half cycle testing are also provided.

Using a residual current device as additional protection

If you use a residual current device as additional protection against direct contact, the value of its residual operating current must not exceed 30 milliamps and must provide a tripping time of 0.04 seconds at a residual fault current of five times I∆N. If your test is unsuccessful, then the duration of the test current is limited to 0.5 seconds. Refer to Regulation 412-06-02 and BS 4293:1983.

Functional testing – what it means

Functional testing is checking out the equipment and accessories installed to ensure they operate as intended. Your checklist could include some of the following items:

- operation of the door entry system,
- check the operation of the cooker, hood and plinth heater,
- confirm that the night storage heaters are operational,
- check out the central heating controls,
- test out the immersion heater and adjust thermostat setting,
- set up/adjust passive infra-red security lighting,

green/yellow lead to the disconnected earth electrode conductor. This lead is usually provided with an insulated crocodile clip. The *brown* and *blue* leads are both fitted with insulated probes which are press-connected to the incoming phase and neutral supply terminals respectively. As this meter is polarised, it is very important to ensure the test leads are correctly positioned. If a mistake is made a red warning light will appear advising that your connections are incorrect.

- Next, press the red 'TEST' button and read off the value in ohms of your earth electrode.

This is by far the easiest and quickest means of testing an earth electrode and is used more frequently than Method one. Remember to record your test results.

- commission and fine tune the intruder alarm system,
- test the automatic mains assisted smoke alarms or communal fire alarm system. Check the decibel level with the requirements of the specification.

There are many more and they will all vary in technical content. It will depend on the type of installation which has been carried out and the demands of the general specification.

Minor works certificate

This type of certificate is issued to a client at the completion of a minor works programme which does not include the provision of a new circuit. For example, use when luminaries are added to an existing lighting arrangement or when a two-way lighting circuit is converted into a two-way and intermediate system.

The certificate is formatted as a five part A4 form in which there are text boxes and data spaces for appropriate information. The information needed to complete this form is:

- The top section is reserved for your client's name and address and a small box is provided for the certificate's in-house serial number.
- Part 1 of the minor works certificate describes the work you have carried out and where and when it was completed. There is also a space reserved for any departures you have made from BS 7671.
- Part 2 concentrates on the installation details – the type of earthing arrangement available; your method of protection against indirect contact; the style and rating of the protective device you have used and any relevant comments on the existing installation.
- Part 3 deals with all the essential tests you have carried out:
  - the value in ohms of your earth continuity conductor,
  - the polarity and the continuity of all circuit conductors,
  - all insulation resistance values you have obtained,
  - the value of the earth-loop impedance of the electrical mains,
  - the residual current rating of the RCD and its operating time in milliseconds,
  - functional testing of the equipment,
  - record of the test instruments used and their serial numbers.
- Part 4 is the *declaration* where you formally sign and declare your position within your company. Please remember you must have sufficient technical knowledge and experience within the electrical installation engineering industry and be totally familiar with all testing and inspection procedures.

  In the space after to 'amended to', please add the latest amendment – for example, No. 2: 2004, No.1: 2008, etc.

The Minor Works Certificate document is based on a model shown in Appendix 6 of BS 7671: 2001. You are welcome to create your own in-house forms based on this design, but if you do, please add a courtesy acknowledgment at the bottom of your document.

The requirements of testing

Sequence of testing

You must carry out your *initial testing* in the following order. This you will find in the *IEE's Guidance Notes Number 3* (GN 3) available at your college library or technical reference point.

- Check the continuity of your protective conductors and your main and supplementary equipotential bonding conductors – Regulation 713-02-01.
- Measure the continuity and value of the final ring circuit conductors – Regulation 713-03-01.
- Test the insulation values of all your conductors – Regulation 713-04.
- Determine the insulation value of any site applied insulation.
- Check out protection afforded by the separation of circuits:
- Check out the protection given by barriers or enclosures provided during the erection of the installation.

- Test the polarity of all your accessories and fixed equipment (with the supply not on) – Regulation 713-09.
- Measure the earth electrode resistance – Regulation 713-10.
- Test the tripping time of any residual current devices and RCBOs you have in circuit circuit. A warning notice should be posted informing your client that that the device is automatic and will switch 'OFF' the supply in the event of a fault to earth developing – see Figure 8.22; Regulation 713-13 confirms this. The IEE recommends that you recheck the polarity when the electrical supply has been switched 'ON' before supplementary testing is carried out using an appropriate volt meter.
- The earth-loop impedance value of the incoming supply.

Please remember that periodic tests are carried out in a slightly different sequence than initial inspection testing procedures. Please check out GN 3 for exact details. An advisory notice must be placed on or near to the distribution centre recommending when the next periodic inspection should take place.

Test instruments – what you should know

Your test instrument's test leads are made to the requirements of GS 38 (General Series 38). Test lead accessories such as crocodile clips, probes and fused probes must either be 'snap-locked' fitted or screw-fitted to the test lead to provide firm contact. Briefly, the GS 38 directive recommends the following for test leads serving test instruments:

- to be well insulated,
- be about 1.2 metre in length,
- not to have any exposed conductive parts other than the probe tips,
- be coloured differently to distinguish one lead from another,
- be strong, yet flexible with a doubly insulated conductor of adequate size,
- be so designed that exposed conductors are not accessible to your fingers if a test lead became detached from a probe or from the instrument whilst working,
- must have the protection of a fuse when used for voltage measurement,
- must not have more than 4 mm of exposed metal at the working end and ideally be spring-loaded,
- ideally, moulded finger barriers should be provided as illustrated in Figure 8.23.

The accuracy of your test instruments

For an analogue instrument, 2% of full scale deflection is adequate and 2% of the higher limit range is OK for digital meters. To accomplish this standard, you should have your test instruments recalibrated at least once a year or after each mechanical mishap such as an accidental static impact (i.e. bumping into

---

YOUR RESIDUAL CURRENT DEVICE

Your installation, or part of it, is protected by a device will automatically switch off the supply if an earth fault develops. Please test quarterly by pressing the button marked 'T'. If the device does not switch off the supply when the button is pressed, seek expert advice.

Figure 8.22 – *A warning notice must be placed near to the distribution unit to inform when a residual current device has been place within a circuit that it will automatically switch 'off' should a fault occur between a phase or neutral conductor and earth*

something) or if dropped from a height, etc. For details of other specialised test instruments requirements, please refer back to the paragraph titled 'Choosing your test instruments' on page 235 in this chapter.

Your health and safety

There are many reasons why accidents happen when using test instruments. Often they are caused due to carelessness or lack of training. Some other reasons are:

- unsuitable or make-shift test leads,
- cracked or broken probes together with moisture content,
- probe tip far too long, causing a short circuit condition to happen,
- too much current drawn through the instrument,
- the test meter selected to the wrong function.

For additional information, please seek advice from the manufacturer's handbook.

The precautions you can take

Try to keep to the following safety rules:

1 Never take risks with electricity – it's far bigger than you are!
2 Keep to the guidelines you have been taught.
3 Avoid using old-fashioned test equipment.
4 If you are unsure of which working range to select, start at the highest and work your way down.

5 Check for damage and sloppy terminal connections before using your instrument.
6 Report defects and shortcomings to your supervisor as soon as possible.

Display scales

Your test instrument will have one of the following types of display scales:

- Figure 8.24 illustrates the plain digital type.
- A digital scale with a bar graph analogue indicator is shown as Figure 8.25.
- Figure 8.26 illustrates a straightforward analogue scale.

Our regulations as applied to testing

- BS 7671, Regulation 711-01-01 demands that every installation is inspected and tested both during production and on completion of the work.

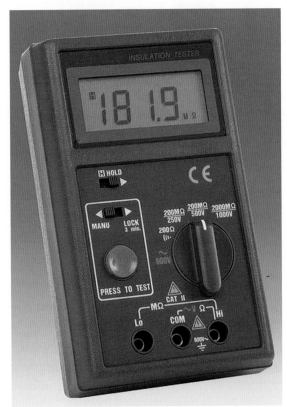

Figure 8.24 – Test meter displays – plain digital

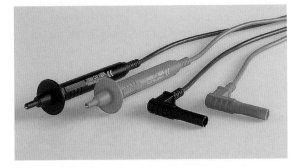

Figure 8.23 – *Moulded finger barriers are essential for test instruments*

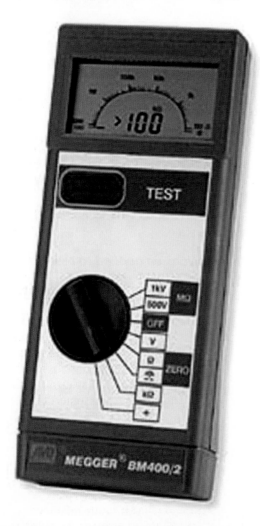

Figure 8.25 – *Test meter displays – digital with a bar graph*

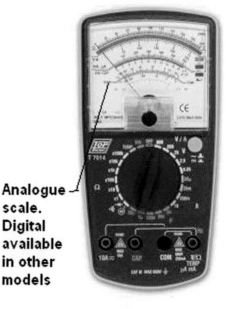

Analogue scale. Digital available in other models

Figure 8.26 – *Test meter displays – analogue*

- You must take preventative measures to avoid damage to both people and property during your testing programme.
- On completion of a successful programme of testing, a signed electrical installation certificate must be given to your customer.
- Inspection must be carried out *before* your testing sequence and is normally done with that part of your completed work disconnected from the supply. Regulation 712-01-01 will confirm.

- Regulation 712-01-02 advises that the purpose of the inspection and testing procedure is to make sure that the equipment and cabling used is of the correct type/colour and fitted in agreement with BS 7671 and is not noticeably broken so as to weaken the safety of the installation.
- BS 7671 demands that relevant items are inspected in the order recommended within the previous paragraphs of this chapter. Please refer to Regulation 712-01-03 for further details.
- Regulation 713-01-01 asks for initial testing procedures to be carried out in the order described in previous paragraphs within this chapter – see page 252.

Section 528 of our regulations advises that we must take into account how close other electrical services of different voltage bands and other non-electrical utilities are near to the installation.

Regulation 528-01-04 reminds us that fire alarm and emergency lighting circuits (for example, lighting from a generator of a central battery system) must be separated from other cables. This follows the spirit of BS 5266 and BS 5839. Band 1 (not exceeding 50 volts AC or 120 volts DC) and Band 2 cabling (from 51 to 1000 volts AC or 120 to 1500 volts DC) must never be mixed within the same wiring system unless physically separated by a barrier or wired with cabling best suited to the highest voltage present. For example, bell wire or telephone cabling must never be mixed with low voltage mains cables.

Testing smoke alarms in private dwellings

Remember to test the smoke alarms you have installed – use a decibel sound meter if required. There must be at least one on each floor and a detector within 7 metres of kitchens and living rooms or anywhere a fire could start. All alarms must be interconnected with one another and be installed at least 300 mm away from any lighting point whether positioned on a ceiling or wall. Originate your circuit, which does not have to be fire retardant cable, from either a distribution centre, supplied from a dedicated over-protection device (mcb or fuse rated at 6/5 amp respectively) or wired from a locally used lighting circuit – the choice is yours!

A smoke alarm circuit must *not* be protected by a 30 mA RCD. If your installation is supplied with a 'TT' earthing system, then the circuit should be connected to the fixed equipment section of the distribution centre where the residual current rating is 100 mA.

Other documentation

Report forms

From time to time your company may ask you to carry out an in-house technical review. This is just an electrical survey targeted on a selective property in which commercial interest is shown. One of the reasons why data is required is often because work is due to commence in the form of a 'design and build' project in the near future. This can be a straightforward means of obtaining the necessary information required. These reviews are not formal or binding as when accepting an electrical periodic inspection but are simply for commercial and technical reference. Forms such as these are in-house produced and could resemble the following, as illustrated in Figure 8.27.

It is a good idea to site-write your report in pencil on a sheet of lined paper. Once edited and presented how you would like it, you can transfer your text and figures onto an in-house form. This will minimise grammar and spelling errors within your document and show off your report in a good light.

Minor works certificates are a type of report which provides information to your client – for details please refer back to page 243.

Hand your in-house report to your supervisor or company manager. He or she will place it within a relevant job file for future reference. This type of data is very important when little is known concerning a potential job.

Providing documentation to your customer

Automatic copies are produced when hand writing an electrical installation certificate. The copy must be separated from the original document and properly filed for your records. Present your certificate well by placing within a suitable 'see-through' pocket and mailing it off with an accompanying letter to your customer.

TARA ELECTRICAL PLC
276 HIGH STREET WHITEPARISH WILTSHIRE SP5 34SP EC
Engineer's in-house report

Operative's name: *Andrea Schiffer*   **Date of inspection:** *14.10.12*

Client: *Dr. A.Y Lindenhurst*   **Address:** *'Zertos High St. Whiteparish*

Earthing arrangements: *'TT'*   **Earth loop impedance:** *10.42 Ω*

COMMENTS
Dr. Lindenhurst has asked for the installation to be rewired. It is now wired in rubber insulated imperial size cables which have very low insulation values – some as low as 0.26 MΩ from phase/neutral conductors to earth.
He has asked for the following:
Outside: *One wall lighting point [back and front] controlled by two PIRS.*
Porch: *One batten holder, one-way controlled and two twin 13 amp socket outlets.*
Kitchen: *One 58 watt fluorescent fitting with a diffuser [two-way controlled] one cooker point, et cetera, et cetera.*

Peter Kingsley.

Figure 8.27 – *An engineer's 'in-house' report form*

Handy hints

- Very long machine screws used to fix accessories to steel back-boxes can slice through the cable's insulation and cause fault problems.

- Switch off all fused connection units and double-poll switches serving remotely sited accessories when checking the continuity of your final ring circuit. If a load/supply wiring error has been made it will immediately show up.

- Use crocodile clips attached to your circuit when insulation testing. It is difficult trying to position the probes correctly whilst pressing the test button at the same time. You often require three hands!

- When an insulation test is carried out on a very long length of mineral insulated cable, always discharge the capacitance which has built up by short circuiting all current carrying conductors to the copper sheath.

- Some types of test buttons serving RCD timing/isolation meters are designed by default to be held down until the value is read. Working with this type of instrument means a casual push on the test button will allow you to glimpse the display for just a fraction of a

Handy hints

second before the screen switches to 'blank-mode'.

- Take care when insulation testing inductive luminaries. Short out your current carrying conductors (between each other and also to earth) to avoid charged capacitance discharging when you least expect it.

- If you are tracing an 'earth fault' which appears to have a very low impedance to earth on either the brown or blue conductor, try working from the ohms scale on your continuity meter. The physically nearer to your problem you are, the lower the insulation value.

- Make a habit of checking twice if the value you obtain is unexpectedly high.

- The Electrical Installation Certificate uses the term '*point*' within one of its text boxes within the 'Circuit details' section of this document. BS 7671 defines this term as 'any independent switch, socket outlet, fused connection unit, lighting point or a termination where fixed wiring yields to a permanently connected current carrying load'. A twin socket outlet is defined as two points whereas a domestic cooker and control switch is described as just one.

- Wear rubber electrical gloves, as illustrated in Figure 8.28, when carrying out earth-electrode resistance field tests.

- Make sure that all neon indicator lamps have been isolated from your circuit before carrying out installation testing. This will prevent false values being obtained.

- Isolate any electronic equipment from circuit such as door entry controls, intruder and fire alarm systems. Such circuits will not tolerate high test voltages and will be severely damaged.

- Make sure, when you are field testing the resistance of an earth electrode, the test spikes are the recommended distance apart. If your spikes are too close to each other their 'earth-resistance areas' will overlap one another and unreliable values will be gained.

- It is best to disconnect the load from a RCD before formally testing. Testing, mechanically by way of the built-in test button within the device, is no substitute for using a RCD test meter which measures the disconnection time in milliseconds.

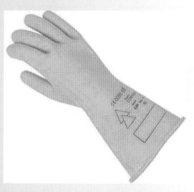

Figure 8.28 – *Wear electrical rubber gloves (made from natural rubber) which are suitable for testing when you are calculating the value of your installation earth electrode*

Summary so far . . .

- A precise detailed inspection and formal test is essential to conform to the requirements of BS 7671.

- Snapped, pinched and nailed conductors can often be the cause of fault conditions.

- Have your test instruments regularly calibrated.

- Carry out a periodic inspection in a way that disruption and disturbance to others will be minimised.

- Check that the equipment and cabling used within the installation conforms to British Standards or an EU equivalent.

- The term 'external influences' can mean many effects including the following: surrounding temperature, an external heat source, the presence of water and humidity, vibration and solar radiation.

- Charts, diagrams and technical information are ideally mounted within large document frames and fitted to the wall next to the electrical mains switch gear.

- Use your earth-fault loop impedance tester to obtain values at any point throughout your installation.

- Expect a prospective short circuit (PSC) value of between 2 and 6.8 kA in industrial urban areas.

- The plug-in socket polarity tester is a useful aid to testing. Not only will it provide polarity details but it will indicate whether wiring problems exist within the socket outlet.

- Always aim higher than the minimum accepted insulation value of 0.5 MΩ.

- The British Standard number for miniature circuit breakers (MCB) is BS EN 60898. MCBs are divided into types, and are used for certain categories of circuit. Types available to date are: 1, 2 and 3, B, C and D.

- Expect a value of between 0.65 and 0.85 Ω when measuring the phase conductor of a final ring circuit from end to end within a two bedroom flat.

- Use a RCD disconnection/timing meter to obtain isolation speeds in milliseconds. You will need to carry out two tests, one at the normal residual current rating of the device, the other at a faster test setting.

- If the values you obtain during insulation testing are below those which are prescribed by BS 7671, a check must be made to find the reason for the fault condition.

- If you use a residual current device as additional protection against direct contact, the value of the residual operating current must not exceed 30 mA and must provide a tripping time of 0.04 s at a residual fault current of 150 mA.

- Functional testing means checking out the equipment and accessories installed to make certain they operate as intended.

- The metal probe tips serving your instrument's test leads must be no longer than 2 mm in length and ideally, spring-loaded.

- BS 7671 demands that you take precautionary measures to avoid damage to both people and property during your testing activities.

- A mains-operated smoke detector must not be served from a 30 mA residual current device.

Review questions

1   List four typical fault conditions you might find on completion of your installation.

2   Briefly describe the 'thermal couple effect' which some times causes nuisance to testing activities.

3   Provide four reasons for carrying out a periodic inspection and test.

4   Describe two 'external effects' you would have to consider when routing cables.

5   Target four areas where you feel you should place labels within your installation.

6   Please list the types of test instruments needed for a formal domestic installation test.

7   State a typical value of earth loop impedance you might expect from a 'TT' earthing arrangement.

8   Briefly describe how you would carry out a prospective short circuit test at the source of the supply.

9   Describe the function of a socket polarity tester.

10 List two problems which can cause higher than expected insulation test values to occur.

11 How often must you carry out an earth electrode test when using supplementary spikes?

12 Briefly describe the meaning of functional testing.

13 Briefly explain what you would use a Minor Works Certificate for.

14 In the recommended sequence of testing, what are the first three tests needed to be carried out?

15 TRUE/FALSE – A domestic mains operated smoke alarm can be incorporated within a local lighting circuit served from a 30 milliamp residual current device.

9 Fault diagnosis and restoration

Introduction

This chapter is all about fault finding on electrical installations and equipment and identifies the principles and methods you will need to adopt. It deals with the safe and formally approved procedures for fault diagnosis and the return to normal operation conditions.

There are nine underpinning knowledge requirements in this chapter, all of which are compulsory.

◻ Identifying electrotechnical systems.

◻ Safe working practices and procedures when undertaking problem solving.

◻ The basic principles of fault finding.

◻ The stages of logical fault finding and making good.

◻ Where faults may occur within an electrical system.

◻ The symptoms of common faults within a system.

◻ The factors which could influence repair or substitution.

◻ The reasons which could affect renovation and generally making good.

◻ When special safety measures should apply.

Practical activities

There are just two formal activities which accompany this chapter and you will have to produce evidence to your assessor that you are able to carry them out to the satisfactory standard.

◻ Develop and make use of safe means for carrying out fault finding procedures.

◻ Resolve faults in electrical installations and equipment.

Single and three phase supplies – installations and equipment

There are many single and three phase electrical systems of which I am sure you have covered in your level 2 studies and during work time activities. Here are just some of them as a refresher:

■ *Power circuits* – single 230 volts in the form of final ring circuits and common radials (cooker, immersion heater, industrial fan, etc.). Three phase circuits will serve site transformers and cement silos, industrial motors, etc. The nominal voltage (the declared voltage) is 230 volts at single phase and 400 volts in a three phase circuit. These voltages are seldom true to their declared value and can vary up to 8% in either direction.

■ *Lighting arrangements* – unless highly specialised, these are single phase arrangements but are often circulated by means of a 400 volt, three phase distribution system. They may be wired in many different switching arrangements such as single way, two way, two way and intermediate and double poled, etc. Extra-low voltage (ELV) systems are popular in modern houses but care must be taken to control the thermal effects – they get very hot!

- *Control systems* – these are found both in industry, commerce and within domestic installations. Control circuitry for electric motors and other industrial applications can be either single or three phase, depending on the design of the equipment. In a domestic setting control circuitry serving central heating systems garden power requirements is single phase.

- *Components* – these are common throughout the electrical installation engineering industry. Look in any electrical wholesaler's catalogue and you will find hundreds! Remember to match up your components with the rest of your installation. It would be unwise to choose the wrong value of capacitance for an electric motor for example. Choose wisely and always used as the manufacturer intended.

Fault finding and safe working procedures

It is essential that the correct procedure is adopted before fault finding commences. Please remember the following – it could stop you getting into trouble!

- Locate the local switching arrangement serving the item you are investigating.

- Switch off and test with a volt meter that there are no conductive parts. Never use a neon test screwdriver or an electronic voltage detector as they are unreliable. This is important as some types of equipment have two sources of energy as with, for example, certain categories of central heating units.

- Ensure your isolation device(s) is/are off – remove the breaker or fuse, lock or physically tape up your switch in the off position.

- It is advisable to place a coloured warning notice near to your isolator or switch – ELECTRICIAN WORKING ON CIRCUIT. DO NOT SWITCH 'ON'.

- Before switching back 'ON' after carrying out your remedial work make sure that everything is as it should be and there is nothing left half done or in a dangerous situation. This is most important especially in houses which are occupied.

Basic principles of fault finding

It is important to know how a problem area operates and functions under normal working conditions. This will help you considerably and stop you adopting a 'try-it-and-see' approach to your problem. Use the expertise and the experience you have gathered together over the years concerning this and other similar systems. If stuck for a solution, use your mobile phone and have a chat with someone who knows – this can really help.

It is important to use a step by step and logical approach when you are carrying out fault diagnosis work and above all be extremely careful if you are forced to complete your investigation with the power switch on.

From fault finding to re-commissioning

This is best described in listed form:

1 If you are not sure of what to expect, collect together as much data as possible and have a list of contact telephone numbers ready in case you require them. Find out the sequence of events which led up the electrical breakdown. It is a good idea to speak to relevant people and to seek out any written reports which might be on hand.

2 Study the evidence you have gathered and use standard tests in order to diagnose the possible cause of your problem. These tests you would have studied in your level 2 course and include voltage testing, continuity, current tests and observation.

3 Come to a logical conclusion – it could be loss of continuity, heavy current drawn from your circuit, component failure or a voltage problem indicating certain fault conditions. Draw on your experience to make a final judgement.

4 Make good your fault condition and check that all is satisfactory and that it will work as intended. Be sure what you have done complies with our current electrical regulations.

Common faults within electrotechnical systems

There are many, so many a book of this size could be easily dedicated to this topic alone. Here is a few which can be browsed through; but problem solving will be made far easier when you know how the system you are working on, works.

- *Misplaced conductors* – control circuits and panels,
- *snapped conductors* – within accessory boxes, etc.,
- *machine screw serving an accessory box piercing the insulation around the conductor* – badly positioned conductors within an accessory box,
- *conductor which miss the target connection area and are screwed up without making connecting* – connecting without due care and attention,
- *loose conductors* – this can cause arcing when your circuit in on load,
- *burnt conductors* – often seen in immersion heaters circuits if the wrong type of flexible cable has been installed,
- *circuit incorrectly wired* – wiring error or lack of knowledge,
- *industrial vibration causing the insulation around your conductors to wear thin into a fault condition* – sometimes experienced within motor terminal housings,
- *dampness* – this will affect older properties and farms, etc.,
- *poor earthing* – can be the cause of electric shocks under a fault condition to earth,
- *dimmer switching and extra-low voltage transformer problems caused by electronic failure* – sometimes caused by over-heating due to close-proximity thermal insulation,
- *worn bearing, burnt-out windings within electric motors and contactor coil failure* – can be caused by age,
- *a low voltage value within your circuit or a low supply voltage can cause problems to your system* – this could be a problem which can only be solved by your supplier,
- *accessories, protective devices, luminaries, components and flexible cord and cables* – can also be the cause of electrical failure.

Typical symptoms of a fault condition

These will vary from fault to fault but some problem areas seem to show similar symptoms. This is where you will have to draw on your past experience to decide on the best possible option to take.

Loss of supply at the origin of your installation

- *Check your supply at your main distribution board* – phase to neutral, phase to phase and finally, phase to earth.
- *Possibly your suppliers fuse is ruptured* – check using a continuity meter.
- *Maybe an area distribution problem* – are any other properties involved? Check the supply at the source of the installation.
- *Your supplier's service fuse could be tested but it might mean breaking the security seal. If this is done, please inform your supplier* – check using a continuity meter.

Local loss of supply

- Your circuit fuse or breaker has switched the circuit to OFF.
- Over-current or overload device brought into play.
- A component, accessory or cabling problem.
- Cable or conductor accidentally severed by the site builders.
- A poorly or loosely connected conductor.
- Insulation pinched or sliced into by an accessory's machine screw.
- Rodent damage – sometimes in country cottages but often within farms and milking parlours.
- Phase fault to earth problem auto disconnecting a residual current device.
- A residual current device which has become over-sensitive or has broken down.
- Site sabotage or out of hours vandalism.

Your overcurrent device repeatedly switches to OFF

- *Too much electrical load* – motor circuit, etc.

- *Your over-current device has become faulty* – test by replacement.
- *A phase to earth fault has occurred* – immersion heater problem, a nail through a cable, etc.
- *A cable problem created in manufacture* – yes it does happen! Test by disconnection and using an insulation tester.
- *Ambient temperature* – when the weather is very hot.
- *The circuit could be incorrectly wired* – recheck with available data.
- *Wrong valued over-current device* – check the design current of the circuit and replace your breaker with a correct one.
- *An incorrect type of over-current device has been fitted* – a type 'B' breaker for site 10 kVA transformer, for example, where a type 'C' should have been installed.
- *A common filament lamp coming to the end of its useful life and rupturing will sometimes auto switch OFF your local over-current device* – confirm and replace the spent lamp.
- *The wrong size semi-enclosed fuse could have been fitted* – make sure and change.
- *Too many appliances connected to one circuit* – this sometimes happens when one final ring circuit serves every room in a house or business premise.

Transient voltages

This is the name given to *short lived high voltages* which have a short life expectancy of milliseconds – these, to all intents and purposes, are just a collection of lethal unwanted energy packages.

If they are valued *less* than twice that of the nominal voltage these momentary energy patterns are called *spikes* but if *greater* than twice the declared voltage they are known as *voltage surges*. Found within AC power lines and communication wiring, these transient voltages can do a great deal of damage.

Transients are often generated internally whenever the line current is interrupted – electric motor circuits which are constantly being switched ON and OFF automatically, photocopiers and fluorescent lighting circuits are examples which will cause this phenomenon. External transient voltages, on the other hand,

are usually cause by lighting strikes on power lines.

Spikes account for about 29% of supply line disturbances whilst transient voltage surges (TVS) amount to some 40% of all other disorderly activity. This voltage instability can and will do a great deal of damage to both equipment and components. Permanent damage can be caused to sensitive electronic equipment such as microprocessors and cause them to malfunction.

The way to stop damage from being caused is to fit a transient voltage surge suppressor in line. This is a device which limits the amount of energy arising from a TVS and protects electrical equipment such as computers and other types of microprocessors from harm. This device can be fitted as a permanent hard wired unit at the source of your supply or presents itsself in the form of a special extension lead which is slightly larger in size than normal. Many home computer terminals have this arrangement.

Insulation failure

This can cause a great deal of trouble within an installation or established equipment. There are many reasons why the insulation around our conductors will fail – here are some of them:

- Pinched or sliced conductors within steel cable carriers such as conduit and trunking or where cables are forced through small holes in walls.
- Vibration causing the insulation to be worn down to conductive parts. This can happen on motor circuits within the terminal housing box.
- Machine screws serving an accessory can pierce the insulation when initially fitted unless care is taken to avoid this happening.
- When a smaller size conductor is used than called for, heat will be generated and the insulation will soften. If the cables are disturbed there is a risk of insulation failure within the circuit most affected. This is very noticeable when the conductors are accommodated in metal clad cable carriers such as steel trunking, accessory boxes and conduit.
- PVC cables do not tolerate ultra-violet (UV) light. They will discolour and crack causing a

breakdown within the insulation. Keep your cables away from direct sunlight.

- Rubber cables will, over a number of years, dry out and crack causing the insulation to flake away from the conductor. This process always happens at the termination points of a rubber insulated cable installation – seldom within the main cable run itself.
- An insulation breakdown can occur within pre-1986 types of mineral insulated cable (MI). If the plastic compound should dry out, moisture will be absorbed in the white compacted insulation. MI cables were first introduced into Britain in 1936 – so there are many years of cable installation which could break down though the passage of moisture passing through dried out cable glands.

Component failure

It is wise to research the reliability of the components you are to use, such as extra-low voltage lighting transformers and dimmer switches, etc., at the pricing stage of a potential contract. Buying cheaply isn't always wise.

Components fail for a number of reasons. Rather than bore you I have prepared a small selection for you to browse through:

- *Extra-low voltage lighting transformers* – cheap ones will not last very long and will fail unexpectedly.
- *Dimmer switches* – they will either refuse to dim your lighting circuit, or not operate at all; buy sensibly.
- Capacitors will breakdown for a number of reasons:
  - plates shorting out due to the movement of silver in damp conditions,
  - bloated due to very high temperature and moisture,
  - short circuit due to mechanical shock, rupture or too high a voltage,
  - intermittent open circuit due to mechanical damage,
  - changeable values due to fault conditions occurring between the dielectric and the capacitor plates.
- Contactor coils and electric motor windings will burn out from time to time. It is helpful to know the value in ohms of the

component/part which has failed. A comparison can then be made for replacement.

- Chokes and ballast units sometimes overheat, especially in the summer months when it can be very hot. This causes the enamelled insulation serving the windings to become soft and melt to expose bare conductive windings. The winding short circuits and the component is destroyed.
- Low-priced equipment served with an electronic component will often fail long before they should. If this happens too many times it can become quite embarrassing to carry out the same remedial work over and over again. Choose all of your components wisely – it will pay off in the long run.

Factors influencing repair or replacement

There are many; a case of 'horses for courses' and also depending on many practical factors. Here are just a few you should familiarise with.

Costs

If the projected total cost proves too uneconomical for repair then it would be considered sensible to apply one of the following options:

- Discontinue the repair.
- Find a compatible cheaper alternative.
- Look for a good second hand matching part.
- Put on hold until it would be considered financially sound to carry out the repair.

The availability of a replacement

If the availability of a genuine component part is estimated to take weeks or even months to arrive on site, then you have the following options to consider:

- Temporarily replace using a safe substitute part until the arrival of the genuine component.
- If manageable, replace the faulty equipment with a hired or borrowed counterpart until such time the new serviceable part arrives for fitting within the original equipment.
- Withdraw from the remedial work.

Downtime under fault conditions

In case some of you are not familiar with this term, 'downtime' is a period of time when something or somebody is not operating or being productive due to an electrical fault condition or a total malfunction of a system. Everything stops!

A breakdown occurring some 150 miles away from your company's centre of operations could be resolved very quickly with a telephone call to a local electrical contractor who could carry out any remedial work on your behalf. You would then compensate them for their troubles. This would save a great deal of time in travelling to and from your remote site. Downtime under fault conditions would be, in effect, completely done away with and should not affect efficient repair or replacement.

Your legal responsibility

Your company has both a legal and contractual responsibility to carry out remedial and after sales services of all work you have carried out. It will not affect repair or replacement as warranties must be honoured by law – so it is wise to price your potential work very precisely at the contractual stage of your proposed agreement.

If you are unable to fulfil any after sales corrective work then your client has every right to employ another company to make good any fault condition and deduct the cost of this service from your financial retainer.

When equipment is past its useful life

It is very easy for a person to say, 'Well, I don't know why you can't repair it – it's been alright for the past forty years!' This, regrettably, is said every now and then – so when a piece of equipment is well and truly beyond economical repair the kindest thing to do is to completely skip it and suggest to your client that he/she buy a new one. Don't waste your time trying to botch up equipment you know you can't repair satisfactorily – it will return to haunt you!

Factors which might affect improvement or making good

Access to an installation or piece of equipment

You may have limited access to your problem area. Talk to your client and agree on a date and time when you can gain access to a property. Try to estimate how long you will be involved with the work you have to carry out and exchange contact numbers in case you have to speak to your client.

Using a standby supply

If your client feels that certain services must be maintained during a planned shutdown, arrange for a small, but adequate, standby generator to be installed to provide power to essential areas. Although cabling will be of a temporary nature, please keep to the spirit of BS 7671.

When a continuous supply is demanded

This is very difficult to achieve within an electrical installation without resorting to temporary mobile circuits supplied from another fixed source or portable generator. Ensure that all short-term installations are in place before the power is switched off and keep to the electrical regulations.

Situations where special precautions should be applied

Antistatic precautions

The following three methods can be used to help reduce the risk of damage to static-sensitive devices used in electronics:

- Workshop areas can be set aside to accommodate special floor and bench mats which are electrically linked together and then grounded to an electrical earth point. Electricians working with static-sensitive devices (SSD) are electrically earthed to the bench work mat using a suitable wrist strap and a light flexible cable. These areas are often referred to as special handling areas where all tools, soldering tips and other technical aids are electrically earthed.

- The use of anti-static handling devices to prevent fingers from physically touching the lead wires of the SSD components.

- By handling only the ceramic or plastic packaging and by never removing the static-sensitive device while 'live' in circuit, otherwise damage to the component may arise.

Do not underestimate the damage which can be caused by static voltages – keep to the guidelines!

Damage to electronic devices by 'over-voltage'

It is very important that the correct voltage is applied to an electronic circuit – in 99.9% of cases it usually is. The real problem is *transient voltages* within our own supply system and these are often created locally within our individual installation.

Most electronic devices such as computers have built-in power supplies which have been designed to cope with problems such as these. These tailor-made supplies can be trusted to deal with over-voltage conditions of up to 10% of the nominal voltage.

Transient voltages are in effect over-voltages. *Spikes* and *surges* are just two examples of these conditions. Spikes, the lesser of the two evils, tend to be very short lived, lasting no more then a few billionths to millionths of one second and rarely do any harm to electronic components. A voltage surge, which can be far greater in strength than a spike, can prove extremely damaging to an electronic circuit mainly because of the duration of the attack which can be measured in thousandths of a second.

The way ahead

What you will need to do is to place a *surge arrestor* in line with the equipment you wish to protect and this could include a telephone line to your computer. There are several different types of surge arrestors; some are used for surges and spikes generated within your own installation whilst others will protect from external lightning strikes which target distribution and power lines. The essential task of a surge arrestor is to divert any damaging spikes or surges around your equipment to earth. How efficient your protection will be will depend on three major factors:

1 The strength of the surge or spike experienced (magnitude/time).

2 The electronic response time of the arrestor in milli/microseconds.

3 The capability/efficiency of the arrestor to disperse the surge to earth safely.

Avoiding embarrassing moments such as when accidentally closing down an IT centre or computer terminal served from your installation

This is very easy to do – especially if you already have a large agenda to cope with and you have been pressurised in doing even more.

It is easy to say that circuits serving a c omputer suite, etc., should be clearly labelled but often, unfortunately, this is not so. We all live in the real world and often IT centres are established long after the original installation has been commissioned by the electrical contractor.

If you are honestly unsure of the circuits which serve your client's IT equipment you must arrange with him/her for an out-of-hour's investigation programme to be carried out when the IT suite can be safely shut down.

Once all the circuits are identified, labels can be fitted advising people working on or around the distribution system not to switch off certain breakers or pull particular fuses – if necessary, the over-current devices can be colour coded to assist recognition.

The risk of high frequency or high capacitive circuits

Be very careful when dealing with high frequency or large capacitive loads. If necessary seek advice for they can responsible for the following:

■ sudden death,

■ fibrillation – when your heart beats very irregularly,

■ electric and physical shock.

Very long mineral insulated cable runs can hold high capacitance after 1000/500 volt insulating testing has been carried out. This could prove dangerous if, for example, you are standing on a ladder or pair of steps and then touch the bare conductors directly after testing. The shock could cause you to react abruptly and fall off onto the ground below. Earth out all the conductors; this will discharge any high capacitance lingering.

Safety precautions with storage batteries

Storage batteries contain many different types of toxic chemicals and elements – they must be respected and handled with a great deal of care.

Please take time to read through the prepared list of precautions you should take when dealing and working with batteries:

- Always use carrying straps when transporting batteries.
- The terminals must never be shorted out.
- Keep the electrolyte within the battery – don't spill it and always keep upright. If spilt – clean up immediately using plenty of water. Do not allow the acid to come into contact with your skin.
- Always wear personal protective equipment in the form of a rubber apron and gloves, together with a face shield when working with batteries.
- Never smoke whilst working with batteries.
- Never place an open flame near to charging batteries. A mixture of hydrogen and oxygen is given off and an explosion could result.
- Since the density of acid is greater than water, splashing will occur if distilled water is poured into the acid. If you need to mix, then add the acid to the distilled water.
- Remove any electrically conductive objects such as rings and watches from your hands and wrists before using metal tools on secondary cells.

If you accidentally splash electrolyte over your skin, please wash it off *immediately* and rinse the affected part with plenty of clean water for between 10 to 15 minutes. Your local medical centre should be informed as soon as possible.

The capacity of a storage battery is measured in amp-hours and is the product of the current in amps and the time in hours during which the battery will supply this current.

Precautions to take with fibre optic cabling

Fibre optic cabling still seems to be a specialised practice and not all companies are willing to take up the challenge in this direction. There are many special precautions that you will need to take when working with fibre optic cable. The following outlines some of the safety measures you will need to put in place:

- Tight bends are *not allowed* – your radius should be about 12 times the diameter of the cable but check using data from the manufacturer.
- Never place fibre optic cabling under any stress. If your cable has to be pulled into long ducts, specialised equipment and processes must be used. This, your college will demonstrate.
- If your cable runs vertically, loops must be inserted in the run every 10 metres or so.
- Place loops every so often within your installation so that accidental damage can be easily repaired.
- If you use cable ties to support your cable, please do not over-tighten as it will cause localised bending/squashing and cable stress within the fibre optic cable.
- If you have to leave cable ends exposed, seal them with a suitable heat shrink capping accessory – this will stop both muck and moisture from attacking the cable.
- Remember to clearly label all your cable runs and ends in accordance with your job specifications – this is important.
- Leave about five metres of cable at each end. This then would be ideal to allow for testing, remedial work and for the positioning of enclosures.
- If you use metal accessories within your installation, these should be earthed in accordance with your job specification. In practice this will require bonding all exposed metallic parts or moisture barriers to earth at one end of your cable run only.

Accurate cable preparation is essential for splicing and connecting – use the correct tools and continue to practice this skill. The more you do, the better you will become and your confidence will grow. Your college tutor will demonstrate these techniques during your practical course time.

Handy hints

- A fault condition within a temporary festoon lighting arrangement could be the result of a shorted-out/smashed lamp.

- Try to be logical when fault finding and do not panic if you are not sure what to do. Seek advice and gather together as much data relevant to your problem as possible. Phone around if this will help.

- A fault on a final ring circuit can be made easier by disconnecting both phase and neutral at your distribution centre and then dividing the ring into two halves at a mid-point socket. Then test each half of your ring to target the fault.

- Never insulation test with a mutimeter – always use a 250/500/1000 volt insulation tester.

- The closer you get to a fault condition, the lower the resistance will be when your insulation tester is switched to the ohms scale. This is ideal when working on a ring circuit if the fault condition is very low and near to zero.

- A new fault found within a new installation will generally be different to a fault appearing within an older system.

- Fine dust can set off domestic smoke detectors – this is not a fault condition. Always, after testing, place a dust cover over the unit. This can then be removed by the homeowner upon moving in.

- Make sure your test instrument calibration certificate is up to date and, if you are using batteries, check they are in good working order before starting work.

- Under no circumstances carry out a high voltage test on electronic appliances – you will do them harm.

- Never dive into murky waters. If you are unsure ask someone you trust for advice.

- Remember that long lengths of cable can hold a large capacitive charge after insulation testing has been carried out. Always short out the conductors and the conductors to earth after you have tested.

- Always handle capacitors with care – they could be fully charged.

- Every capacitor has a safe working voltage. Do not exceed this value when testing or checking.

- Before you switch your distribution centre off for testing or fault diagnosis work, please make sure there are no computers on line – switching them off can cause very loud voices!

- Take time to do your fault finding properly; rushed jobs are seldom good jobs.

Summary so far . . .

- Single phase circuits are supplied with 230 volts AC and comprise the following: radials, final ring circuits, power and lighting arrangements. Three phase installations use 400 volts and can be found in industry and commerce. Circuits include motors, heating equipment, ovens and machinery, etc.

- Control circuitry is common throughout the electrical installation engineering industry and can even be found within domestic installations these days. Both manual and automatic controls are available for many different applications.

- It is important that the correct procedures are adopted before fault finding commences. It is essential to know where to isolate your circuit and always check out with a volt meter that your circuit really is dead.

- Before undertaking fault diagnosis, place warning notices in places you know that will matter and always carry out relevant safety and functional checks before switching the supply back on.

- The basic principle of undertaking fault finding is being able to understand how the relevant system operates and to bring your own work-time experiences into play.

- The use of a logical approach is essential when looking for a problem. Above all, be calm and do not panic.

- Before you start your fault finding programme, gather together as much data as you are able to.

- Analyse the evidence before you and use standard tests to locate your problem area.

- Check out your repair with functional tests and ensure that the remedial work you have carried out meets with our current regulations.

- Faults may occur within the following areas: wiring, cable terminations within switches, protective devices and lighting fittings. Flexible cable and components can also be troublesome.

- Typical fault symptoms include complete loss of supply at the origin, insulation and component failure.

- Transient voltages are very high voltages lasting a mere fraction of a second. They can damage microprocessor circuits if not properly protected.

- Over-current and protective devices can also be mechanically troublesome from time to time causing a localised loss of supply.

- The total cost of remedial work could influence whether a fault was repaired or not.

- Whether a replacement is available, the amount of 'down time' you will experience, and your legal responsibilities, are just three factors which could influence a repair or replacement.

- Obtaining safe access to a faulty system could affect whether it can be repaired or not.

- The availability of a suitable emergency/ standby system or the introduction of a temporary, but continuous supply could be two factors which might affect remedial work and making good.

- The installation of fibre optical cabling requires special training. Read through the recommended precautions needed when carrying out this type of work and take on board what your tutor has to say.

- Electronic equipment can be damaged by spikes and voltage surges. Fit surge arrestors to your circuit to help to eliminate this problem. Be very careful when you switch off breakers from your distribution centre – they could be serving IT equipment. Ask first!

Review questions

1 Provide examples of the following:
   - a lighting circuit,
   - a power circuit,
   - a common component,
   - a control system

2 List the safe working procedures you must apply before carrying out fault diagnosis work.

3 What do you consider to be the basic principles of successful fault finding?

4 Please compile a list briefly outlining the stages of fault finding and rectification.

5 List four fault conditions you could discover within a new or well established electrical installation.

6 Briefly describe transient voltages within a supply.

7 How can you guard against damaging spikes and power surges?

8 What can cause insulation failure within your installation?

9 When would you not spend time to repair and make good a piece of equipment?

10 Briefly describe the term 'down-time under fault conditions'.

11 List four major factors which could influence repair or replacement.

12 State the issues which could affect whether your fault is to be made good.

13 List the precautions you must keep in mind when working with storage batteries.

14 What are precautions to take when working with fibre optical cable?

15 Why should you not over-tighten fibre optical cable when using cable ties for support?

Introduction

This is the last chapter in this book and deals with the following ten issues:

◻ Collecting sources of information useful for commissioning your work.

◻ The requirements of testing and commissioning,

◻ Report forms and dealing with your clients,

◻ Selecting instruments suitable for fault finding and making good.

◻ Verification of test instruments, calibration and the use of documentary evidence.

◻ Carrying out functional checks to confirm the soundness of your work.

◻ Complying with test values and what to do in the case of unsatisfactory results.

◻ Advising your customer of the need for additional refurbishment/repair work to his/her building fabric.

◻ How to get rid of your waste material from site.

◻ After your work is completed – processes you will need to initiate.

Practical activities

There are just four formal activities accompanying this chapter. As a candidate, you will be asked to provide practical evidence learnt within this chapter. The activities are as follows:

◻ Show that you are able made good electrical installations and equipment which were once in fault condition.

◻ Agree on suitable measures for re-commissioning electrical installations, components and equipment which were once in fault condition.

◻ Demonstrate safe and efficient practices to your assessor.

◻ Show your assessor that you are able to restore installations, equipment and components to a suitable working order.

Gathering together information to make re-commissioning easier

There are many sources of information which will help you to re-commission the system you have restored to a good working order. Here is a short list to glance through and maybe provide topics for discussion within the classroom, if time allows.

■ *The Wiring Regulations (BS 7671)* – an up to date database of all aspects concerning work within our industry.

■ *Your job specifications* – this will remind you of the requirements of the installation or equipment you have repaired. A handy tool if you have had little to do with the project you are dealing with.

■ *Distribution arrangements and wiring diagrams* – ideal if you are unsure of how things should be and how certain items of equipment have been designed to interact with one another.

■ *Manufacturer's instructions* – often thrown away; but don't. These instructions are very useful when you are unfamiliar with the work you are undertaking. This information will provide all you wish to know concerning the product you are dealing with.

- *Appropriate legal commands* – our lives seem to be under the control of numerous statutory Acts. If you are unsure of anything which has a legal bearing on your work, please find out by logging into an appropriate web site or asking someone who will know.

The requirements of testing and commissioning

Types of test equipment

Let us suppose you are carrying out tests to restore an installation or an item of electrical equipment to work, as intended, again. The instruments you need will depend on the nature of your problem but it is wise to have at hand a selection of test instruments – just in case you require them. Please browse through the following suggestions:

1 *A continuity meter capable of measuring up to 500 ohms and as low as 0.1 ohm* – it is best to select a purpose-made test meter which, in addition, has a built-in buzzer; not a multi-meter: they are far too general.

2 *A 250/500/1000 volt insulation tester for probing into problem areas* – digital models are easy to work with but do be careful to make certain where the decimal point is in relation to the figures displayed.

3 *An all-in-one voltmeter/clamp meter which is auto-ranging and capable of recording either AC or DC voltages automatically* – this two-in-one test instrument is a very useful member of your testing kit.

4 *Maybe a portable appliance tester* – please remember not to carry out a high voltage test on electronic equipment; it could cause damage.

5 You might require an earth-loop impedance/PSC meter if you have an earthing problem to sort out – remember that some types of ELI meters will trip out a residual current device in the circuit under test when the test button is pressed.

6 If you are having trouble with how quickly your residual current device (RCD) responds to a fault condition, then an RCD timing meter would be very useful to you. Your timing will depend on the type of installation you are working on; please check BS 7671 for the maximum value which is accepted.

The sequence and procedures for testing

The sequence and procedures for testing really depend on the size and the nature of your problem. A logical approach to fault diagnosis is needed at every stage and the following must be considered:

- You might require a safety permit to work – this can be checked out before you go any further.

- Advise co-workers of what you are doing and if necessary ask them to leave the immediate work area.

- Remove unwanted material and equipment from where you are working – think safety.

- De-energise your faulty equipment or installation including any capacitors in circuit. This can be done by making an initial survey of all energy sources by means of a combination of drawings and equipment instruction manuals. This might have to be carried out with the co-operation of others – for example, the owner or if you are working within a factory, the maintenance department.

- Once you have established where everything is located, 'LOCK-OUT' notices or tags must be placed on all relative switches and disruptors and these 'locked off'. Accident prevention notices should provide the reason why the system has been temporary closed down together with 'contact details' of the person placing the notice. Remember to date/time your notice – this is very important! Main switches usually have a suitable hole in which a padlock can be fitted. If the switch has been incorporated within a metal box, then the metal box must be locked. Likewise if a fuse is removed to de-energise the system, the distribution centre has to be also locked. If no provision has been allowed to lock the centre, a suitable hasp could be welded or riveted to the metal door together with a lock staple.

A padlock could then be fitted. The key must be kept with the job holder at all times.

- Make sure that no one is operating machinery prior to closing down the system as sudden loss of power could cause an accident.
- If you are working in a factory do ensure that any device or item under load or pressure is released.
- Do test that the electricity has been safely switched 'OFF' by means of a suitable meter before starting fault diagnosis work.
- If you have to carry out some tests when the system is energised then switch the system to 'OFF' after each test and de-energise any capacitors within your circuits.
- Find your problem with the aid of relevant test instruments and available data.

Remembering our regulations

Any remedial work carried out in the course of your fault diagnosis programme must be well-matched with BS 7671. Our regulations are changed every so often so it is important we keep up to scratch with them. Once it was within the spirit of the electrical regulations to allow two socket outlets to be spurred from a mainline final ring circuit socket. Now we only allow one – so if a problem exists within the spur of an old type circuit we would have to stick strictly to the newer regulations and make redundant one of the spurred sockets.

If you have any doubt, please check out BS 7671 – it is important you get it right!

Procedures

Report forms and documentation

Technical report forms and documentation must be written in a plain and straightforward manner. Watch your spelling and grammar and, if possible, use a computer to add a splash of professionalism. If, for example, you are writing a report concerning an item of equipment, please consider the following:

- *The general condition of the equipment* – good, bad, past its useful life, etc.
- *The model number of the equipment* – often to be found within the handbook.

- *A copy of any accompanying handbook* – this could be photocopied.
- Manufacturer's circuit diagrams, etc.
- Lists of components and sub-components.
- A general assembly drawing with accompanying photographs.
- Description of how to operate the equipment.
- A list of standards covering the equipment and safety regulations.
- How safety requirement and standards have been applied.
- Your report of electrical tests carried out.
- Any relevant comment you may have involving the report.

From time to time your company may ask you to undertake a formal electrical assessment concerning the safety of a property before any remedial work is started. This you will have to do both visually and instrumentally and your results formally logged for future review.

If you are not used to this sort of work, it might be worthwhile to record your comments using a pencil and notepad when making observations and carrying out tests. This will make it easier for you to complete a formal paper electronically, which is both clear and readable. Filling out report forms is not everyone's cup of tea but it plays an important role within our industry.

Dealing with your customer

Adopt a simple personal code of practice when dealing with your customer – a good image and a high standard of professionalism is essential. It is no longer sufficient just to provide the required expertise and good workmanship demanded by your client. You will not be judged by this alone. What is required these days is old fashioned protocol blended with a few social graces. This is often enough to generate the necessary confidence to win a customer over and to form a good working relationship.

Always explain any cause of disruption to your customer's activity such as total or partial power loss or when you will have to work in areas which are occupied – it's good manners and it all helps!

Listed in random order are points for your consideration when dealing with your customer. Please take on board the following:

- personal appearance/hygiene and courteousness.

- showing a positive attitude, together with a good image,
- being tactful and discrete,
- respect your customer's property and any rules he/she might have imposed,
- respond to requests for information promptly and try to develop a positive relationship at all levels of contact,
- when information is unavailable prompt action must be taken,
- avoid speaking to your customer in jargon (the language we use to talk to each other). The technically unaware will just turn off – they will not know what you are talking about!
- keep confidential material to yourself – avoid gossip,
- identify with your customer's needs accurately and always provide advice concerning health and safety matters,
- discuss the range of services available from your company and recommend any improvements or remedial work which could be carried out.

It is a good idea to find out how technically aware your clients are and then act accordingly. It will save a lot of wasted words!

Instruments for fault finding and making good

There are many instruments and we are, in truth, spoilt for choice. Firstly, you must have a full understanding of how your test instruments are used and their limitations – for example, you couldn't expect a multi-meter to carry out the same tasks as your insulation tester. It really is 'horses for courses'. A selection of instruments we use for fault diagnosis work follows:

Insulation tester

There are three types: analogue scaled, digital scaled and the old fashioned mechanical hand-cranked model. They can be used as a continuity tester or a cable insulation tester. When the instrument is used for continuity the voltage delivered is very small compared to an output voltage of 250, 500 or 1000 volts DC used to measure cable insulation values, etc. On the continuity scale all values are measured in ohms or decimal parts of an ohm whereas when switched to the higher voltage range the values are directly displayed in megohms. When using a digital instrument for insulation testing, values will appear on screen as megohms or parts of one megohm. A value of 250 000 ohms will appear as 0.25 megohm whereas half a million megohms will read as 0.5 megohms.

The accuracy factor for an analogue instrument is ± 1.5% of the total scale and slightly better when using a digital meter. A combination continuity and insulation tester can be seen in Chapter 8 – illustrated as Figure 8.5.

Multi-meter

This type of instrument can be either digital or analogue and can have many test functions. Briefly, these include:

- AC and DC voltages up to 999 volts,
- AC and DC current up to 400 milliamps and 20 amps,
- resistance up to 5000 ohms,
- battery power check,
- frequency of your supply,
- capacitance from 2000 picofarads to 20 microfarads.

Not all multi-meters have such a range of functions – it will depend on the cost of the instrument. The accuracy factor is between 0.3 and 1.8% of the recorded value and this will rely on both the make and design of the multi-meter. A site multi-meter is illustrated as Figure 8.9.

Low ohm continuity tester

This type of instrument is used to measure the continuity of a conductor or its precise resistance. The measurement will show if the conductor has been broken or is very loose which could cause a problem whilst on load. As a practical example, consider checking out a workshop machine with a suspected damaged load conductor. For this exercise, consider that all safety precautions have been carried out prior to testing as described in previous paragraphs within this chapter.

- Check the health of the battery serving your continuity tester.
- Check that no electrical supply is present at the machine.
- Disconnect the suspect load conductor from the machine's isolator.

- Open the terminal box serving the machine and disconnect the supply conductor.
- Switch your instrument to the continuity range and 'null' the test leads together to produce a value of 000 ohms.
- Connect your test leads to each end of your suspect conductor and then press the test button. A sound conductor will produce a very low ohmic value but if the conductor is broken or a loose connection exists somewhere, a much higher reading can be expected. If your conductor has a clean break, an infinite reading will be displayed.

Some continuity testers are fitted with an internal buzzer. This will help you when only continuity requires to be measured instead of the actual resistance of the conductor.

A compact voltage, continuity/resistance/current tester is illustrated in Figure 8.14.

Voltmeters and ammeters

It is best to choose a digital model supplied with a heavy duty rubberised holster for site work and this could mean you buying a multi-meter. Many standalone instruments seem to be less strong than the rubberised site equivalent and tend to adopt a bench or 'panel model' profile. Current measured with this type of ammeter have probes connected in series formation with the circuit.

A clamp meter is an ideal way to take the value of current flowing within your circuit – it's so simple to use. To test, open the hooded insulated jaws and place them around a *single* conductor. The electromagnetic induction produced by the circuit will be collected by the jaws of the instrument. The induced current is then processed and then turned into a readable value.

Most clamp meters have a facility for reading voltage when auxiliary probe leads are fitted to the bottom of the instrument. Voltage is measured in parallel with the circuit.

Verifying your test instruments

Take time to make the correct choice of selecting your test instruments when carrying out fault diagnosis work – this is extremely important but there are other things to consider. Please ask yourself the following 10 questions:

1   Is there sufficient insulation on the test probes?

2   Have you checked out your meter and probes for damage?

3   Has each probe finger barriers to guard against accidental hand contact with live conductive parts?

4   Is there good all round insulation apart from a maximum of 2 mm exposed metal forming the probe tip?

5   Are the leads coloured brown and blue so that they can be identified from each other?

6   Are the service leads a suitable length – ideally about 1.2 metre?

7   If the leads come away from the probe, will the person who is testing be safe from exposed conductive parts?

8   Are the leads flexible?

9   Is there a built-in service fuse in the body of the probe?

10 Has the instrument been switched to the correct setting before testing starts?

If you have answered 'YES' to all ten questions, then the demands of *Guidance Note GS 38* from the Health and Safety Executive have been satisfied.

Why bother to calibrate test instruments regularly?

When you use your test instruments on a regular basis their reliability and accuracy steadily weakens. Regular calibration will compensate for inaccuracies which have built up over the months you have used them. It is similar to having a test instrument MOT – it returns the instrument to a dependable and acceptable level. Doing this will maintain the meter's high performance and keep you within the law.

Most electrical wholesale companies will accept test instruments for calibration. They will either carry out the work in-house or contract out to a specialist company like Robin Electronics or Fluke Instruments Limited. A certificate of calibration is always issued by the company and covers the instrument in question.

The use of documentary evidence

The use of documentary evidence plays an increasing role within our industry today.

It verifies that what we have carried out or what we are going to agree to is both correct and satisfactory and within the spirit of our electrical regulations and standards. Listed are a selection of work related activities where documentary evidence is asked for:

- *A permit to work document* – asked for by many companies where hazardous activities are common place.
- *Portable appliance testing* – a certificate is produced to confirm the electrical health of the appliance and dated 'PASS' or 'FAIL' labels are attached to the appliance.
- *Company vehicle/private car parking permits* – to keep check within a restricted site parking area.
- *Domestic/industrial electrical installation certificates and periodic installation certificates* – a great deal of technical information can be extracted from these and will formally verify that the installation is sound and free from fault.
- *Part 'P' from the Building Regulations* – confirmation in the form of documentation is required when work is carried out within a dwelling, especially when undertaken by a DIY enthusiast.
- *Manuals, test certificates, line diagrams and installation layout drawings* – these will all help when carrying out fault diagnosis work.
- *Installation 'as wired drawings' are very useful aids for fault finding* – they are often just company layout drawings which the installation electrician has pencilled in cable routes from one accessory to another.
- Documents for permission to proceed with work which has to be carried out in an explosive atmosphere
- *Vocational certificates and apprenticeship papers* – this are very important in the electrical installation engineering industry.
- *Prices, reports, contracts, etc.* – all vital documents for the smooth running of an efficient company.

There are many more uses of documentary evidence that we all use in our professional lives. The listed ten are just a few examples; maybe you can add to this list or use the topics as a base for a classroom discussion.

Checking your work after making good from fault finding

Checking continuity

For this you will need a low scaled continuity tester to carry out the following procedure:

- First, test the health of the batteries serving your continuity meter.
- Place the leads together to 'null' the internal resistance of the probe leads. This will provide a 000 value within the window of your test instrument.
- If the conductor you are testing is short, place a test probe at either end, then press the test button. The value you will obtain will be in milliohms or decimal parts of one ohm. If a larger value is obtained a check must be made to establish the reason why. When checking out a *long* conductor, extend one of your probe leads to a suitable workable length but remember to 'null' your newly formed test leads as described within the last paragraph before carrying out any tests.

Another way to do this is to place the phase conductor of the circuit you are attending directly onto the earth bar serving the distribution centre. Use your continuity meter to check out your various accessories probing from phase to circuit protective conductor. The value you receive will be zero or near to zero.

Insulation resistance

Test that the insulation surrounding your conductors is sound by using an insulation tester set to the 500 volt scale.

When testing an installation after remedial work has been carried out a check must be made at the distribution centre concerned. The main switch must be off and all breakers and fuses prepared, current consuming equipment must be disconnected, all lamps removed and control switches closed. If lighting fittings and equipment prove impractical to remove, then BS 7671 will allow local switches controlling these problem areas to be opened.

It is a good idea to disconnect any capacitors or neon indicator lamps which are in circuit as these can cause confusing test results. If the circuit you are testing contains electronic devices such as dimmer switches, passive infra-red detectors and sensi-touch switches, etc., please disconnect them before carrying out your insulation test if permanent damage to the circuit is to be avoided.

Place/connect one probe to the earth terminal whilst connecting the other to a phase and neutral combination. The value you record must be greater than 0.5 megohms but if your value is below 2.0 megohms you must investigate as to why this is. If your joint phase/neutral to earth value is well below 0.5 megohms then test both phase and neutral separately to find out which conductor is troubled.

Your final test should be taken between both the phase and neutral conductors and you should resort to similar values. Look at Regulation 713-04-04 for further details.

Polarity

The polarity of your conductor amounts to whether it is phase or neutral, positive or negative.

The phase conductor throughout the EU is brown, the neutral blue. The brown phase conductor will always serve your switch – never the blue neutral conductor. On a machine or an appliance you can check the polarity of your conductors by the use of a suitable voltmeter, attaching one probe to an established neutral whilst the other is free to probe conductors of your choice.

Polarity testing within an installation is different and tests must be carried out to make sure of the following:

- Switches, over-current devices and any single pole controlling device within your circuit are served with a phase conductor.
- That socket outlets have been correctly wired with the phase conductor connected to the terminal marked 'L'.
- The centre connection point serving older types of Edison lamp holders must be connected to the phase conductor; the threaded annulus is reserved for your neutral conductor. Modern Edison type lamp holders don't have this problem.

Testing procedure

Polarity tests must be carried out with all switches placed in the ON position. All lamps must be removed and equipment disconnected.

Fuses and circuit breakers must also be withdrawn and the neutral conductor serving the circuit under test disconnected. The outgoing circuit phase conductor must then be removed from the top of your over-current device and temporarily connected to the earth/bonding bar serving your distribution centre. By use of a continuity meter or combined bell and battery the polarity of all relevant accessories and equipment can be tested. Just place your probe lead onto earth and the other onto the brown phase conductor. If wired correctly the bell will ring or you will obtain a value in ohms/milliohms from your test meter. If you use the bell and battery method, please remember that a volt drop will occur and it might affect the performance of your bell.

After you have finished your testing, remove your temporary placed phase conductor from the earth terminal and reconnect within the over-current device. Lastly, reconnect your neutral conductor into the neutral block in the correct sequence order.

Earth-fault loop impedance

You will occasionally have to carry out an earth-loop impedance test on remedial work you have done. Your earth-loop impedance meter has a set of three coloured probes. These probes are fitted into a three-pin instrument plug on the side of the meter. Switch off the main switch which serves your distribution centre, disconnect the earthing conductor from the main earthing terminal and then connect the green/yellow lead firmly onto the principal earthing point at the source of your supply. The brown coloured probe is firmly pushed into the phase conductor at the start of your supply whilst the blue neutral coloured probe is pressed onto the main neutral terminal. Pressing the test button will display the supply voltage followed by the earth-loop impedance value (Ze) ohms. This is the external earth-loop impedance of your supply and represents the route back to the community transformer.

If you wish to check out the earth-loop impedance value of a circuit (this is the Zs value) – for example a cooker point which has a socket incorporated within the main control switch – just insert a supplementary test lead which incorporates a 13 amp plug into your meter. Plug into your cooker point, switch ON and press the test button. You can then read off the value in ohms of your earth-loop impedance. If the wiring/polarity to an appliance is wrong a red light will appear on your instrument advising you not to proceed any further. Please note that both earthing and bonding conductors should not be removed during this test.

When your installation is supplied by means of 'TT' earthing arrangement you will find the value of your impedance to be very high when compared with a TN-CS or a TN-S system.

Measuring the value of an earth electrode

The easiest way of measuring your earth electrode is to use an earth-loop impedance meter. Your meter will have available a set of three coloured probe conductors which have been described in the previous paragraphs.

Firstly, disconnect the green/yellow conductor serving the earth electrode at the distribution centre so that it is completely free from the principal earth bar then firmly connect the green/yellow lead to it. Alternatively you might be able to disconnect the link serving the earth bar which is easier.

- Press the blue neutral probe onto the incoming supply neutral (this in effect will be at the top of your main double pole switch).
- The brown probe lead must then be pressed onto the incoming phase conductor.
- If you have locked your test button on 'test mode' you will then have an instant delivery in ohms of the value of your earth electrode.

You will find that the value will vary depending on whether the soil is dry, damp or saturated and how far you are away from the community transformer.

As the instrument is polarised it is important to connect the correct probe to the right terminal. If you make a mistake a red warning light will appear advising you not to proceed with your test.

Testing your residual current device (RCD)

Once an RCD has been installed it is good practice to test it regularly by pressing the test button; often marked with a capital 'T'. This button will only work when the device is energised and will

confirm that both the electrical and mechanical components of the RCD are working as intended. It will not check out the continuity of the circuit protective conductor or principal earthing conductor, earth electrode or any other elements concerning the installation earthing system. These will have to done with an earth-loop impedance meter.

Your test will be made on the load side of your residual current device between the brown phase conductor and the related green/yellow circuit protective conductor – but remember to switch off the load during the test.

Pressing the button unbalances the circuitry within the device and it will automatically go into abort mode – switching off the circuit it is monitoring. In domestic installations it is important that the prescribed switch-off time of 0.2 seconds is observed for circuits which incorporate socket outlets.

To formally test you must use an RCD timing meter. This is an instrument which has two sets of independent leads comprising, one, a simple plug which can be fitted into a 13 amp socket and the other in the form of a three pronged probe coloured brown, blue and green/yellow. The lead formed into a standard plug is for testing sockets whilst the other one is used for testing the response of other circuitry such as lighting arrangements, etc.

To test

Clip the earth probe onto the circuit protective conductor (cpc) serving the circuit you wish to test whilst placing the tip of the brown coloured probe securely onto the phase conductor and the blue onto the neutral. Press the test button firmly down until you have noted the disconnection value in milliseconds. To test the disconnection time of an RCD serving a circuit of sockets, just place your 13 amp socket lead into your instrument and press the test button until you receive the information you require. Different models and makes of residual current devices produce different disconnection times. It also depends on how far you are away from your community transformer and the conductivity of the sub-soil.

When an earth leakage current of 50% of the rated current is flowing the device should remain closed but when increased and equal to 100% of the rated tripping current, then the device should automatically switch off in less than 200 milliseconds.

An RCD protected socket outlet to BS 7288 must automatically switch off within 200 milliseconds when an earth leakage current is flowing equivalent to the tripping current of the RCD.

Functional testing

Functional testing after a repair or completion of an installation is essential both to you and your customer as it will identify problem areas earlier and set aside failures.

When functional testing within an installation it would be wise to ask yourself the following question: 'Is this appliance or circuit doing what is intended?' If YES, it can be ticked off. If NO, then investigate why failure has occurred and carry out simple remedial work if this is practical or wise – but remember you could harm the terms of the product guarantee if radical remedial work is undertaken.

The functional testing of industrial/ commercial appliances and installations will be much different from working within the domestic arena. Load and supply voltage measurements will have to be taken into consideration, safety circuits checked. Complicated equipment is usually commissioned by experts who will often employ software programs or specialist test equipment to complete their functional testing.

Again, as with domestic installations, please ask the same basic question: 'Is this doing what it intended to do?' and if not, why not.

It is very important to carry out functional testing on fire alarm and intruder alarm systems. This is often considered to be highly specialised work and is usually treated as such.

Other functional tests will include the following:

- *Residual current devices* – check that it will do as intended and within the specified time frame.
- Any mechanical interlock mechanism.
- All your switch gear.
- *Controls and control circuits* – both automatic and manual.
- Check also that your installation is physically secure and will not collapse at the slightest touch.

- Check that all your over-current circuit breakers are firmly tightened both top and bottom.

Complying with your recorded test values

Earth-loop impedance test

Each circuit has a maximum permitted impedance which will efficiently allow your over-current device to operate under an earth-fault condition. These values are based on the rating of your over-current protection device and the type number/letter of the breaker concerned. For example, a B32 breaker has a maximum permitted impedance of 1.2 ohms whereas a B16 has a value of 2.4 ohms. The values we support when formally testing are usually about 80% of the maximum limits specified in BS 7671 at an ambient temperature of 20°C. This 'err' on the side of caution ensures that the protective devices will automatically switch off within the permitted time.

If, for example, your final ring circuit has a maximum impedance of 4.6 ohms, is supplied from a B32 breaker and served from a TN-S earthing arrangement, your circuit will have departed from the spirit of the regulations. To remedy this problem you would have to remove the B32 breaker and position an RCBO in its place. Doing this you will be allowed a maximum of 1666 ohms impedance.

Cable size in relation to current flow

For this exercise please ignore any cable correction factors which might spring to mind.

It is important that the cross sectional area of your conductor reflects the design current of your circuit. Installing under-rated conductors in relation to current flow is a recipe for countless problems in the form of volt drop and over-heating, etc. If your accompanying over-current device is over-rated it will just compound your difficulties.

Undesired heat generated as a by-product within under-rated cables can be very troublesome. A conductor's ability to lose heat energy efficiently will depend on how and where it is installed. An example of this can be drawn from conductors grouped together within cable trunking or conduit. Their ability to dissipate heat safely would be restricted and somewhat inefficient. This would result in serious damage to the insulation of the circuit concerned and to other neighbouring conductors using the same cable carrier.

Insulation values

When your insulation test has shown an unsatisfactory result the circuit must be corrected. An insulation resistance value of above 500 000 ohms does act in accordance with BS 7671, but if the value you record is less than 2 megohm your circuit may have a hidden problem. This potential troublespot must be addressed and, if possible, your initial test value increased in value to above 2 megohms.

The minimum value of insulation resistance in SELV and PELV circuits is 0.25 megohms at a test voltage of 250 volts DC.

Sometimes general site dampness is to blame for bad values. When this is the cause you will find the insulation value on your instrument will creep up the scale whilst applying pressure on the test button.

Polarity

You need to comply with test values when carrying out polarity testing – failure to do so could prove very dangerous. Please confirm the following and, if found to be wrong, put right immediately:

- That the polarities of all socket outlets are correct with the phase conductor placed into the terminal marked 'L'.
- Other than Edison screw lamp holders, types E14 and E27 to BS EN 60238, the centre terminal serving the original type of ES lamp holder must be connected to the phase conductor while the neutral is strictly reserved for connection to the threaded section of the lamp holder.
- All switches and single phase controls are served with a phase conductor.
- All over-current devices such as fuses and miniature circuit breakers, etc., must be supplied with a phase conductor – never a neutral.

Earth electrode resistance (RA)

The electrical resistance of an earth electrode can vary quite radically from season to season depending how dry the soil is. Test using an earth-loop impedance meter as described in previous paragraphs – but please remember to take care to avoid shock hazard, both to yourself or others working along side of you. Always replace the test link on your common earth terminal bar once you have completed your test.

Your earth electrode resistance when served within a 'TT' system should not exceed 200 ohms in value or, to put it another way, the value of your earth electrode multiplied by the operating current of your protective device ($I\Delta N$) must not rise above 50 volts.

A practical example

If the measured resistance of your earth electrode (RA) is 200 ohms then your maximum residual operating current must not exceed 250 milliamps.

If you discover that your earth electrode resistance exceeds 200 ohms then a longer or possibly several additional earth electrodes made common with each other will be required over a larger area. Re-test after each amendment or improvement is carried out until you receive the value you require.

Prospective fault current values (PSC)

You must measure the value of the prospective fault current of the installation for which you are responsible for the following reasons:

- All protective circuit breakers, for example, miniature circuit breakers and fuses are rated with what is known as a 'breaking capacity'.
- The value of this breaking capacity tells us the amount of fault current the device is able to handle without blowing to pieces.
- If, for example a type 'B' MCB has a rating of 6000 amps it will safely disconnect a fault current of up to 6000 amps.
- If, supposing the fault current were to be a staggering 10 000 amps or more, it would be sufficiently high to skip past this breaker and arc its way into the circuitry causing potential damage to the system.

By measuring the prospective fault current you will be able to choose the short circuit rating of your breakers, making certain that all protective devices fitted have a suitably high kiloamp rating to safely disconnect the supply should a serious fault condition occur.

If, for example, your PSC value is 2.34 kiloamps then it would be safe to install 6 kA breakers throughout your installation – they would faithfully isolate the supply in the event of a fault condition. On the other hand, if the PSC value was found to be a massive 10.4 kA it would be foolish to fit breakers rated at 6 kA – it would be a recipe for potential trouble. Sometimes a protective device is not always clearly marked as to its rated short circuit current characteristics. Usually they have a 'boxed' number 6000 or 15 000, etc., printed on its side. Other times an 'M' number will be displayed on the breaker such as M6, M10 or M15 – 'M' being the Roman numeral for one thousand.

Please keep in mind that if the supply cabling is ever up-graded to your installation the PCS value at the distribution centre will alter. The value of your new fault current would be the recorded voltage divide by the new earth-loop impedance value and this could mean upgrading all the breakers should this happen.

It is wise to check out the earth-loop impedance/supply arrangements *before* any design decision has been made concerning the over-current protection. Table 10.1 explains the relationship between an assortment of common protective devices and their short circuit ratings in kiloamps.

Measurements for PSC values must be made at the source of the supply (the distribution centre) with the principal earth conductor and bonding conductors in place. Use a PCS/earth-loop impedance meter to carry out this test. The instrument is equipped with three coloured probes – one to serve the phase conductor, one the neutral and, last, one to serve the principal earth conductor. Connect the probes as instructed and press the test button. The values obtained will be in kiloamps (kA). Rural installations are often surprisingly low having a PSC value of 0.6 to 1.2 kA whereas industrial urban sites served by TN-S or TN-C-S earthing arrangements can be as much as 3.5 kA or more.

Non-electrical remedial work

When carrying out your customer's electrical work you might come upon areas which could benefit from non-electrical restoration work: damage to the building fabric which could have been caused by the work you have done or the work which has been carried out by other trades. This could include redundant holes, gaps within your client's ceiling caused by lack of care, cable access routes which require filling, etce. Other spoilt areas could be the following:

■ The fair face brick work – *damaged, loose and missing bricks.*

■ *The plaster on the ceiling and walls* – damage caused by forming holes for accessory boxes out or fitting luminaries to the ceiling.

■ The decorations (paintwork, wall covering, emulsion paint, etc.) – *scratch marks, dirty hand marks and missed paintwork, etc.*

■ Fitments; for example pieces of furniture or equipment which have been site designed for a particular space or room – *fabric damage and damage to electrical accessories caused by other trades.*

Don't wait for your client to speak to you about it – discuss any problem areas with him/her at once and come to a helpful and friendly agreement between you.

Table 10.1 – *The short circuit ratings of common protective devices*

Protective device	Description	kA rating of the device
BS 1361	Cartridge fuse	From 16 to 33
BS 3036	Semi-enclosed fuse	Between 1 and 4
BS EN 60898	Miniature circuit breaker	From 1.5 up to 15
BS 88	Cartridge fuse	From 50 to 80
BS 3871	Moulded case circuit breaker (MCCB)	Between 1 and 9

Site waste disposal

You must dispose of your waste sensibly and leave your site in a safe and clean condition. One of the simplest solutions is to hire a skip from a waste disposal company and ensure that the skip is large enough to accommodate all of your site rubbish and litter.

A word of warning! The new *Hazardous Waste (England and Wales) Regulations* were put into operation on 16 July 2005. One of the many rules is that producers of hazardous waste must pre-register before any waste is collected from their premises. That will involve you and a possibility that you will require two skips or a skip with a dividing fillet.

Surprisingly, there are many items which can be found on a typical site that are now considered hazardous – here are just a few of them:

- *Industrial paint* – especially very old tins which are pre-1960 which contain both lead and mercury.
- Company van motor oil, gasoline and antifreeze leftovers.
- *Fluorescent lamps* – never break these; just keep them in one piece.
- Different types of spent high pressure discharge lamps.
- Batteries, both dry and lead acid.
- *Old redundant overload pot oil* – this could contain cancer-causing agents.
- *Ionisation type smoke detectors* – especially the very old ones which have a higher percentage of radioactivity present.
- *Old transformer oil* – some of which could be carcinogenic.

If you are in any doubt, telephone your waste disposal company who will advise you professionally.

After you have finished your remedial work

The responsibilities of various people within your organisation play an important role once site remedial work has been completed.

Your testing department

Your work might have to be formally tested and the results presented to your customer on an official test sheet after your remedial work has been carried out. This formal test might be in addition to any site testing which has been performed by you or your staff.

Your line manager

His/her role is to inspect your work and to be responsible for producing all necessary technical documentation, any day work or additional work which has been carried out.

A service manual will have to be assembled with all relative details concerning the work which has been done.

Your client's account will have to be processed and given to your accounts department. Finally, there will be someone within your company who will write to your customer to advise that all works have been completed.

A service contract could be offered to your client with your terms and conditions simply explained.

Accounts department

Their roll is to formally present their account to their customer and to complete any documentation work.

Handy hints

- Never become a minor monument of carelessness. Always think first!

- Check your electrical test instruments long before testing is carried out – calibration may be overdue and batteries low.

- Use electrically rated rubber gloves when carrying out 'field earth electrode testing' to guard against electric shock.

- Never voltage test or fault find with a neon screwdriver or an electronic voltage indicator – they are not reliable.

- Keep electronic, extra-low voltage lighting transformers free from thermal loft insulation – they will over-heat and become faulty very quickly.

- A pulsating magnetic 'growl' (single phasing) from a three phase motor indicates a missing supply phase – check out the protective devices serving the motor. These will probably be fuses of some sort.

- A loose terminal is sufficient to cause a three phase motor to single phase and to activate the over-current protection devices serving the motor.

- A 'chattering' switch contactor will, if left unattended, lead to the contactor coil burning out.

- A 'chattering' contactor is often caused by an over-sensitive probe, temperature device or has a too finely adjusted thermal differential.

- A simple method of removing a rust seized fan from a motor spindle is to play a gentle flame around the neck of the fan blade assembly. Rotating the blades will ensure that the heat is applied evenly. Remove by gently tapping a copper drift around the neck of the fan once the fan screws have been loosened. This method is not suitable when the fan blades are made from plastic. Take care not to damage the bearings!

- A capacitor can develop a fault condition for many difference reasons:
  - too hot an environment,
  - damp conditions,
  - general contamination,
  - prolonged and frequent starting,
  - too high a working voltage,
  - mechanical damage.

- Two wire-fuse elements wired in parallel formation will reduce their joint current carrying rating by about 8% and three by 15%.

- A good magnifying glass and a torch are handy tools for reading small or faded print in poor lighting conditions.

- PVC electrical tape can often become hard and brittle during the colder months but, if placed in an overall pocket, will absorb sufficient body warmth to keep it supple and in a usable condition.

- A smear of petroleum jelly or grease over the threads of a PVC solvent tin will prevent the lid from sticking and provide an efficient way of removing the lid.

Summary so far ...

- Sources of information which will help you with your re-commissioning programme include the following:
  - the Wiring Regulations (BS 7671),
  - your contract specification,
  - distribution and wiring diagrams,
  - paperwork from the manufacturer (information/diagram, etc.),
  - legal regulations relevant to the work in hand.

- Only reliable and trusted test instruments with a valid calibration date should be used. These will include an insulation tester, continuity meter, an earth-loop impedance/PSC instrument and a residual current timing tester, etc.

- Periodic testing need not be in any formal order – initial and formal testing must.
  Refer to this book or the IEE's 'On Site Guide' for details.

- Check out BS 7671 to help with your commissioning requirements.

- Report forms must be accurate and written in a plain and straightforward way.

- Adopt a good code of practice when dealing with your customers and avoid using jargon if the person to whom you are speaking is technically unaware.

- Instruments suitable for fault diagnosis include the following:
  - multi-meter,
  - a low ohm scale continuity meter,
  - an insulation tester with a 250, 500 and 1000 volt DC test range,
  - a combination volt/ammeter.

- You must make the correct choice when selecting your test instruments for fault diagnosis work – it's 'horses for courses'.

- Regular calibration of your test instruments is essential to avoid false values.

- Use documentary evidence to help with fault finding – very important and helpful if you are not quite sure how to deal with your problem.

- To check for polarity, remove your phase conductor from its breaker and place onto the common earth bar whilst disconnecting your circuit neutral from the circuit. Use a continuity meter or bell and battery to check the polarity between the phase and earth point at each accessory and piece of electrical equipment. Reinstate your conductors when your test is completed.

- An earth electrode resistance can be measured with an earth-loop impedance meter. Disconnect the earth bar link first before testing. Reinstate afterwards.

- Use a continuity tester to check the soundness of your conductors by placing a probe at the end of each wire you wish to test.

- Functional test after each repair and ask yourself if what you have done is what was intended.

- Each circuit has a maximum permitted impedance which will efficiently allow your over-current device to operate under a fault condition.

- The cross sectional area of your conductor reflects the value of the design current of your circuit – always keep the two well-matched!

- Often you will need to consult your customer regarding the need for additional restoration work to the brickwork, plastering and decorating once your job has been completed.

- Dispose of your site rubbish carefully and remember that hazardous waste must be pre-registered before a collection is made from site.

- Your line manager could be the person who technically processes your client's account before passing it to the accounts department for invoicing.

- After the close and completion of your work, a service contract could be suggested to your customer.

Review questions

1 List the basic instruments you will require for testing an installation.

2 Outline the sources of information which could help you re-commission an installation or an item of repaired equipment.

3 What type of instruments would you use for fault diagnosis and rectification work?

4 Why is it essential to regularly calibrate your test instruments?

5 Why is polarity testing carried out?

6 What is the kA (kiloamp) rating of a common BS EN 60898 miniature circuit breaker?

7 What course of action would you take if the breakers installed within your installation were of the 6 kA rating range and the prospective short circuit value recorded was 8.34 kA?

8 What difference would you expect to see between earth-loop impedance values taken from a TT earthing arrangement and one taken from a TN-S earthing arrangement installation?

9 Prepare a list of four hazardous waste products you could possibly find on site.

10 What would you consider to be a good and well-organised way of disposing of rubbish from site?

11 Your installation is served from a TN-S earthing arrangement and the final ring circuit originates from a type 'B' (BS EN 60898) circuit breaker rated at 32 amps. When calculated, the total value of $R_1 + R_2$ and the external earth-loop impedance for this circuit amounted to 2.3 ohms whilst the total impedance permitted is only 1.2 ohm. Describe what course of action you would take to correct this departure from BS 7671.

12 Your earth electrode resistance serving a TT earthing arrangement must not exceed one of the following values:

(a) 1666 ohms,
(b) 200 ohms,
(c) 1.2 ohms,
(d) 30 milliohms.

13 Describe a way of decreasing the value in ohms of your earth electrode to a level which is acceptable

14 Describe two problem areas which will occur when the current flow is too high for the conductors installed.

15 State why old redundant ionised smoke detectors should never be placed in a site skip or thrown out with domestic waste.

Answers to Review questions

Chapter 1

1   Your safety officer.

2   There are many – please be guided along these lines: PPE, forming a safety committee, fire fighting equipment, protection from machinery, a place where meals can be taken, messroom heating arrangements, a drying room, fresh water for drinking, a means to boil water, washing facilities, good levels of lighting within the working environment and the maintenance of ladders and steps.

3   (c)

4   Asbestosis, occupational asthma, cancer and dermatitis, etc.

5   Sack-trucks or a fork lift truck, if trained to drive and operate one.

6   When five or more people are employed.

7   A formal talk outlining the rules and site regulations and a means to highlight the risks and dangers involved with construction site working.

8   A statement of intent, organisational matters and company arrangements (training, accident reporting and fire safety, etc.).

9   An EHO will look into objections made by members of the public concerning environmental issues – smoke, fumes, smells and noise pollution, etc.

10 Training, stopping extended overtime, stricter supervision and good lighting arrangements are just four suggestions.

11 A double skinned vessel with a drip tray.

12 (d)

13 Briefly: must be suitable for the purpose required. If not the disposable type – it must be cleaned and must have a good seal: no leakage. Used where conditions are hazardous to breathing.

14 Please accept a format similar to the following: an unexpected occurrence resulting in personal injury, ill health or death.

It also covers loss of property and business prospects.

15 A legally binding written agreement between two people.

Chapter 2

1   A formal written system where permits are issued to control certain types of work which could be potentially dangerous or hazardous.

2   Provides assurance and guarantees that the circuit to be worked on will be completely isolated from the electrical mains.

3   Please be guided on the following or similar lines: restricted spaces, deep trenches, moving machinery, working from a tall ladder, and carrying out work on or very near live bus-bar chambers.

4   Your company accident book.

5   Drilling walls and ceilings, working in deep trenches, using an angle grinder, battery acid work, chasing walls and floors, making/ forming large holes and welding and brazing, etc.

6   Move or drag the victim away from the source of current with anything which is insulated – a dry wooden broom, a waterless hose pipe or a length of nylon rope.

7   Seven suggestions: never fool around with an insulation tester; always wear the correct protective clothing. Work in *pairs* on or near to live bus-bar chambers. Never use 230 volt equipment on a construction site, remove personal jewellery when working on live conductive parts or with secondary cells. Regularly test portable appliances in your site equipment and always handle fully charged capacitors with respect.

8   There are many; here are four as suggestions. Avoid fatigue as tiredness can quickly lead to carelessness. Check the status of conductors – never assume they are 'dead'. Use only approved test equipment. Short out long runs

of insulation tested cable to collapse the capacitance created.

9   No recommendations have been offered by the HSE – it is assumed that PAT inspections will form part of a works/site planned maintenance programme.

10   10 kVA

11   Many, including the following random samples: video presentation, group discussion with colourful illustrations, health and safety policy, site traffic routes, external car parking facilities, working times, welfare amenities and fire procedures and information. Finally, company rules and regulations.

12   Asphyxia is when oxygen is prevented from reaching the lungs. You are unable to breath and often gasping for breath. Causes – choking on food, drowning, suffocation, strangulation and a general lack of oxygen in the air within a confined space.

13   Here are six of many: never wedge fire doors open, chip pans must not be over-filled with oil in a site canteen, use intrinsically safe 110 volt power tools within flammable areas. Waste products must never be burnt on site, stub your cigarette out sensibly – do not throw it on the floor. Keep a suitable fire extinguisher by your side when carrying out 'hot work'.

14   Battery or mains operated fire detection points/alarms, break-glass points in passageways, audio alarms, heat detectors in works canteens, auto magnetic site fire doors and mechanical sprinkler heads.

15   A fire blanket made from glass fibre fabric.

Chapter 3

1   A college course, learn with a postal 'direct learning' organisation (a correspondence school), an on-line computer course and hands-on experience with your workplace overseen by an experienced operative.

2   A registered apprenticeship, an NVQ award at Level 3 in electrical installation engineering, be at least 22 years of age, work without supervision and have knowledge of BS 7671, British Standards and various codes of practice serving our industry.

3   Storming, forming, norming and performing.

4   (c)

5   (c)

6   FALSE

7   When treatment to a person of ethnic origin is less favourable than that offered to others.

8   When an employer is far less favourable to a man or woman because of his/her gender.

9   Life, a fair trial, the respect for privacy and family life, freedom of thought and expression, not to be discriminated against and the protection of an individual's property.

10   In order to be free from other trades within a chosen area where work has been planned.

11   An international organisation of national standards.

12   This standard advances and cultivates workable schemes which are both practical and sound in order to develop both productivity and performance within the workplace.

13   Postal questionnaires, telephone interviews, e-mail surveys, face to face interviews, employing others to research for you and customer satisfaction reviewed by means of your web site.

14   Fit global tracking devices on all company vehicles and plan journeys to avoid traffic problems.

15   Your company will be added to a builder's list of preferred companies they are confident to do business with. Although you will not win every contract you are asked to tender for, you will win some.

Chapter 4

1   This is a measurement of a material's ability to oppose the flow of an electric current.

2   What the conductor is made from, the cross sectional area of the conductor, ambient temperature and the length of the conductor.

3   In parallel – 25 ohms; in series – 400 ohms.

4   An invisible field of magnetic force which is found around any load bearing conductor or the invisible field around a permanent magnet. The first law of magnetism is: unlike poles attract whilst like poles will oppose one another.

5 When a body, which possess magnetic properties, can cause or induce a voltage in another body without direct hard-wire contact.

6 The property of a circuit that contains capacitance and/or inductance which jointly, with any resistive component, makes up the impedance of the circuit. Symbol, X; measures in ohms (Ω).

7 Doping.

8 The non-conductive part of a capacitor which is sandwiched between the plates. It may also be described as a non-conductor where an applied electric charge creates an electrical field which causes electrical deformity within the atoms serving the dielectric.

9 The electrical pressure between the plates of a capacitor produces an oval warp within the orbits of the electrons serving the dielectric. This is electrical stress. Expressed in volts per millimetres.

10 Can keep their charge for long periods. Will cause heart fibrillations if shocked by one. Some older types of metal capacitors may have toxic oil. The oil from older metal-can capacitors is known to be cancer producing.

11 As a switch and as an amplifier.

12 Insulated wire, a step-down transformer, a capacitor, four diodes and possible a resistor.

13 A semiconductor diode will only allow the flow of an electric current in one direction. It is used in rectification circuits.

14 Optical fibre cables are used in the transmission of data, telephone channels and telephone programmes, etc. Advantages include the following: free from electrical interference, lighter and easier to install, improved security, greater transmission distances can be achieved and more information can be carried compared to conventional means.

15 Ninety degrees.

Chapter 5

1 Wind turbine, wave motion, hydro, oil–diesel alternators, gas turbine and nuclear energy.

2 The current flow within the cables is greatly reduced. This enables much smaller cables to be employed for power transmission

3 The secondary winding is in 'star' formation and the centre point of the star is earthed. The neutral conductor is then taken from this centre point to provide for a single phase supply.

4 (d)

5 Transformers carrying small loads, fluorescent lighting, any luminaries with a ballast unit and any type of discharge light.

6 In industrial installations where many principal main switches have to be employed.

7 Eddy currents occur within the steel core of a transformer or an electric motor. They are generated by mutual induction caused by the ever-changing magnetic field. Once the lines of magnetic flux cut across the core of the transformer, currents are induced in it.

8 Magnetic fluxes generated within the primary winding induce an emf into the secondary winding. The windings are completely independent from each other.

9 As the auto transformer is made from a single length of wire, the primary and secondary windings are in series with each other. A direct connect exists between both 'IN' and 'OUT' supplies.

10 FALSE

11 (a) A site generator; (b) town (older installations); (c) usually town – not used very often now; (d) in rural installations.

12 (a) General domestic use; (b) transformers; (c) electric motors and large site transformers.

13 Basically, a false earth is created by the connection of an earthing conductor to the supply neutral. (a) PME earthing arrangement; (b) mainly found in urban installations, sometimes new rural installations will support this type of earthing arrangement.

14 Responds to both over-current and phase or neutral to earth fault conditions. (a) Where $R_1 + R_2$ values are too high to sustain a 'B' type miniature circuit breaker; (b) wrong test values will be recorded.

15 A sensi-touch switch or any type of semiconductor device

Chapter 6

1 Frequency, the number of pairs of poles, the applied voltage and the strength of the magnetic field strength generated.

2 A recessed copper squirrel cage, all of which are common; coils are let into soft iron slots – this forms the stator. Connecting a triple phase supply to the winding will produce a rotating magnetic field.

3 The number of pairs of poles the motor has.

4 Synchronous; induction and commutator motor.

5 TRUE

6 Where high starting torque is required; large shearing machines; water pumps and large refrigeration plants.

7 Cost.

8 By linear resistance, through electronics means and by the use of an auto transformer.

9 A slower speed would alter the status of the centrifugal switch.

10 Swap over winding lead *Z1* with *Z2*.

11 Miniature bimetal overload *with* an integral heater and a miniature bimetal overload *without* an integral heater.

12 Advantages: reduced starting torque, torque directly proportional to the voltage applied and a smooth starting sequence.

Disadvantages: high installation costs, regular maintenance required and only suitable for certain types of motors.

13 Rotor resistance speed control mechanism.

14 A wound commutator rotor placed in series formation with the stator.

15 Bearings can wear through over-heating; the grease could melt and leak into the windings of the motor causing damage.

Chapter 7

1 A site office, a canteen/drying room and a male and a female toilet.

2 To oversee progress, to discipline, to advise and inform and to cater for site material requirements.

3 A useful office tool for keeping site tasks on schedule. May be formatted as a bar chart.

4 (b)

5 ELCB voltage, RCD and RCBO.

6 *Switching*: the safe disconnection of the supply when on load. *Isolating*: switching off a supply when off load when, for example, maintenance work is to be carried out.

7 Enclosures made from PVC-u, plastic cable cariers and fittings and materials made from glass reinforced polyester.

8 Fire alarm circuits, central battery emergency lighting circuits and any other sensitive circuit requiring protection.

9 By the use of a reliable milliohm meter.

10 For example: never mix aluminium fittings or accessories with copper, paint bare copper earthing links with a suitable finish, use covered MI cable for outside applications and take care to use the correct soldering flux when making a joint

11 The factors drawn from relevant tables allow the size of conduit needed for the installation to be calculated.

12 By use of an RCBO in circuit; protection afforded by a residual current device; means of isolation; checking everything before your final connection is made.

13 As a means of protecting a large site transformer and protection afforded to the circuit of a large electric motor.

14 Conduit with a diameter of 25 millimetres.

15 *Earthing*: this is the return path back to the secondary winding's 'star-point' of the local community transformer. The star point is grounded to the general mass of earth. *Bonding*: this brings the potential difference to zero when metal infrastructure and pipe work, etc., are bonded together as one. A potential difference could never be maintained under these conditions.

Chapter 8

1 Examples follow: scuffed insulation to bare conductor; trapped or pinched conductors; conductors pierced by an accessory screw; snapped conductors resulting from using side cutters instead of wire strippers, and wood

screws or masonry nails driven through a cable.

2 Tiny electrical currents generated by temperature differences.

3 After damage caused by fire, water or any other mechanical means; a change of use for the building; when extensive alterations are made to the original installation; a change of ownership; after a qualifying period of time and a request made from an insurance broker.

4 Six examples from many follow: the ambient temperature; any external heat sources; water and high humidity; stress and vibration; presence of animals or/and plants and the effects of solar radiation.

5 Warning or advisory labelling on or near to distribution centres; notices and diagrams on 'mains room' walls; warning labels on the terminal/connection box serving a three phase 400 volt motor; on three phase control equipment; within a caravan site; written circuit destinations to serve a distribution centre; on or near to an auto transformer and the labelling of main switches and isolators.

6 A 500 volt insulation tester, a low scale continuity meter, an earth-loop impedance meter, a prospective short circuit meter and a residual current device timing meter. These examples are all standalone.

7 Anything from 0.9 to 20, 30 or 40 ohms depending on the dampness of the soil and how far the local community transformer is sited from the installation.

8 There are three independent test leads. Clip the *green/yellow* earth lead onto the principal earthing point, the *brown* phase test lead must be firmly attached to the source of the supply whilst the *blue* neutral test lead must be placed on the incoming supply neutral conductor. Select the highest earth-loop impedance scale and progress downwards if necessary. Press the test button and read off the value. The value recorded will depend on the applied voltage and the earth-loop impedance.

9 (I refer to post-2005 models here.) This modern plug-in device will not trip a residual current device or RCBO, will check the applied voltage, any phase faults and the polarity of the socket. The device will also indicate

wrongly placed conductors within the socket outlet and whether an open earth or neutral conductor fault conditions exists. Some will check out earth-loop impedance parameters.

10 Seven examples of many: test meter problems; low battery value; contamination within an accessory; loose connection; plaster-buried joints and accessories; site damaged cables; and broken conductors.

11 Three successive tests.

12 Making sure that everything works as intended – hobs, hoods, final ring circuits, machinery, etc.

13 Any electrical work carried out which does not include the provision of a new circuit.

14 *First test*: continuity of protective conductors and the main supplementary bonding conductor. *Second test*: measuring the continuity of the final ring circuit(s). *Third test*: insulation values recorded for all circuits.

15 FALSE.

Chapter 9

1 (a) A two-way and intermediate switching arrangement, for example.

(b) Final ring circuit, radial and an industrial motor circuit are examples of many.

(c) Dimmer switch, capacitor, inductor and an ELV lighting transformer for example.

(d) Controls serving a central heating system, a thermal sensor controlling an underfloor heating arrangement and so on.

2 Firstly, locate your local switch and switch off the circuit.

Check, using a voltmeter, that there is no supply present.

Remove the breaker or fuse from the distribution board.

Secure this action by locking if possible.

Fit a warning notice by your isolator that work is being done.

After completion, ensure that all is as intended and test.

3 Do not use the 'try-it-and-see' approach.

Apply your expertise and experience.

Ask for advice when needed.

Use a logical step by step approach to your problem.

Be very careful if you have to work on live equipment.

4   Collect together all relevant data.

List helpful telephone contacts should you require their help.

Enquire of the sequence of events leading to the fault appearing.

Apply standard tests.

Come to a technical conclusion.

Make good your problem and re-test.

5   Missed place conductors, snapped conductors. Squashed or pierced conductors to earth. Loose wiring, component problems and burnt conductors, etc.

6   Found in AC power lines and communication wiring. A short lived high voltage phenomenon – spikes are twice the declared voltage but, if greater, are known as surges.

7   Fit a transient voltage surge suppressor in line or use a special surge suppressor extension lead

8   Pinched or sliced conductors grounded to earth.

Vibration causing insulation wear over time.

Over-current within small conductors.

Ultra-violet light (UV).

Perished rubber cabling over a period of time.

Dampness within older types of mineral insulated cables (MI).

9   When a replacement is not available.

When past its useful life.

10 A period of time when something or somebody is not operating or being productive.

11 Too costly.

A replacement is not available.

Too old to repair.

Excessive 'downtime' involved.

12 Whether access can be safely gained to an installation or an item of equipment, whether a standby generator can be used or when a continuous supply is demanded.

13 Use carrying straps when transporting batteries.

Never short out the terminals.

Avoid accidental spillage of acid on your skin.

Wear personal protective equipment (PPE).

Never smoke when working in a battery charging room.

Keep naked fames away from charging batteries.

Remove rings and watches when handing batteries.

14 Tight bends must be avoided.

Never place a cable under stress.

Insert a cable loop every 10 metres or so.

If securing with cable ties, do not over-tighten.

Seal exposed cable ends with heat-shrink caps.

Leave about 5 metres of cable at either end.

Earth metal accessories in accordance with your specification.

15 Over-tightening creates pressure within the cable and causes it to bend and squeeze out of shape.

Chapter 10

1   A continuity meter.

An insulation tester which will deliver 250, 500 and 1000 volts.

Voltmeter/clamp meter.

May be an earth-loop impedance/PSC tester.

An RCD timing meter.

Possible a portable appliance tester.

2   BS 7671.

Your job specifications.

Wiring and circuit diagrams.

Manufacturer's instructions and advice.

Legal requirements.

3   Insulation tester with three DC test voltages.

A multi-meter.

A low ohm continuity tester.

A volt meter and amp meter.

4   A test meter's reliability weakens over a period of time.

5 Switches, over-current devices or any single pole controlling device must be served with a phase conductor – never a neutral. If reversed, an installation could be potentially dangerous as an unswitched phase conductor would be present where normally a neutral conductor would be.

6 1.5 kA to 15 kA; many used for domestic installations are 6 kA.

7 Upgrade the breaker to a more suitable short circuit rating.

8 The earth loop impedance value for a 'TT' earthing arrangement would be lower in value than the 'TN-S' arrangement.

9 Old industrial paint.
Motor oil, gasoline and antifreeze.
Fluorescent lamps and high pressure lamps.
All types of batteries.

Transformer – especially very old transformer oil.
Old ionisation type smoke detectors.

10 Hire a large enough skip from a waste disposal service company. Advise them if you have any hazardous waste products you wish to dispose of.

11 Remove the Type 'B' breaker and put a 32 amp RCBO in its place.

12 (b) 200 ohms.

13 By installing a longer earth electrode or joining several linked together to serve a larger area.

14 Heat generated within the under-rated conductors. Volt drop will occur.

15 Older types of ionisation smoke detectors have a much higher level of radioactivity present than the one we use today.

Index

AC *see* alternating current
accidents
    accident books 40
    causes 26–7
    definition 28
    employer's responsibility 12–13
    investigation forms 12–13
    prevention 26–7
    reporting 12–13, 20, 27–8, 40
    working practices and procedures 20
accommodation planning 166–7
accounts department 282
air capacitors 91
alarms 54–6, 255, 258, 268
alphanumerical identification 104
alternating current (AC)
    circuits 94–5, 109
    electrolytic capacitors 91
    motors 137–8, 147–52, 161
    star/delta starters 157–8
alternators 138
aluminium 34, 108, 191
ammeters 275
amp per metre unit 83
amplifiers 103–4
antimony 97, 98
antistatic precautions 265–6
apparent power 94–5, 124
applied voltage 81–2, 85, 100–1, 241
apprentices 66
approved electricians 66
aqueous film-forming foam fire extinguishers 60
arms, injury prevention 42
arrestors 266, 269
arsenic 97
asphyxia 53–4, 63
assembly points 54–5
atoms 95–9
auto transformers 125, 162
automatic fire fighting equipment 59–61, 63

balanced loads 181–2
bar charts 172, 225
barriers 17–18
basket systems 201
batteries 266–7
bearings 143, 160
belts 161
bimetallic devices 155–6, 162, 220, 221
blankets for fire fighting 58, 60
block diagrams 170
block terminations 205
bonding installations 213–15
bonds, semiconductors 96
boron 98
boxing out 165
bridge rectifiers 103, 110
British Standards 30, 230, 258
    BS 7671 Regulations 169, 227, 254, 271, 274
    BSEN ISO 14001 30
Building Regulations [1991], role 169
bullet *see* pin and bullet cable terminations
busbar terminations 204
busbar trunking 118

cable carriers 194–203
cable factors 210–12
cable trays 201, 202
cable trunking 118
cables 118–19
    channelling 165
    handy hints 34
    routing 231, 258
    size factors 279–80
    terminals 191–2, 225
    terminations 191–2, 203–7
calibration 228, 242, 275, 284
capacitance 89, 90–1, 93–5
capacitors 89–93, 105, 109
    capacitor-start motors 151–2
    dangers 92–3, 109
    faults 283
    safety precautions 266, 268

    single phase AC motors 149–52
    types 91–2
car starter motors 139
carbon dioxide fire extinguishers 60
career options 66–7, 77
career patterns 65
cartridge fuses 132, 185, 186, 219
causes of accidents 26–7
centrifugal switches 150, 161
certificates 243–6, 251, 257
change, causes of 31–2
charge 90–1, 96–100
chattering contactors 283
circuit breakers 128, 135, 185–8, 205, 219–20
circuits
    design currents 207–9
    diagrams 171
    isolation 38–9
    protective devices 218–22
    test certificates 244–5
clamp meters 236, 272
Clean Air Act [1993] 31
clothing 2, 11, 16–17
code of practice 5, 18, 46
coefficient of resistance 95
colour coding 104, 109
COMAH *see* Control of Major Accident and Hazards Regulations
commissioning installations 227
communication 18, 177–8
commutators 138, 160, 161, 162
competition 32
compiling test certificates 243–6
completing installations 163–225
    bonding 213–15
    cable carriers 194–203
    cable terminals 191–2, 225
    cable terminations 191–2, 203–7

circuit protective devices 218–22
contracts 171–6
corrosion prevention 206–7, 224, 225
dangers 189–94
distribution voltages 178–80
drawings 168–71
earth fault protection devices 186–9
earthing 182–3, 212–13, 215–18, 225
enclosures 194–203
fuses 185, 186, 218–19, 225
harmonizing with other trades 176
isolation 183–4
over-current protection 184–6, 218–22
planning 166–8
reality of variations 177
regulatory requirements 168, 169
relationship maintenance 177–8
safe working areas 163–4
single phase supplies 182–3
specifications and drawings 168–71
supply voltages 178–80
switch gear 183–4, 222–3, 225
terminals 191–2, 225
testing 230–5
three-phase distributions 180–2
voltages for supply/distribution 178–80
wiring systems 194–203
components *see* electronic components
compound motors 140
condensation 224
conductors
  handy hints 35
  resistance 79–81
  size factors 207–10, 225, 231
conduit 118, 199–200, 210–12, 225
confined space working 19
confirmation procedures 228

Construction (Health, Safety and Welfare) Regulations [1996] 60
consultancy 67
continuity testing 229, 256
  electric motors 149–51, 161
  meters 235, 241, 272
  methods 246–7
  working order restorations 272, 274–5, 276, 284
contracting, career options 66
contracts 33–4, 171–6
Control of Major Accident and Hazards Regulations (COMAH) [1999] 6
Control of Substances Hazardous to Health Regulations (COSHH) [1999] 4
control systems 261, 269
Controlled Waste Regulations [1998] 31
controlling fires 56–61
correction factors 207–8, 225
corrosion prevention 206–7, 224, 225
COSHH *see* Control of Substances Hazardous to Health Regulations
costs, fault diagnosis 264, 269
covalent bonds 96
CPN *see* critical path networks
crimped terminations 203, 204, 205
critical path networks (CPN) 172–3
critical voltages 101
crocodile clips 256
cross sectional area 79–80, 81
crystal lattices 95–9
current 3
  cable size 279–80
  relays 150–1
  three phase systems 115–17
  transformers 122–5
  voltage–torque relationship 159
customer relations 177–8
  expectations 32–4
  satisfaction 74, 77, 78
  working order restorations 273–4, 284

Dangerous Substances and Preparation and Chemical Regulations [2000] 31
dangers 189–94
  asphyxia 53–4, 63
  capacitors 92–3
  dangerous areas 19
  dangerous occurrences 54, 63
  visual inspections 233, 234
  see also electric shock
Data Protection Act [1998] 68, 77
DC *see* direct current
decay, fires 56, 63
delta starters 157–8
delta/star connections 114–16, 134, 180–1, 182
demand, changes 31–2
depletion layers 99–100
detecting fires 59, 61, 63
diagrams 168–71, 175, 233–4, 271
dial type kilowatt-hour meters 239–40
diaries 174
dielectrics 90, 109
digital display kilowatt-hour meters 240–1
diodes 99–101, 106, 109–10
direct contact protection 53, 63, 190, 209, 231
direct current (DC) motors 138, 139–43, 157, 160
direct-on-line (DOL) starters 153–4
Disability Discrimination Act [1995] 68–9, 77
disconnection procedures 52, 210
discrimination acts 68–70, 77
display scales 253–4
disruption avoidance 167
dissipated power 82
distribution 111–20, 178–81
  four wire systems 180–2, 225
  neutral systems 116–17, 180–2, 225
  three phase systems 180–2, 225, 260–1, 269
  voltages 178–80, 225
  working order restorations 271

documentation
  handy hints 34
  testing 242–6, 251, 255–6
  working order restorations
    271–3, 275, 284
DOL *see* direct-on-line
doping 97–9
double cage induction motors
  143, 145–6, 161
downtime 265, 269
drawings and specifications
  168–71
drilling and notching 165
dualing 118
duct/ducting systems 201
dynamic inductance 85, 109
dynamo motors 137–8

earth electrode resistance tests
  229, 249–50, 257
  working order restorations
    278, 280, 284
earth-fault loop impedance
  testing 229
  meters 237
  test certificates 244–5
  testing methods 248, 257,
    258
  working order restorations
    272, 277–8, 279, 284
earth-fault protection devices
  186–9
earth-leakage circuit breakers
  189
earthing
  completing installations
    182–3, 212–13, 215–18,
    225
  electrical supply systems
    130–2, 135
eddy currents 121, 154
education, career options 67
effective working practices
  65–78
  employment regulations
    67–70
  improvement benefits 74–6,
    77
  performance 65–7
  team working methods 67,
    77
efficacy, lamps 34

efficiency
  transformers 122, 134–5
  work procedures 70–4, 77
EHO *see* environmental health
  officers
electric fields 90
electric motors 137–62
  auto transformers 162
  bimetallic devices 155–6, 162
  direct-on-line starters 153–4
  function 137–43
  miniature bimetallic devices
    155–6, 162
  over-current protection 140,
    154–5, 161
  overload devices 156
  series wound 139, 147–8
  shunt type 140, 157, 161
  speed 144, 147–8, 159–62
  standard enclosures 152
  starters 153–9
  synchronous 138–9
  three-phase induction motors
    143–6
  transformers 162
electric shock 44–53, 62–3
  avoidance 51–3
  capacitor dangers 92, 109
  conductor size factors 209–10
  dealing with 42–4
  portable appliances 45–7
  prevention 44–5
  risks 44–5
  visual inspections 231
electric stress 90
electric tools 62
electrical drawings 133
electrical inductance 84–9
electrical labourers 66
electrical regulations 169, 227,
  254, 271, 274
electrical resistance 79–81
electrical rubber gloves 16, 257,
  283
Electrical Safety Act [2002] 18
electrical supply systems 111–35
  distribution systems 111–20,
    178–82, 225, 260–1, 269
  earthing systems 130–2, 135
  generation 114–17, 134
  industrial distribution systems
    117–20

  protection 132–3
  switch gear 125–30
  transformers 114–16, 120–5,
    134–5
  voltages 125–30, 178–80
electrician grading systems 66
Electricity at Work Regulations
  [1989] 3
electrolytic attack 207
electrolytic capacitors 91
electromechanical devices 157,
  186–7
electronic components
  capacitors 89–93, 105, 109
  fault diagnosis 261, 264
  identifying 104
  location dangers 193–4
  resistors 79–82, 105
  semiconductors 95–103,
    105–6
  transistors 103–4, 105, 107,
    110
  use 105–8
electronic RCDs 187
electrons 90, 96–100
electrotechnical systems 262
emergency procedures 19, 61,
  184
emf 85, 138
employer's responsibility 7–15
employment regulations 67–70
Employment Rights Act [1996]
  67–8, 77, 78
EMS *see* environmental
  management systems
enclosures 152, 194–203
end shields 143
environmental health officers
  (EHO) 25–6
Environmental (Information)
  Regulations [1992] 169
environmental management
  systems (EMS) 30
Environmental Protection Act
  [1990] 31
equipment
  erection/assembly 233, 235,
    258
  location dangers 193–4
  location diagrams 171, 175
  safety 8
equipotential bonding 213–14

evacuation procedures 54–6, 61
external influences 232, 258
extinguishers 57–61, 63
eyes 16, 41

fault diagnosis 260–9
  basic principles 261
  capacitors 283
  electrotechnical systems
    262
  improvement influencing
    factors 265
  legal responsibility 265
  making good influencing
    factors 265
  precautions 265–7
  re-commissioning 261
  repair influencing factors
    264–5
  replacement influencing
    factors 264–5
  restoration 261
  safe working procedures 261
  safety precautions 265–7
  shut down 266
  single phase supplies 260–1,
    269
  supply losses 262
  symptoms 262–4
  three phase supplies 260–1,
    269
  working order restorations
    274–9, 284
  see also testing
ferrules 199
fibre optics 108, 110, 267
fibrillations 92
field strength 83
fires
  barriers 231
  causes 56
  controlling 56–61, 63
  discovery 54
  doors 56
  equipment signs 23
  evacuation procedures 54–6,
    61
  fire fighting equipment
    57–61, 63
  Fire Marshals 55
  fire officers 30
  fire retardant cables 198

prevention 56–61, 63
protection rules and
    regulations 60–1, 63
first aid 40–1
first-fix cables 165
fixing accessory boxes 165
flexible cable terminals 191–2,
    225
floor boards 165
fluxes 83, 84, 87, 89
foam fire extinguishers 60
footwear 16
forces, magnetism 83–4, 88
forming 67
forward bias voltage 100
four wire three phase systems
    180–2, 225
FP 200 cables 119
free electrons 96–100
full-wave rectification 101,
    102
functional testing 230, 251,
    279, 284
funnelling effect 62
fuses 125, 132, 185–6, 196,
    218–19, 225

gas meter bonding 214
gates 103–4
generation 114–17, 134
generators 137–43
germanium 95, 98
gloves 16, 257, 283
grading systems 66
growth periods, fires 56, 63

half-wave rectification 101,
    102
halon fire extinguishers 60, 63
hand injury prevention 41–2
handling methods 11–12
harmonic currents 134
harmonizing with other trades
    176
harnesses 30
hazardous malfunction 54
hazardous substances 20
hazardous waste disposal 282,
    284
hazards 29–30, 92
HBC *see* high breaking capacity
    fuses

health and safety
  capacitor dangers 92
  executives 5, 25, 60
  test instruments 253
Health and Safety at Work Act
    [1974] 2–3
Health and Safety at Work
    Regulations [1974] 169
Health and Safety Inspectors 2,
    25
heating appliance thermal links
    220, 222
hep-tape 199
high breaking capacity fuses
    (HBC) 185, 186, 196, 218
high frequency risks 266
high voltage systems 112–15,
    126–7, 132, 134–5
holes (electrons) 96–100
horsepower 134
housing caps 160
Human Rights Act [1998] 70, 77
hysteresis losses 121–2

identifying your tools 224
ignition, fires 56, 63
impedance 93–5
  see also earth-fault loop
    impedance testing
improvement influencing factors
    265
impurities 97–9
In-Service Inspection and
    Testing of Electrical
    Equipment 46
indirect contact protection 183,
    190–1, 209–10, 231
indium 98
inductance 84–9, 93–5, 109
induction
  motors 143–6, 149–52, 160,
    161
  safety measures 49, 50–1
  stickers 223
industrial distribution systems
    117–20
infra-red source/detection
    devices 107
initial confirmation procedures
    228
injury prevention 41–2
input losses 121–2

inspection boxes 199–200
installation engineering regulatory requirements 168, 169
installation planning 166–8
instruction, employers responsibility 13–15
instrumental inspections 227
insulation
    fault diagnosis 263–4, 268
    resistance 245, 247–8, 272, 274, 276–8, 280
    testers 229, 235–6, 241, 256, 257
insurance 133
integrated circuits 105, 110
International Organisation for Standardisation (ISO) 30, 71–2, 77
Investors in People 73–4, 77
iron filings 84
ISO see International Organisation for Standardisation
isolation 38–9, 40, 51–3, 183–4
isolators 129–30, 190, 231–2, 257
IT earthing systems 131, 135

joint box terminations 192
Joint Industry Board 65
joists 34

kilovolt amp reactive power (kVAR) 116
kilovolt amps (kVA) 116, 124–5, 134
kilowatt-hour meters 239–41
kilowatts (kW) 116, 134, 239–41
kVA see kilovolt amps
kVAR see kilovolt amp reactive power
kW see kilowatts

labelling 232
ladder systems 201, 203
lagging current 94
lamps, efficacy 34
laser beam temperature measurers 107
lattices 95–9

leading current 94
leads 104, 252–3, 258
learning related performance 65–7, 78
leg injury prevention 42
legal obligations 33–4, 265
length factors 79–80, 81
lighting 34, 260, 268
line current 124–5
line managers 282, 284
lines of force 83–4
list preparation 168, 170
location dangers 193–4
location diagrams 171, 175
lock-off procedures 38–9, 62
losses
    fault diagnosis 262
    power 112–15
    transformers 121–2
low ohm continuity testers 274–5
low resistance continuity meters 235
Low Smoke and Fire (LSF) cables 119
low voltage systems 125–33, 135
LSF see Low Smoke and Fire
lug terminations 204–5

magnetism 83–9, 109
    fluxes 83, 84, 87, 89
    inductance 84–9
    magnetic circuits 83–5
    magnetic fields 83–4, 88, 109, 120, 141–3, 161
    magnetic growl 283
    three phase stators 141–3, 161
    transformers 120
main equipotential bonding 213–14
mains layouts 223
maintenance, career options 66
management, career options 67
Management of Health and Safety Regulations [1998] 7
marketing 32
material composition 79, 177

MCB see miniature circuit breakers
MCCB see moulded case circuit breakers
mechanical starters 156–7
medium voltage switch gear 126–7, 135
melting points 133
method statements 19
MI see mineral insulated cables
mica capacitors 91–2
microfarads 89
mineral insulated (MI) cables 119, 196, 197, 198
miniature bimetallic devices 155–6, 162, 220, 221
miniature circuit breakers (MCB) 128, 135, 185–6, 205, 219–20, 258
minor works certificates 251
mobile telephones 224
modular panel boards 126
monitoring 172–6, 232
motors see electric motors
moulded case circuit breakers (MCCB) 18–65, 128–9, 135, 186, 219, 221
mouth to mouth resuscitation 43–4
multimeters 237–8, 241, 274
multiplate capacitors 92
multivoltage transformers 124
mutual induction 88–9

n-type semiconductors 97–8
National Grid 111–12, 178–80
negative coefficient of resistance 95
neutral three phase systems 116–17, 180–2
Noise Act [1985] 6
Noise and Statutory Nuisance Act [1993] 6
non-electrical remedial work 281–2
norming 67
notching 165
notices
    electric shock 49, 52
    health and safety 2
    lock-off procedures 39

three phase supplies 269
visual inspections 233, 234
working practices and
procedures 20
nuisance tripping 188

ohm continuity testers 274–5
ohm impedances 244–5
ohm/metre 80
Ohm's law 81, 82
on-site training 14–15
operator safety requirements
16–18
optical fibres 108, 110, 267
optocouplers 107
output losses 121–2
over-current protection
completing installations
184–6, 218–22
electric motors 140, 154–5,
161
fault diagnosis 262–3, 269
starters 154–5
overhead collectors 118
overloads 135, 156–7, 220, 222
overvoltage protection 266
oxygen starvation 53–4, 63

p-n-p transistors 104
p-type semiconductors 98–9
paper insulated capacitors 91
Paper Insulated Lead
Covered/Steel Wire
Armoured (PILC/SWA)
cables 119
parallel circuits 82
passive infrared detectors 129
PAT *see* portable appliance
testing
pd *see* potential difference
peak inverse voltages (PIV) 101
performance 65–7
performing, definition 67
periodic inspections 229–30,
258
Permit-to-Work systems 38, 62
personal performance 65–7
personal protective equipment
(PPE) 2, 11, 16–17
personnel, safety policies 21
petroleum jelly 224, 283
phase values 115–16

phones 224
phosphorus 97
photo cells 106
photodiodes 106
phototransistors 107
picofarads 89
PILC/SWA *see* Paper Insulated
Lead Covered . . .
pin and bullet cable
terminations 203, 205
PIV *see* peak inverse voltages
plant safety 8
plaster/plastering 34, 165
plug fuses 218
PME (protective multiple
earthing) *see* TN-C-S
earthing
point-to-point wiring diagrams
170
polarity
test certificates 245
testing 229, 231, 241, 248,
258
working order restorations
277, 280, 284
policies 15, 18–21
pollution 19, 31, 206–7, 225
Pollution Prevention and Control
Act [1999] 31
portable appliance testing (PAT)
5, 34, 45–7, 272
potential difference (pd) 90
powder fire extinguishers 60
power
AC circuits 94–5
circuit fault diagnosis 260
dissipation 82
factors 94–5, 109, 116, 124
losses 112–15
single-phase transformers 48
tools 46
transformers 123–4
powers, responsibilities and
roles 23–6
PPE *see* personal protective
equipment
practices, carrying out 18–21
prevention
accidents 26–7
electric shock 44–5
fires 56–61
primary windings 120

probes 252–3, 258
procedures, carrying out 18–21
productivity 32, 74–5, 77
profitability 75–6, 78
prohibition signs 22
prospective short circuit current
testers 238–9, 258, 272,
280–1
protection
electrical supply systems
132–3
low voltage systems 132–3
protective equipment 3
rules and regulations 60–1, 63
protective multiple earthing
(PME) *see* TN-C-S earthing
systems
provision, first aid treatment
40–1
Provision and use of Work
Equipment Regulations
(PUWER) [1998] 4–5
publications 230
pure inductance 109
PUWER *see* Provision and use of
Work Equipment
Regulations
PVC
cables 119, 197, 200, 202
conduit 199–200
trunking 200, 202

quality
checks 282
competition 32
standards 72, 77

Race Relations Act 69, 77
Radioactive Substances Act
[1993] 31
ratings 124, 134, 147–52
RCBO *see* residual current
breaker with overload
protection
RCCB *see* residual current circuit
breakers
RCD *see* residual current devices
re-commissioning 271–3
reactance 85, 93–4, 109
reading kilowatt-hour meters
240
records 174

rectifier circuits 99–103, 110
reduced voltage 47–9
regulations
    BS 7671 Regulations 169, 227, 254, 271, 274
    fire prevention 60–1
    requirements 168, 169
    test instruments 254–5
    testing 227, 254–5, 273
    working order restorations 271, 273, 274
relationship maintenance *see* customer relations
religious orders 34
remedial work 271–84
repair influencing factors 264–5
replacement influencing factors 264–5
Reportable Diseases and Dangerous Occurrences Regulations (RIDDOR) [1985] 6–7
reporting accidents 12–13, 20, 27–8, 40
reporting hazards 29–30
reports
    forms 255–6, 273, 284
    risk assessments 28–9
rescue teams 30
residual current breaker with overload protection (RCBO) 133, 135, 185–6, 188, 196, 219–20
residual current circuit breakers (RCCB) 187–8
residual current devices (RCD)
    completing installations 186–7
    testing 238, 245, 250–2, 256–8, 272
    working order restorations 272, 278–9
resistance
    coefficients 95
    definition 79–81, 109
    inductance/capacitance/ impedance 93–5
    meters 249–50
    test certificates 244–5
resistivity 80–1, 109
resistors 79–82, 105

resources, working practices 75, 78
respiratory protective equipment (RPE) 17
responsibilities
    employment regulations 67–70, 77, 78
    roles and powers 23–6
restoration 260, 261
resuscitation 43–4
reverse bias voltage 100–1
reverse rotation 147, 148, 151, 152, 161
revolutions per minute, rotor speeds 141
RIDDOR *see* Reportable Diseases and Dangerous Occurrences Regulations
rising mains 119
risk areas 21
risk assessments 19, 20–1, 28–9
risks 44–5, 190–1
road safety 19
road traffic accidents (RTA) 12–13
roles, responsibilities and powers 23–6
rotating magnetic fields 141–3, 161
rotors 141, 143–4, 145–6
RPE *see* respiratory protective equipment
RTA *see* road traffic accidents
rubber gloves 16, 257, 283
rules, fire prevention 60–1
rust 283

safe methods in handling, storage and transport 11–12
safety
    electric shock 44–53, 62–3
    employer's responsibility 7–15
    fault diagnosis 261, 265–7
    gloves 16, 257, 283
    harnesses 16
    officers 23–4, 29–30
    operators 16–18
    policies 15, 20–1
    procedures 7–11, 18–20, 38–40, 70–1, 77, 163–4, 261

regulation awareness 1–7
representatives 24–5, 29–30
signs 22–3
training 49, 50–1
working areas 9–11, 163–4
working environment provision 9–11
working practices 7–11, 18–20, 38–40, 70–1, 77, 163–4, 261
sales agreements/contracts 33–4
scaffolding 19, 34
scale of competition 32
screwdrivers 283
secondary windings 120
secure areas 17–18
self induction 87–8
semi-enclosed fuses 186, 219, 225
semiconductors 95–103, 105–6, 109–10
    manufacture 95–9
    rectifier circuits 99–103, 110
    types 99–103
series circuits 81–2
series wound motors 139, 147–8
Sex Discrimination Act [1975] 69–70, 77
sexual harassment 70
shape factors 104
shock *see* electric shock
short circuit current testers 238–9, 258, 272, 280–1
shunt motors 140, 157, 161
shut down 266, 268
signs 20, 22–3, 49, 51, 56–8
silicon semiconductors 95, 101
silver mica capacitors 91–2
single-phase motors 147–52, 156–7, 159
single-phase supply 114–17, 182–3, 260–1, 269
single-phase transformers 48
site accommodation 166–7
site diaries 174
site emergency alarms 54–6
site evacuation 54–6, 61
site induction 49, 50–1, 223, 224
site records 174
site variation orders 174–6, 177
site waste disposal 282, 284

size factors 207–12, 225, 231, 279–80
slip 144, 161
slip rings 145–6, 160–2
smoke alarms/detectors 255, 258, 268
socket polarity testers 241
soldered cable-lug terminations 205
solid state temperature devices 107
space factors 202
specialised electrical engineering 66
specifications 168–71, 271
speed 144, 147–8, 159–62
spikes 266, 269
split phase motors 148
sprinklers 56
squirrel cages 143–6, 160, 162
standard motor enclosures 152
standard multimeters 237–8
standards
  Investors in People 73–4, 77
  ISO 30, 71–2, 77
  working practices 71–4, 77
  see also British Standards
star connections 114–16, 134, 180–1, 182
star/delta starters 157–8
starters 153–9
  direct current motors 157
  motor over-current protection 154–5
  single-phase motors 156–7
  star/delta starters 157–8
static inductance 86–7
stators 138, 141–3, 161
steel conduit/trunking 200
Steel Wire Armoured (SWA) cables 119, 134, 197–8
step-up transformers 125
storage 11–12, 167
storage batteries 266–7
storage heater bricks 62, 133
storming 67
stress 90
supervision/supervisors 13–15, 29
supplementary bonding 214–15
supply *see* electrical supply systems

surges 263, 266, 268, 269
SWA *see* Steel Wire Armoured cables
switches 39, 103, 128–9
  switch fuses 125
  switch gear 125–30, 222–3, 225, 232–3
  switching arrangements 183–4, 222–3, 225, 231–2
synchronous motors 138–9

tag terminations 205
team working 67, 77
technical career options 67
technicians 66
temperature 80
terminals 191–2, 225
terminating cables 191–2, 203–7, 225
termination point size 206
terms of reference 54
test certificates 243–6
test leads 252–3
testing 227–58
  bimetallic devices 156
  BS 7671 Regulations 227, 254
  completed installations 230–5
  confirmation procedures 228
  documentary evidence 242–6, 251, 255–6
  faults 227–8
  frequency 5
  important issues 242–3
  initial confirmation procedures 228
  methods 246–51
  periodic inspections 229–30
  procedures 272–3
  regulations 227, 254–5, 273
  report forms 255–6
  requirements 252
  sequences 252, 272–3
  smoke alarms 255, 258
  test instruments 235–43, 252–5, 272
    accuracy 253
    calibration 242, 275, 284
    important issues 242–3
    verifying 275
    working order restorations 274–5
  visual inspections 227, 230–5

wiring regulations 230
working order restorations 272–3
thermal effects 231
thermal links 220, 222
thermal overloads 154–5
thermistors 105, 107, 110
third harmonic currents 134
three phase systems 225, 260–1, 269
  four wire systems 114–17, 180–2
  generators 114–17
  magnetic fields 141–3, 161
  motors 143–6, 160
thyristors 101
time factors 167–8, 177, 238
timing meters 238
TN earthing 212, 215–18
TN-C earthing 131, 217–18
TN-C-S earthing 131, 135, 216–17
TN-S earthing 130–1, 135, 182–3, 216
tong testers 236, 272
tool identification 224
toroidal transformers 125
torque 159
toxicity 92
training 13–15, 49, 50–1
transformers 120–5, 134–5
  electric motors 162
  single-phase transformer power requirements 48
  windings 114–16
transient voltages 263, 266, 269
transistors 103–4, 105, 107, 110
transmission 112–15, 134
transport 11–12
triacs 106, 110
triboelectricity 86
trimmer capacitors 91
triple pole and neutral distribution 181
tripping 188
true power 124
trunking 118, 200, 202
TT earthing 130, 135, 212, 215
tungsten halogen lamps 134
turns of wire 123–4

under-floor dualing 118
under voltage protection 232
units, electricity 240
universal motors 147–8

valency shells 96
variable capacitors 92
variation orders 174–6, 177
vaseline 224, 283
vibration 108
visual inspections 227, 230–5
voltage
    distribution and supply
        178–80, 225
    drop 81, 208–9
    earthing systems 131–2
    insulation meters 272
    protection 132–4, 135, 232
    reduction 111–12
    semiconductors 100–1
    supply and distribution
        178–80, 225
    surges 263, 266, 268, 269
    switch gear 125–30
    testers 241, 272, 275
    three phase systems 115–17
    torque 159
    transmission 112–15, 134
voltmeters 272, 275

warning signs 22, 49, 51
waste disposal 31, 282, 284
water, bonding 214
water fire extinguishers 60
watts 123
waveforms 101–3
welfare 20
windings 114–16, 120, 123–4,
    180–2
wiring
    completing installations
        194–203
    diagrams 168, 169–70
    direct-on-line starters 153–4
    regulations 230, 271, 274
Work Time Regulations [1998] 7
working area safety 163–4
working drawings 168, 169–70
working environment safety
    9–11, 40, 163–4
    see also health . . .; safety
working in isolation 40
working order restorations
    271–84
    documentation 271–3, 275
    fault diagnosis 274–5, 276–9,
        284
    insulation resistance 272,
        274, 276–8, 280

non-electrical remedial work
    281–2
    procedures 273–4
    quality checks 282
    test instruments 272, 274–5,
        284
    testing 272–3, 274–5, 284
working practices
    effectiveness 65–78
    efficient work procedures
        70–1, 77
    employer's responsibility
        7–9
    employment regulations
        67–70, 77, 78
    improvement benefits 74–6,
        77
    performance 65–7
    safe work procedures 7–11,
        18–20, 38–40, 70–1, 77,
        163–4, 261
    standards 71–4, 77
    team working methods
        67
workplace fire drills 58–9, 63
works certificates 251
    see also certificates
works rescue teams 30
wound rotors 145–6